METHODS IN MOLECULAR BIOLOGY

Series Editor
John M. Walker
School of Life and Medical Sciences
University of Hertfordshire
Hatfield, Hertfordshire, AL10 9AB, UK

For further volumes:
http://www.springer.com/series/7651

Skeletal Muscle Regeneration in the Mouse

Methods and Protocols

Edited by

Michael Kyba

Lillehei Heart Institute, University of Minnesota, Minneapolis, MN, USA

Humana Press

Editor
Michael Kyba
Lillehei Heart Institute
University of Minnesota
Minneapolis, MN, USA

ISSN 1064-3745 ISSN 1940-6029 (electronic)
Methods in Molecular Biology
ISBN 978-1-4939-3808-7 ISBN 978-1-4939-3810-0 (eBook)
DOI 10.1007/978-1-4939-3810-0

Library of Congress Control Number: 2016948837

Printed on acid-free paper

This Springer imprint is published by Springer Nature
The registered company is Springer Science+Business Media LLC New York

Preface

Although skeletal muscle is the most abundant tissue in the body, playing key roles in metabolism, heat generation, and macromolecule storage in addition to the obvious role in locomotion, our understanding of the mechanisms underlying the growth, damage, degeneration, and repair of this tissue is still very limited. The past decade has seen rapid advances in our understanding of the development and cell biology of skeletal muscle progenitor cells and a burgeoning of laboratory research on skeletal muscle regeneration. These advances rest in large part on novel technologies and methods that, combined with new genetic tools available in the mouse model, have reshaped the field. As the field has swelled with many new investigators focused on the mouse model, there seemed a need for a volume that collects these critical protocols, optimized for mouse but in most cases easily adaptable to rat or other mammalian models. I thank my many wonderful colleagues for their contributions to this collection. Methods have been grouped into three general areas: inducing damage, analysis of the progenitor cells of skeletal muscle, and measuring overall muscle function. It is our hope that this volume will serve as a comprehensive laboratory reference for research on skeletal muscle growth, damage, repair, degeneration, and regenerative therapy in the mouse model system.

Minneapolis, MN, USA *Michael Kyba*

Contents

Part III Muscle Functional Assays

Contributors

MARIE LUE ANTONY • *Lillehei Heart Institute, University of Minnesota, Minneapolis, MN, USA*

ROBERT W. ARPKE • *Department of Pediatrics, Lillehei Heart Institute and Department of Pediatrics, University of Minnesota, Minneapolis, MN, USA*

ATSUSHI ASAKURA • *Stem Cell Institute, University of Minnesota Medical School, Minneapolis, MN, USA; Paul & Sheila Wellstone Muscular Dystrophy Center, University of Minnesota Medical School, Minneapolis, MN, USA; Department of Neurology, University of Minnesota Medical School, Minneapolis, MN, USA*

YOKO ASAKURA • *Stem Cell Institute, University of Minnesota Medical School, Minneapolis, MN, USA; Paul & Sheila Wellstone Muscular Dystrophy Center, University of Minnesota Medical School, Minneapolis, MN, USA; Department of Neurology, University of Minnesota Medical School, Minneapolis, MN, USA*

ARPITA A. BISWAS • *Department of Molecular & Cell Biology, University of Connecticut Stem Cell Institute, University of Connecticut, Storrs, CT, USA*

HELEN M. BLAU • *Department of Microbiology & Immunology, Baxter Laboratory for Stem Cell Biology, Stanford University School of Medicine, Stanford, CA, USA*

SUSAN V. BROOKS • *Department of Biomedical Engineering, University of Michigan Medical School, Ann Arbor, MI, USA; Department of Molecular & Integrative Physiology, University of Michigan Medical School, Ann Arbor, MI, USA*

ADAM B. CADWALLADER • *Department of Molecular, Cellular and Developmental Biology, University of Colorado at Boulder, Boulder, CO, USA*

JARROD A. CALL • *Department of Kinesiology, University of Georgia, Athens, GA, USA; Regenerative Bioscience Center, University of Georgia, Athens, GA, USA*

DENNIS R. CLAFLIN • *Department of Surgery, Section of Plastic Surgery, University of Michigan Medical School, Ann Arbor, MI, USA; Department of Biomedical Engineering, University of Michigan Medical School, Ann Arbor, MI, USA*

BENJAMIN T. CORONA • *Extremity Trauma and Regenerative Medicine, United States Army Institute of Surgical Research, Fort Sam Houston, TX, USA*

TIFFANY L. ELSTON • *Department of Molecular, Cellular and Developmental Biology, University of Colorado at Boulder, Boulder, CO, USA*

CHRISTOPHER S. FRY • *Department of Nutrition and Metabolism, School of Health Professions, University of Texas Medical Branch, Galveston, TX, USA*

KATHERINE E. GADEK • *Department of Molecular, Cellular and Developmental Biology, University of Colorado at Boulder, Boulder, CO, USA*

LAURA GARCÍA-PRAT • *Cell Biology Group, Department of Experimental and Health Sciences, Pompeu Fabra University (UPF), CIBER on Neurodegenerative Diseases (CIBERNED), Barcelona, Spain*

DANIEL J. GARRY • *Lillehei Heart Institute, University of Minnesota, Minneapolis, MN, USA*

GLYNNIS A. GARRY • *UT Southwestern Medical Center, Dallas, TX, USA*

DAVID J. GOLDHAMER • *Department of Molecular & Cell Biology, University of Connecticut Stem Cell Institute, University of Connecticut, Storrs, CT, USA*

KATSUMASA GOTO • *Department of Physiology, Graduate School of Health Sciences, Toyohashi SOZO University, Aichi, Japan*

ROBERT W. GRANGE • *Department of Human Nutrition, Foods and Exercise, Corporate Research Center, Virginia Tech University, Blacksburg, VA, USA*

SARAH M. GREISING • *Department of Physiology and Biomedical Engineering, Mayo Clinic, Rochester, MN, USA*

JONATHAN P. GUMUCIO • *Department of Orthopaedic Surgery, University of Michigan Medical School, Ann Arbor, MI, USA; Department of Molecular & Integrative Physiology, University of Michigan Medical School, Ann Arbor, MI, USA*

DENIS C. GUTTRIDGE • *Department of Molecular Virology, Immunology, and Medical Genetics, The Ohio State University, Columbus, OH, USA; Center for Muscle Health and Neuromuscular Disorders, The Ohio State University, Columbus, OH, USA*

ERICK O. HERNÁNDEZ-OCHOA • *Department of Biochemistry & Molecular Biology, University of Maryland School of Medicine, Baltimore, MD, USA*

ANDREW T.V. HO • *Department of Microbiology & Immunology, Baxter Laboratory for Stem Cell Biology, Stanford University School of Medicine, Stanford, CA, USA*

TANIA INCITTI • *Department of Medicine, Lillehei Heart Institute, University of Minnesota, Minneapolis, MN, USA*

SHAMA R. IYER • *Department of Orthopaedics, University of Maryland School of Medicine, Baltimore, MD, USA*

TAKEHIRO KASAI • *Department of Orthopedic Surgery, Nagoya University Graduate School of Medicine, Aichi, Japan*

TYLER J. KIRBY • *Center for Muscle Biology, University of Kentucky, Lexington, KY, USA; Department of Physiology, College of Medicine, University of Kentucky, Lexington, KY, USA*

MICHAEL KYBA • *Lillehei Heart Institute and Department of Pediatrics, University of Minnesota, Minneapolis, MN, USA*

GENGYUN LE • *Department of Physical Medicine and Rehabilitation, University of Minnesota, Minneapolis, MN, USA*

RICHARD M. LOVERING • *Department of Orthopaedics, University of Maryland School of Medicine, Baltimore, MD, USA*

DAWN A. LOWE • *Programs in Physical Therapy and Rehabilitation Science, Department of Physical Medicine and Rehabilitation Medical School, University of Minnesota, Minneapolis, MN, USA*

ALESSANDRO MAGLI • *Department of Medicine, Lillehei Heart Institute, University of Minnesota, Minneapolis, MN, USA*

CARLOS B. MANTILLA • *Department of Physiology and Biomedical Engineering, Mayo Clinic, Rochester, MN, USA; Department of Anesthesiology, Mayo Clinic, Rochester, MN, USA*

MARTA MARTÍNEZ-VICENTE • *Neurodegenerative Diseases Research Group, Vall d'Hebron Research Institute-CIBERNED, Barcelona, Spain*

JOHN J. MCCARTHY • *Center for Muscle Biology, University of Kentucky, Lexington, KY, USA; Department of Physiology, College of Medicine, University of Kentucky, Lexington, KY, USA*

CHRISTOPHER L. MENDIAS • *Department of Orthopaedic Surgery, University of Michigan Medical School, Ann Arbor, MI, USA; Department of Molecular & Integrative Physiology, University of Michigan Medical School, Ann Arbor, MI, USA*

PURA MUÑOZ-CÁNOVES • *Cell Biology Group, Department of Experimental and Health Sciences, Pompeu Fabra University (UPF), CIBER on Neurodegenerative Diseases (CIBERNED), Barcelona, Spain; Institució Catalana de Recerca i Estudis Avançats (ICREA), Barcelona, Spain*

BRADLEY B. OLWIN • *Department of Molecular, Cellular and Developmental Biology, University of Colorado at Boulder, Boulder, CO, USA*

HUASCAR P. ORTUSTE QUIROGA • *Randall Division of Cell and Molecular Biophysics, King's College London, London, UK*

MAURA H. PARKER • *Clinical Research Division, Fred Hutchinson Cancer Research Center, Seattle, WA, USA; Department of Medicine, University of Washington, Seattle, WA, USA*

RITA C.R. PERLINGEIRO • *Department of Medicine, Lillehei Heart Institute, University of Minnesota, Minneapolis, MN, USA*

PATRIZIA PESSINA • *Cell Biology Group, Department of Experimental and Health Sciences, Pompeu Fabra University (UPF), CIBER on Neurodegenerative Diseases (CIBERNED), Barcelona, Spain*

CHARLOTTE A. PETERSON • *Center for Muscle Biology, University of Kentucky, Lexington, KY, USA; Department of Rehabilitation Sciences, College of Health Sciences, University of Kentucky, Lexington, KY, USA*

BETH E. POLLOT • *Extremity Trauma and Regenerative Medicine, United States Army Institute of Surgical Research, Fort Sam Houston, TX, USA*

STUART M. ROCHE • *Department of Orthopaedic Surgery, University of Michigan Medical School, Ann Arbor, MI, USA*

BHAVANI SR MURKONDA • *Stem Cell Institute, University of Minnesota Medical School, Minneapolis, MN, USA; Paul & Sheila Wellstone Muscular Dystrophy Center, University of Minnesota Medical School, Minneapolis, MN, USA; Department of Neurology, University of Minnesota Medical School, Minneapolis, MN, USA*

ALESSANDRA SACCO • *Development, Aging and Regeneration Program, Sanford Children's Health Research Center, Sanford Burnham Prebys Medical Discovery Institute, La Jolla, CA, USA*

JONATHAN SHINTAKU • *Department of Molecular Virology, Immunology, and Medical Genetics, The Ohio State University, Columbus, OH, USA; Center for Muscle Health and Neuromuscular Disorders, The Ohio State University, Columbus, OH, USA; Molecular, Cellular and Developmental Biology Graduate Program, The Ohio State University, Columbus, OH, USA*

GARY C. SIECK • *Department of Physiology and Biomedical Engineering, Mayo Clinic, Rochester, MN, USA; Department of Anesthesiology, Mayo Clinic, Rochester, MN, USA*

JUSTIN E. SPERRINGER • *Department of Human Nutrition, Foods and Exercise, Corporate Research Center, Virginia Tech University, Blacksburg, VA, USA*

PASCAL STUELSATZ • *Department of Biological Structure, School of Medicine, University of Washington, Seattle, WA, USA*

MATTHEW T. TIERNEY • *Graduate School of Biomedical Sciences, Sanford Burnham Prebys Medical Discovery Institute, La Jolla, CA, USA; Development, Aging and Regeneration Program, Sanford Children's Health Research Center, Sanford Burnham Prebys Medical Discovery Institute, La Jolla, CA, USA*

Kunihiro Tsuchida • *Division for Therapies Against Intractable Diseases, Institute for Comprehensive Medical Science, Fujita Health University, Aichi, Japan*
Akiyoshi Uezumi • *Division for Therapies Against Intractable Diseases, Institute for Comprehensive Medical Science, Fujita Health University, Aichi, Japan*
Ana P. Valencia • *Department of Orthopaedics, University of Maryland School of Medicine, Baltimore, MD, USA*
Mayank Verma • *Medical Scientist Training Program, University of Minnesota Medical School, Minneapolis, MN, USA; Stem Cell Institute, University of Minnesota Medical School, Minneapolis, MN, USA; Paul & Sheila Wellstone Muscular Dystrophy Center, University of Minnesota Medical School, Minneapolis, MN, USA; Department of Neurology, University of Minnesota Medical School, Minneapolis, MN, USA*
Thomas O. Vogler • *Department of Molecular, Cellular and Developmental Biology, University of Colorado at Boulder, Boulder, CO, USA*
Zipora Yablonka-Reuveni • *Department of Biological Structure, School of Medicine, University of Washington, Seattle, WA, USA*
Peter S. Zammit • *Randall Division of Cell and Molecular Biophysics, King's College London, London, UK*

Part I

Injury Models

Chapter 1

Eccentric Contraction-Induced Muscle Injury: Reproducible, Quantitative, Physiological Models to Impair Skeletal Muscle's Capacity to Generate Force

Jarrod A. Call and Dawn A. Lowe

Abstract

In order to investigate the molecular and cellular mechanisms of muscle regeneration an experimental injury model is required. Advantages of eccentric contraction-induced injury are that it is a controllable, reproducible, and physiologically relevant model to cause muscle injury, with injury being defined as a loss of force generating capacity. While eccentric contractions can be incorporated into conscious animal study designs such as downhill treadmill running, electrophysiological approaches to elicit eccentric contractions and examine muscle contractility, for example before and after the injurious eccentric contractions, allows researchers to circumvent common issues in determining muscle function in a conscious animal (e.g., unwillingness to participate). Herein, we describe in vitro and in vivo methods that are reliable, repeatable, and truly maximal because the muscle contractions are evoked in a controlled, quantifiable manner independent of subject motivation. Both methods can be used to initiate eccentric contraction-induced injury and are suitable for monitoring functional muscle regeneration hours to days to weeks post-injury.

Key words Force drop, Lengthening contraction, Muscle damage

1 Introduction

Eccentric contractions are contractions in which the external load or resistance placed on an activated muscle is greater than the force generated by that muscle, and subsequently the muscle is lengthened while it is active. There is an immediate loss of strength following the performance of eccentric contractions that is attributed to disruption of the excitation–contraction coupling process and/or frank damage to force-generating or -transmitting structures within the muscle [1–3]. Forces generated during maximal eccentric contractions exceed those produced during isometric and concentric contractions by as much as 80%, and injury resulting from eccentric contractions has been largely attributed to these high forces [4, 5]. The extent of the strength loss can vary depending on age and disease of the subject. For example, acute

Michael Kyba (ed.), *Skeletal Muscle Regeneration in the Mouse: Methods and Protocols*, Methods in Molecular Biology, vol. 1460, DOI 10.1007/978-1-4939-3810-0_1, © Springer Science+Business Media New York 2016

strength loss can be as high as 50% in muscles of healthy C57Bl mice, but in excess of 75% in an *mdx* mouse (model for Duchenne muscular dystrophy).

In vitro and in vivo methods are ideal for executing the eccentric contraction-induced injury model because the severity of muscle injury can be monitored in real time and controlled precisely by altering the number of eccentric contractions performed, the distance the muscle is lengthened, and/or the velocity at which the lengthening occurs. Additionally, these methods provide a level of consistency as far as injury at the fiber level because all motor units are activated which is in sharp contrast to fully conscious experimental models, e.g., downhill treadmill running.

1.1 In Vitro Eccentric Contractions

Live, isolated muscle preparations are utilized to assess contractile capacity of those organs. Small muscles of mice, such as the soleus and extensor digitorum longus (EDL) muscles, are suitable and most commonly used because oxygen diffusion to fibers in the core of muscle is adequate [6, 7]. These muscles also have well-defined tendons making dissection and attachment to force transducers straightforward. Interfaced to equipment as described by Sperringer and Grange in Chapter 19 of this volume [8], a host of contractile tests can be performed. The in vitro section of this chapter will focus on eccentric contractions by isolated muscles and the consequent contraction-induced injury. In the muscular dystrophy field, the term “force drop” is becoming common. This term is used to describe the outcome of eccentric contraction testing in EDL muscle lacking dystrophin as these muscles are highly susceptible to injury, as measured by the loss of force-generating capacity during and immediately following eccentric contractions. Before detailing methods for an in vitro eccentric contraction-injury protocol, a description of EDL muscle dissection is presented in this chapter because careful dissection is important in order to study contraction-induced injury to the muscle as opposed to dissection-induced injury.

1.2 In Vivo Eccentric Contractions

We have developed a modified in vivo muscle testing apparatus and protocol based on the original in vivo system reported by Ashton-Miller et al. [9], and similar to that described by Iyer et al., in Chapter 21 of this volume [10]. Briefly, peak isometric contractility of either the ankle dorsi-flexors (tibialis anterior, extensor digitorum longus, extensor hallicus longus muscles) or plantar-flexors (gastrocnemius, soleus, plantaris muscles) in anesthetized mice is determined using percutaneous electrodes to stimulate specific nerves innervating those muscle groups and specialized equipment to record the contractile output. Then an eccentric contraction-induced injury protocol is performed to cause muscle injury as immediately measurable by decrements in torque. Below we detail our eccentric contraction-induced injury model including internal controls and expected outcomes for healthy C57Bl6 mice and some examples from dystrophic *mdx* mice. Finally, a real advantage of the relatively noninvasive muscle analysis in vivo is that muscle contractile function can be measured repeatedly which can be ideal for monitoring ongoing muscle regeneration. Accordingly, we also briefly describe the recovery of muscle strength after eccentric contraction-induced injury.

2 Materials

In Vitro

2.1 Equipment and Surgical Instruments for Delicate Muscle Dissections

1. Dissecting microscope (e.g., Leica S6D).
2. Fiber-Lite Dual Gooseneck Light (this is needed if there is not a light ring on dissecting scope).
3. Halstead Mosquito Forceps (~hemostat), curved, delicate 5′ length (e.g., George Tiemann #105-1107).
4. Dressing and Tissue Forceps, delicate with teeth (e.g., George Tiemann #105-205-1).
5. Extra fine Graefe Forceps, curved, finely grated tips (Fine Science #11151-10).
6. Dumont Forceps, pointed tip (e.g., Fine Science #11295-10).
7. Tissue Scissors, delicate and straight 3 3/4″ (e.g., George Tiemann #105-421).
8. Student Vannas Spring Scissors, sharp and non-serrated tips (e.g., Fine Science #91500-09).
9. McPherson Vannas Scissors, very finely serrated and delicate tips (George Tiemann #160-140).
10. Digital Caliper (e.g., World Precision Instrument).
11. 5-0 braided silk suture, non-absorbable; cut in 5 in. pieces.
12. Gel type cyanoacrylate (i.e., super glue) and 25 G needle for applying the glue.

2.2 In Vitro System for Eccentric Contractions

1. 300C-LR, Dual-Mode Lever System, Aurora Scientific.
2. S48 Stimulator with SUI5 Stimulus Isolation Unit, Grass Technologies.
3. Refrigerated/heating circulating water bath, controllable to 0.1°.
4. Organ bath, volume 1.2 ml; custom made by glass blower (Fig. 1b) or large organ baths available commercially.
5. Thermocouple and probe to measure/confirm temperature of buffer in organ bath (e.g., BAT-12 Microprobe Thermometer with IT-18 Flexible Microprobe).
6. Platinum wire electrodes (1.3 mm diameter, 99.95 pure platinum) (Fig. 1b).

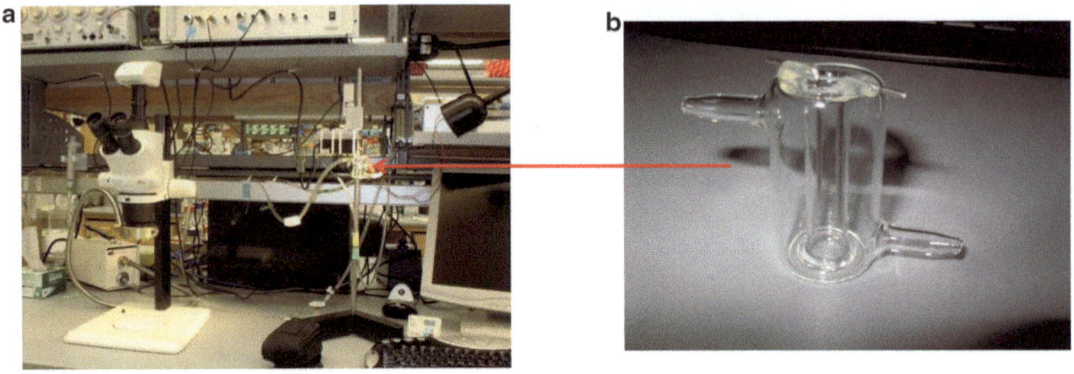

Fig. 1 (**a**) In vitro muscle physiology system and (**b**) custom, 1.2 ml organ bath with electrodes sized for mouse soleus or EDL muscle. The in vitro system described here is similar to that described by Sperringer and Grange in Chapter 19 of this volume [8], particularly in regard to the lever system

7. 95 % O_2, 5 % CO_2 tank with corresponding gas regulator.
8. Flow meter, 5.8 ml/min air with high resolution valve (e.g., Cole-Parmer, WZ-32044-00).
9. Computer, A/D board, and software.

In Vivo

2.3 Equipment for In Vivo Eccentric Contractions

1. Dual-Mode Lever System (Aurora Scientific Inc., 300C-LR).
2. 16-bit National Instruments M series A/D card (Aurora, 603C).
3. Signal interface (Aurora, 604A), Grass stimulator and stimulation isolation unit (Grass s48 and SIU5) or high power follow stimulator (Aurora 701C).
4. Dynamic muscle Control/Dynamic Muscle Analysis-High Throughput Software Suite (Aurora, 615B) or similar data acquisition and analysis software (e.g., TestPoint).
5. Fuzzy logic PID temperature controller (Cole-Parmer C-89810-04).
6. Self-adhesive probe (Cole-Parmer C-08519-54).
7. Kapton heater pad (Cole-Parmer C-36060-50).
8. Extension adapter cord (Cole-Parmer C-03122-71).
9. Platinum-iridium electrodes (Grass Technologies E2-12).
10. Soldering helping hands, adhesive tape, and computer with Windows OS.

A plexiglass platform and mouse footplate were both custom manufactured by the university machine shop. An alternative platform and footplate are available through Aurora Scientific (809B testing apparatus). A heated circulating water bath will be necessary if using Aurora Scientific testing apparatus (809A).

3 Methods

In Vitro

3.1 EDL Muscle Dissection

1. Anesthetize mouse with an intraperitoneal injection of sodium pentobarbital at 100 mg/kg body mass using a 3/10 cc insulin syringe (e.g., for a 30 g mouse use 60 μL of a 50 mg/ml pentobarbital stock) (*see* **Note 1**).
2. Pull skin back from posterior hind limb up to the hamstrings using dressing and tissue forceps with teeth.
3. From the anterior hind limb, working on big toe side of the tibialis anterior muscle (TA) to stay away from the EDL muscle, pull the skin up past the knee and down to the ankle.
4. Pull the skin back over the heel being careful to not break the ankle.
5. From this point forward, it is highly recommended to use a dissecting microscope.
6. Remove fascia from TA using Dumont pointed forceps to puncture fascia at the mid-belly of TA.
7. Tear back fascia towards proximal end of the muscle and then towards distal end, again working on big toe side to avoid poking the EDL especially at the distal end.
8. Separate the distal EDL and TA tendons by inserting Dumont pointed forceps between the two tendons (that is, forceps should be under the TA tendon and over the EDL tendon; Fig. 2a).
9. Slide the forceps (still under the TA) up to the knee to completely separate the TA and EDL muscles. This is only possible when fascia is completely removed.
10. Insert one blade of the McPherson Vannas scissors under the distal TA tendon and cut as close to the foot as possible.
11. Stabilize the foot with the hemostat and then pull back TA towards the knee using Graefe forceps.
12. Pull the TA to the side and get one blade of the McPherson Vannas scissors under the TA fascia, parallel to EDL fibers, and cut a small slit to expose the proximal EDL tendon at the knee (Fig. 2b); a hamstring muscle inserting at the knee also lies perpendicular, over the proximal EDL tendon, and will be severed with this cut.
13. Use tissue scissors to cut off the TA at the knee.
14. Locate and pull out the extensor hallucis longus muscle (the big toe muscle).

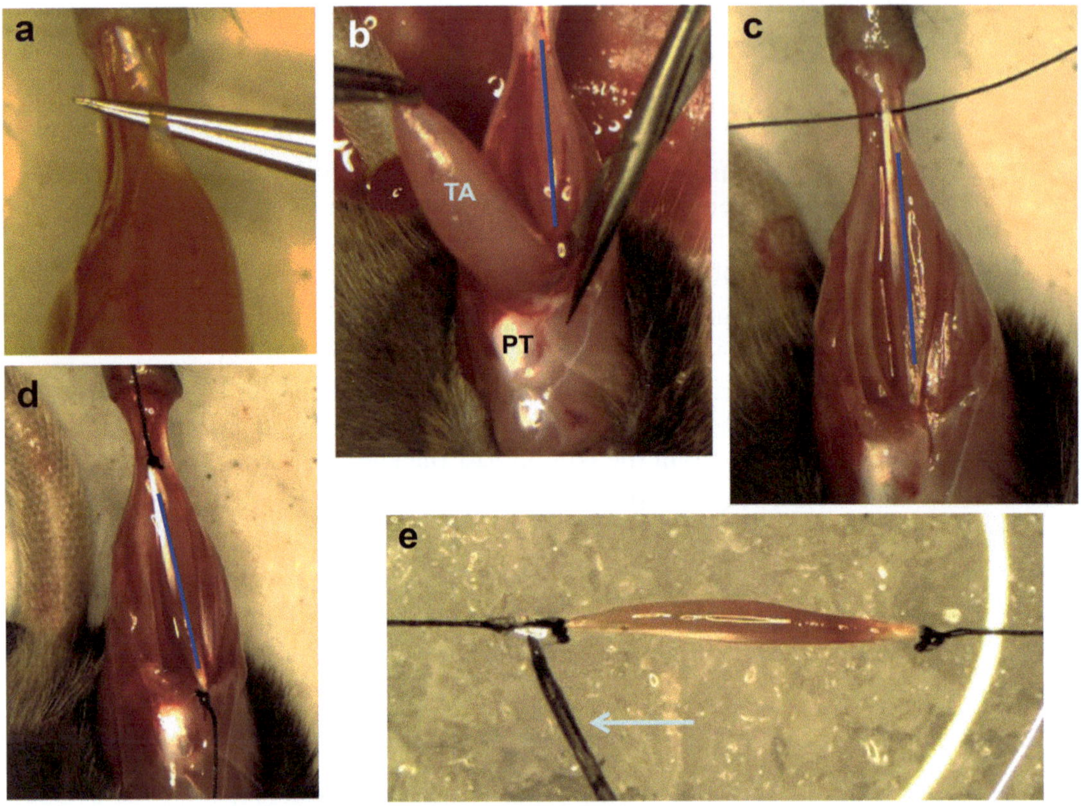

Fig. 2 Dissection of mouse EDL muscle. Orientation is such that in (**a**)–(**d**) the foot of the mouse is at the top of the picture. (**a**) Distal tendon of the tibialis anterior (TA) muscle is separated from the distal EDL tendon by forceps. (**b**) Distal TA tendon is cut exposing EDL muscle. Blade of Vannas scissors is shown cutting proximal TA muscle to expose medial EDL tendon. *Blue line* demarks midline of the EDL muscle. PT = patellar tendon. (**c**) Suture placed under distal EDL tendon, precisely at myotendinous junction. Note that tendon has been cleared of connective tissue and fat. (**d**) Both distal tendon (*top*) and medial tendon (*bottom*) of the EDL muscle are secured by suture. We prefer to tie sutures on the muscle in vivo to keep muscle at its anatomical length as long as possible during the dissection and to minimize handling of the muscle ex vivo. (**e**) Once completely excised, gel type cyanoacrylate adhesive is carefully applied with a needle (*arrow*) to cover tendons and knots of the suture to avoid slippage during eccentric contractions

15. Carefully remove any connective tissue or fat around the distal tendon of the EDL (*see* **Note 2**).
16. Insert Dumont pointed forceps under the distal tendon of the EDL muscle and pull suture through, under the tendon. Position suture precisely at the myotendinous junction (MTJ) making sure there is no connective tissue or fat between the suture and the tendon before tying knots (Fig. 2c). Also, be certain that the suture is around all four slips of the tendon, which are often distinguishable in atrophied EDL muscles.
17. Tie three knots, with the first being perpendicular to the fibers and the final parallel to fibers so that the transmission of force to the transducer is in series with fibers.

18. Using Student Vannas scissors, cut retinaculum of the foot parallel to the EDL tendon to expose about 2 mm of the EDL tendon distal to the MTJ and then cut the tendon (*see* **Notes 3** and **4**).
19. Using Graefe forceps to hold the tendon, gently pull back the EDL muscle toward the knee; if "slit" at the knee (**step 12**) was done correctly when removing the TA, the proximal EDL tendon will be exposed and easily pops out.
20. Place the second suture under proximal tendon, precisely at the MTJ, and tie three knots as described for the distal tendon; before doing this make sure the muscle is not twisted (you should be able to see the distal tendon on the top surface of the muscle; Fig. 2d).
21. Make small cut in knee capsule to get longer tendon by cutting parallel to tendon; the length of the proximal tendon is shorter than the distal tendon but strive to get 0.5 mm.
22. Place excised EDL muscle in a petri dish with chilled, oxygenated Krebs buffer (for example recipe *see* Sperringer and Grange in this volume [8]); use a hemostat on each suture to keep slight tension on the muscle (*see* **Note 5**).
23. Raise muscle out of buffer, blot off excess buffer on tendons and knots, and carefully apply super glue with a needle to each knot with the tendon positioned parallel with the suture (Fig. 2e) (*see* **Note 6**).
24. Move muscle to the organ bath of the in vitro system and attach suture to lever arm, add Krebs buffer and apply O_2; occasionally confirm that temperature of the buffer in the organ bath is precisely as desired using the microprobe thermometer.
25. Let the muscle equilibrate at least 5 min, set the muscle to its anatomical optimal muscle length (L_o), and then measure that length, from MTJ to MTJ, using the calipers (*see* **Note 7**).

3.2 Eccentric Contraction Protocol: In Vitro

Determine maximal isometric tetanic force (P_o) to establish pre-injury force generating capacity

1. With our system, maximal tetanic contraction of EDL muscle is elicited with 200 ms of 0.5-ms pulses at 180 Hz with stimulator set to maximal 150 V (*see* **Note 8**).
2. Repeat tetanic contractions with 3 min of rest in between (to avoid fatigue) until minimal increase in force occurs between two consecutive contractions (<0.5 g); this typically occurs at three contractions for EDL muscle and 5–8 contractions for soleus muscle with our set up at 25 °C (*see* **Note 9**).
3. In between these "warm up" tetanic contractions, optimize L_o; we do this by adjusting resting tension to 0.4 g as it will drift down (refer to **Note** 7 for justification).

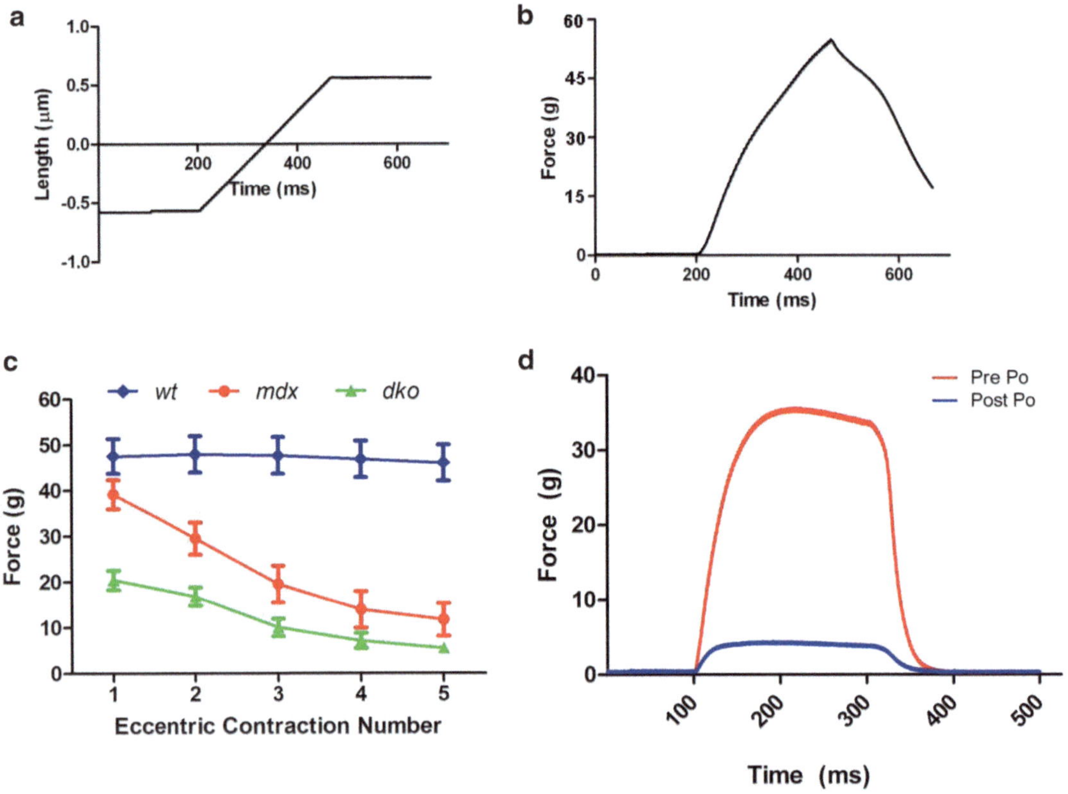

Fig. 3 (**a**) Representative length-time and (**b**) force-time tracings of a single in vitro eccentric contraction by a mouse EDL muscle. (**c**) A series of five eccentric contractions can be titered by using only 10 % L_o length change at a slow velocity of 0.75 L_o/s to result in no force loss in healthy muscle (*wt*) but substantial force drop in muscle lacking dystrophin (*mdx*) and muscle lacking both dystrophin and utrophin (*dko*). This establishes an optimal situation to detect therapeutic improvement in the disease models. (d) In vitro contraction-induced injury shown as a loss in maximal isometric tetanic force (Po) for *mdx* mouse muscle following eccentric contractions (Post) as compared to before those injurious contractions (Pre). Note, though the force is typically measured in grams (g), it should be converted to newton (N), which is the SI unit of force for reporting. For example, an 8 mg EDL mouse muscle should generate ~35 g, i.e., 343 mN, of force. Notice that the peak-eccentric force (in **b**) is 150–175 % of peak-isometric force (Pre in **d**)

Eccentric contractions

4. Calculate 20 % of the L_o measured with calipers; this will be the length change during the eccentric contraction (e.g., 2.4 mm is 20 % of an EDL muscle with a L_o of 12 mm).
5. Passively shorten the muscle 10 % of L_o (e.g., 1.2 mm) (*see* **Note 10** and Fig. 3a).
6. Initiate a 133 ms contraction at 180 Hz while simultaneously lengthening the muscle 20 % of L_o (e.g., 2.4 mm) at a rate of 1.5 L_o/s (*see* **Notes 11** and **12**).
7. Passively move the muscle back to the starting position, i.e., L_o.
8. Repeat eccentric contractions every 3 min up to 19 times.

Measure post-injury P_o to determine extent of injury, i.e., loss of force generating capacity

9. Upon completion of the final eccentric contraction, reset L_o.
10. Three minutes after the final eccentric contraction, re-measure P_o.
11. Calculate force loss (aka force drop)

$$\% \text{ eccentric force loss} = [(\text{ECC}\mathit{last} \quad \text{ECC}\mathit{first}) / \text{ECC}\mathit{first}] \times 100$$

$$\% \text{ isometric force loss} = [(\text{post}P_o \quad \text{pre}P_o) / \text{pre}P_o] \times 100$$

Expected outcomes:

12. Re-injury P_o of an 8 mg EDL muscle from a wild type mouse should generate about 35 mg of isometric force (which should be converted to SI unit of force = 343 mN for reporting).
13. To calculate specific force, fiber to muscle length must be known; *see* Table 1.
14. The first eccentric contraction should generate force that is >150% of P_o; this is critical because it is the high force generation that causes injury to the muscle [4, 5] (Fig. 3b and d).
15. A fewer number of eccentric contractions are typically performed on *mdx* muscles due to their higher susceptibility to injury (Fig. 3c).
16. The % of eccentric force loss should be similar in extent to that calculated for % isometric force loss.

In Vivo

Table 1
Reported fiber length to muscle length ratios across mouse background and age

Fiber length–muscle length ratio					Animal model			
TA	**EDL**	**Soleus**	**L Gastroc**	**M Gastroc**	**Species**	**Sex**	**Age**	**Ref.**
0.61	0.51	0.72	0.46	0.45	HSD mice	Female	Adult	[11]
	0.45	0.71			C57Bl/6 mice	Male	3–4 months	[12]
	0.44	0.69			C57Bl/6 mice	Male	9–10 month	
	0.45	0.69			C57Bl/6 mice	Male	26–27 months	
	0.45*	0.71*			C57Bl/6 mice	Male	3–4 months	
	0.47*	0.68*			C57Bl/6 mice	Male	9–10 month	
	0.44*	0.70*			C57Bl/6 mice	Male	26–27 months	
	0.44				Albino mice	Female	4–6 weeks	[13]

Ratios are fiber length over muscle length. Harlan Sprague Dawley (HSD). *Asterisk* indicates fiber length was determined after using nitric acid digestion (otherwise determined by microdissection)

3.3 In Vivo System for Eccentric Contractions

The in vivo system is very similar to that described by Iyer et al. in Chapter 21 of this volume [10], with only slight modifications (Fig. 4a).

1. The in vivo apparatus is setup so that the animal lies on its side with the left hind limb secured to the footplate and force transducer (Fig. 4b). This provides best access to the common peroneal and sciatic nerve for percutaneous stimulation with the platinum-iridium electrode needles.
2. In lieu of a platform connected to a circulating water bath, place a Kapton heater pad on the testing platform and the self-adhesive probe is positioned so that it will rest beneath the animal during testing. The probe provides animal body temperature feedback to the temperature controller which can adjust the temperature of the heating pad appropriately (*see* **Note 13**).

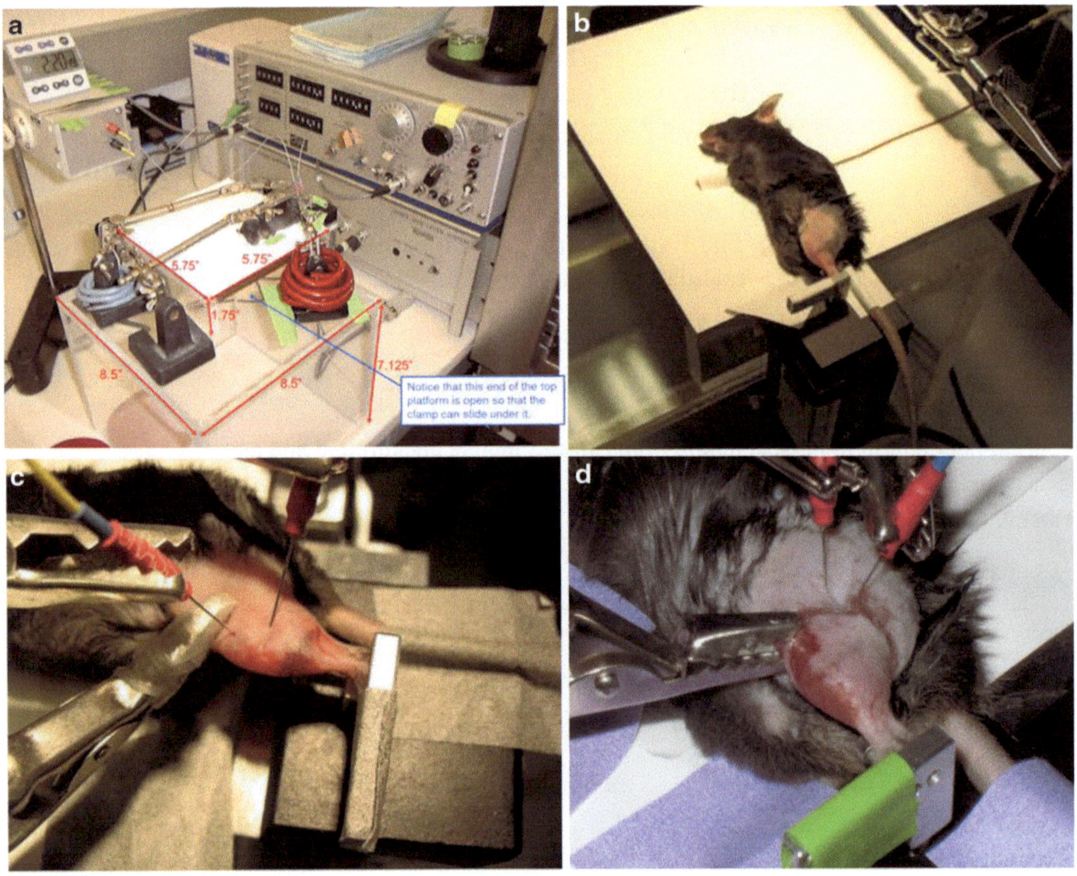

Fig. 4 (**a**) The custom in vivo platform with dimensions. (**b**) Optimal placement of the mouse on the platform. The mouse position is secured with adhesive tape, and the self-adhesive temperature probe lies underneath the animal. (**c**) Appropriate electrode placement on either side of the peroneal nerve for dorsiflexion. (**d**) Appropriate electrode placement on either side of the sciatic nerve for plantarflexion

3. Two soldering helping hands will grasp the needle electrodes and hold them in place during testing (Fig. 4c). A third soldering helping hand will serve as a knee clamp and maintain proper knee-ankle joint alignment and resting torque during testing.

3.4 Animal Preparation for In Vivo Eccentric Contractions

1. Mice anesthetized as described by Iyer et al. in this volume [10] (*see* **Note 13**).
2. The left hind limb from ankle to hip is shaved, hair removed with depilatory cream, and disinfected with repeated washes of betadine and 70 % ethanol.
3. The animal is secured to the pre-warmed (37 °C) in vivo platform (*see* **Note 14**), and knee secured with the knee clamp so that the ankle and knee joints are at 90° (Fig. 4c).
4. Pure platinum-iridium (Pt-Ir) electrodes are placed on either side of the peroneal nerve that innervates the dorsi-flexors (Fig. 4c) (*see* **Note 15**).
5. To injure the plantar-flexors, the peroneal nerve is severed and then Pt-Ir electrodes are placed on either side of the sciatic nerve (Fig. 4d). The sciatic nerve branches into the tibial nerve that innervates the plantar-flexors and the peroneal nerve that innervates the dorsi-flexors. Direct stimulation of the tibial branch nerve is not possible, and that is why the sciatic nerve is stimulated instead. The peroneal nerve must be severed as to not co-contract the antagonist dorsi-flexors (*see* **Notes 16** and **17**).

3.5 Eccentric Contraction Protocol: In Vivo

1. We recommend that isometric contractions are used to optimize current or voltage applied to the nerve (in lieu of twitches).
2. A starting current of 0.5 Amps (A) or voltage (V) of 2 V is recommended. Perform a 150 ms isometric contraction (250 Hz) every 45 s adjusting the stimulation until a peak-isometric torque is achieved. Increments of 0.5 A or 0.2 V are typical.
3. Resting torque values should be accounted for when calculating pre-injury peak-isometric torque.
4. Peak-isometric torques for healthy C57Bl mice are 95–105 N mm per kg body mass (N mm/kg) prior to injury. Peak-isometric torques for dystrophic *mdx* mice can range from 60 to 85 N mm/kg depending on the age of the *mdx* mouse, as disease severity tends to be more severe between 3–8 weeks (*see* **Notes 18** and **19**).
5. After optimizing the stimulation parameters, proceed to the eccentric contraction-induced injury protocol described below:

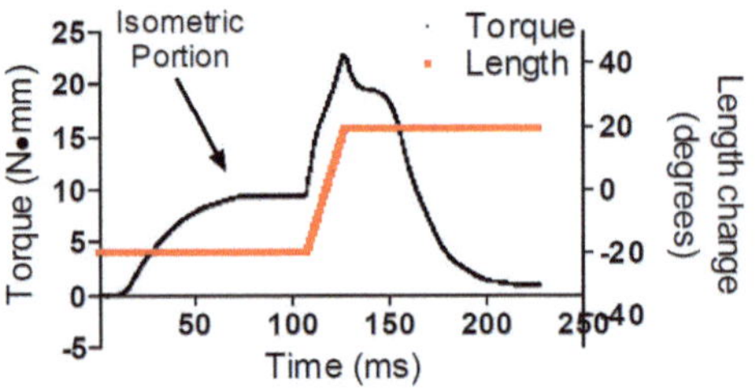

Fig. 5 Representative torque-time tracing of a single in vivo eccentric contraction performed by the plantar-flexors. The length change reflects the foot being moved from 20° plantarflexion to 20° dorsiflexion during the electrical stimulus for contraction. The torque tracing demonstrates an ~100 % increase in torque production during the eccentric portion of the contraction compared to the isometric portion

Eccentric contraction protocol (dorsi-flexors) in vivo

6. The ankle joint should be at a starting position of 90°.
7. Passively (no stimulation) rotate the foot 20° toward dorsiflexion (~1 s).
8. Initiate a 150 ms contraction at maximal stimulation frequency. The first 100 ms will be an isometric contraction at the dorsiflexion position. During the final 50 ms, the ankle joint should move 40° toward plantarflexion at an angular velocity of 800°/s (*see* **Note 20**).
9. Passively (no stimulation) rotate the foot 20° toward dorsiflexion and back to the original starting position (~1 s).
10. Repeat eccentric contractions every 10 s up to 149 times.

Expected outcomes

11. Peak-eccentric torque should be 50–100 % greater than peak-isometric torque pre-injury (Fig. 5). This is a good internal control to ensure proper eccentric contraction protocol.
12. To execute an eccentric contraction-induced injury of the plantar-flexors the above protocol is the same with the exception that the foot is first passively moved toward plantarflexion and then rotated toward dorsiflexion during the eccentric contraction.
13. Figure 6a and b demonstrates the effectiveness of the eccentric contraction-induced injury protocol for initiating muscle damage and strength loss of the dorsi- and plantar-flexors of wild type (WT) and dystrophic mice (*mdx*, het, Fiona).
14. Real time monitoring of injury can be done by analyzing the isometric portion (first 100 ms) of the eccentric contraction and/or the decrease in peak eccentric torque. Figure 6c demonstrates how live monitoring of the strength loss can be used to reach equal injuries in different animal models (wild type vs. *mdx*).

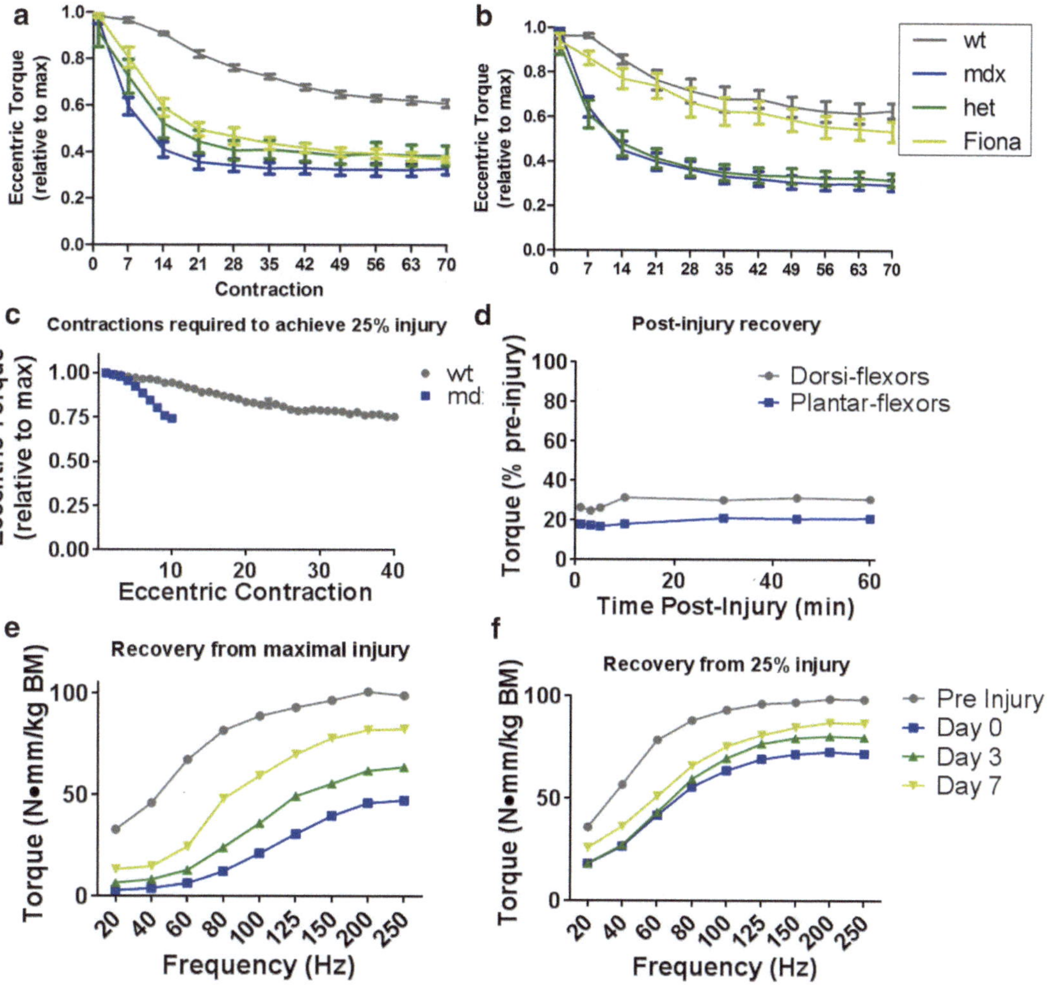

Fig. 6 In vivo contraction-induced injury by the dorsi-flexors (**a**) and plantar-flexors (**b**) among C57Bl controls (*wt*), dystrophin deficient (*mdx*), dystrophin deficient and utrophin heterozygous (*het*), and dystrophin deficient and utrophin overexpression (*Fiona*) mice [15]. (**c**) In vivo contraction-induced injury controlled to a 25 % loss of torque in *wt* and *mdx* muscles. (**d**) Lack of recovery of dorsiflexion and plantarflexion torque during the hour immediate post-injury demonstrates injury as opposed to fatigue. (**e**) Longitudinal recovery of dorsiflexion torque as a function of stimulation frequency after a severe eccentric contraction-induced injury. (**f**) Longitudinal recovery of dorsiflexion torque as a function of stimulation frequency after a relatively mild eccentric contraction-induced injury

15. At the conclusion of the eccentric contraction-induced injury protocol, wait 5–10 min before assessing peak-isometric torque for the immediate post-injury time point. Figure 6d demonstrates the typical recovery of strength during the first hour post-injury. Measurements made within the first 5 min may be variable due to a small fatigue component of the eccentric contraction induced injury protocol.
16. The in vivo system also represents an excellent approach for monitoring muscle regeneration after injury, specifically the

recovery of strength. Figure 6e and f show the recovery of torque at different stimulation frequencies for muscles that were maximally injured (performed 150 eccentric contractions) compared to muscles that were injured to 25 % loss of eccentric torque.

4 Notes

In Vitro

1. For *mdx* mice, we find backing down the pentobarbital dose to 75 mg/kg body mass is preferable to keep mouse deeply anesthetized but still alive during the dissection.
2. We find this critical when doing eccentric contractions so that the suture does not slip when muscle is lengthened.
3. This is a second critical point of dissection when doing eccentric contractions, again to avoid the tendon from slipping.
4. Do not use delicate and expensive McPherson Vannas scissors to cut through the retinaculum or else you will ruin them!
5. We add insulin and branch chained amino acids to the Krebs buffer to improve viability of the preparation [6].
6. The third critical point to avoid tendon slipping out of suture when doing eccentric contractions is applying super glue to cover knots and tendons (but avoiding muscle fibers!).
7. We prefer to set anatomical instead of physiological L_o for reasons delineated in [16, 17]; in practice this is done by adjusting the length of the muscle such that resting tension is 0.4 g for EDL (0.5 g for soleus). Others report resting tensions near 1.0 g (*see* Chapter 19 by Sperringer and Grange in this volume [8]); this should be optimized in each individual lab set up.
8. The pulse duration should be optimized for each set up as not all systems will be optimal at 0.5-ms pulses. Alternatively, a series of twitches are used to optimize contraction parameters (*see* Sperringer and Grange [8]).
9. 25 °C is the most common temperature used because (a) it facilitates lowered muscle metabolism ensuring adequate O_2 diffusion and ATP production keeping muscle viable [6] and (b) extent of injury to muscle in terms of force loss is highly temperature sensitive [17].
10. Our software is programmed to split the total length change around L_o with the goal to keep the length excursion within anatomical range, whereas beginning at L_o and lengthening a full 20 % may reach physiological and anatomical in vivo range for a muscle.

11. In vitro eccentric contractions by other groups are commonly started with the muscle contracting isometrically and then a length change is imposed during the final ~200 ms of the stimulation; eccentric force generation and subsequent force loss are similar for two approaches (e.g., [18]).
12. We use a length change of 10 % and a rate of change of 0.75 L_o/s for *mdx* EDL muscle (as opposed to 20% and 1.5 L_o/s to induce injury to wild type muscle).

In Vivo

13. There are alternatives to isoflurane for anesthesia during in vivo testing, and attention should always be paid to the effect of depth and duration of anesthesia on muscle strength [19].
14. Maintaining the body temperature at 37 °C is critical for the eccentric contraction induced injury. If body temperature is not regulated, the injury will not be as severe.
15. For dorsiflexion, the tibialis anterior muscle contributes approximately 89 % of the torque [20].
16. For plantarflexion, the gastrocnemius muscle makes up ~78 % of the cross-sectional area of the plantar-flexors [21] but contributions to torque have not been thoroughly assessed.
17. Alternative models of eccentric contraction-induced injury have been reported, e.g., stimulating the sciatic nerve before it branches out to the common peroneal and tibial nerves in the absence of a dual-mode lever system. Due to the greater mass of the collective plantar-flexors relative to the dorsi-flexors, a co-contraction would result in a concentric contraction by the plantar-flexors and an eccentric contraction by the dorsi-flexors. These co-contractions have previously been used to injure the dorsi-flexors [22]. This is not a precise, quantifiable model for eccentric contraction-induced injury as initial torque or the subsequent loss of torque generating capacity is not measured or controlled.
18. The common cause of low torque values is improper electrode placement. Muscle contractions should be observed especially for the dorsi-flexors to ensure that the antagonistic plantar-flexors are not also being stimulated.
19. The common cause of extremely high torque values is improper system calibration.
20. The 40° rotation during the eccentric contraction protocol is the excursion limit for the Aurora 300C-LR motor, but the company can accommodate for investigators interested in a larger excursion limit (e.g., 90° rotation). With a greater excursion, muscle lengthening will be greater and fewer eccentric contractions are required to elicit maximal strength loss (*see* Iyer et al., Chapter 21 in this volume [10], as well as [14]).

Acknowledgements

This work was supported by the NIH National Institute on Aging (R01 AG031743).

References

1. Friden J, Lieber RL (1992) Structural and mechanical basis of exercise-induced muscle injury. Med Sci Sports Exerc 24(5):521–530
2. Warren GL, Ingalls CP, Lowe DA, Armstrong RB (2002) What mechanisms contribute to the strength loss that occurs during and in the recovery from skeletal muscle injury? J Orthop Sports Phys Ther 32(2):58–64. doi:10.2519/jospt.2002.32.2.58
3. Faulkner JA, Brooks SV, Opiteck JA (1993) Injury to skeletal muscle fibers during contractions: conditions of occurrence and prevention. Phys Ther 73(12):911–921
4. McCully KK, Faulkner JA (1985) Injury to skeletal muscle fibers of mice following lengthening contractions. J Appl Physiol 59(1): 119–126
5. Warren GL, Hayes DA, Lowe DA, Armstrong RB (1993) Mechanical factors in the initiation of eccentric contraction-induced injury in rat soleus muscle. J Physiol 464:457–475
6. Bonen A, Clark MG, Henriksen EJ (1994) Experimental approaches in muscle metabolism: hindlimb perfusion and isolated muscle incubations. Am J Physiol 266(1 Pt 1): E1–E16
7. Segal SS, Faulkner JA (1985) Temperature-dependent physiological stability of rat skeletal muscle in vitro. Am J Physiol 248(3 Pt 1): C265–C270
8. Sperringer JE, Grange RW (2016) In vitro assays to determine skeletal muscle physiological function. In: Kyba M (ed)
9. Ashton-Miller JA, He Y, Kadhiresan VA, McCubbrey DA, Faulkner JA (1992) An apparatus to measure in vivo biomechanical behavior of dorsi- and plantarflexors of mouse ankle. J Appl Physiol 72(3):1205–1211
10. Iyer SR, Valencia AP, Hernandez-Ochoa EO, Lovering RM (2016) In vivo assessment of muscle contractility in animal studies. In: Kyba M (ed)
11. Burkholder TJ, Fingado B, Baron S, Lieber RL (1994) Relationship between muscle fiber types and sizes and muscle architectural properties in the mouse hindlimb. J Morphol 221(2): 177–190. doi:10.1002/jmor.1052210207
12. Brooks SV, Faulkner JA (1988) Contractile properties of skeletal muscles from young, adult and aged mice. J Physiol 404:71–82
13. McCully KK, Faulkner JA (1986) Characteristics of lengthening contractions associated with injury to skeletal muscle fibers. J Appl Physiol 61(1):293–299
14. Lovering RM, Roche JA, Bloch RJ, De Deyne PG (2007) Recovery of function in skeletal muscle following 2 different contraction-induced injuries. Arch Phys Med Rehabil 88(5):617–625. doi:10.1016/j.apmr.2007.02.010
15. Belanto JJ, Mader TL, Eckhoff MD, Strandjord DM, Banks GB, Gardner MK, Lowe DA, Ervasti JM (2014) Microtubule binding distinguishes dystrophin from utrophin. Proc Natl Acad Sci U S A 111(15):5723–5728. doi:10.1073/pnas.1323842111
16. Warren GL, Hayes DA, Lowe DA, Williams JH, Armstrong RB (1994) Eccentric contraction-induced injury in normal and hindlimb-suspended mouse soleus and EDL muscles. J Appl Physiol 77(3):1421–1430
17. Warren GL, Ingalls CP, Armstrong RB (2002) Temperature dependency of force loss and Ca(2+) homeostasis in mouse EDL muscle after eccentric contractions. Am J Physiol Regul Integr Comp Physiol 282(4):R1122–R1132. doi:10.1152/ajpregu.00671.2001
18. Petrof BJ, Shrager JB, Stedman HH, Kelly AM, Sweeney HL (1993) Dystrophin protects the sarcolemma from stresses developed during muscle contraction. Proc Natl Acad Sci U S A 90(8):3710–3714
19. Ingalls CP, Warren GL, Lowe DA, Boorstein DB, Armstrong RB (1996) Differential effects of anesthetics on in vivo skeletal muscle contractile function in the mouse. J Appl Physiol 80(1):332–340
20. Warren GL, Ingalls CP, Shah SJ, Armstrong RB (1999) Uncoupling of in vivo torque production from EMG in mouse muscles injured by eccentric contractions. J Physiol 515(Pt 2): 609–619
21. Armstrong RB, Taylor CR (1982) Relationship between muscle force and muscle area showing glycogen loss during locomotion. J Exp Biol 97:411–420
22. Baar K, Esser K (1999) Phosphorylation of p70(S6k) correlates with increased skeletal muscle mass following resistance exercise. Am J Physiol 276(1 Pt 1):C120–C127

Chapter 2

Volumetric Muscle Loss

Beth E. Pollot and Benjamin T. Corona

Abstract

Volumetric muscle loss (VML) injury is prevalent in severe extremity trauma and is an emerging focus area among orthopedic and regenerative medicine fields. VML injuries are the result of an abrupt, frank loss of tissue and therefore of different etiology from other standard rodent injury models to include eccentric contraction, ischemia reperfusion, crush, and freeze injury. The current focus of many VML-related research efforts is to regenerate the lost muscle tissue and thereby improve muscle strength. Herein, we describe a VML model in the anterior compartment of the hindlimb that is permissible to repeated neuromuscular strength assessments and is validated in mouse, rat, and pig.

Key words Musculoskeletal trauma, Skeletal muscle injury, Orthopedic, Regenerative medicine, Force, Torque, Mouse, Rat, Pig

1 Introduction

Severe orthopedic limb injuries often present a volumetric loss of skeletal muscle that results in chronic muscle weakness, impaired limb function, and disability [1–3]. This type of injury may result from either the initial trauma or secondarily from debridement or evacuation of injured/contaminated musculature. However, only recently has volumetric muscle loss (VML) injury become a musculoskeletal trauma research emphasis, primarily in response to the high prevalence of VML injuries among casualties in recent wars [1, 4]. Given the nature of the injury, a frank loss of muscle tissue that outstrips the regenerative capacity of the remaining musculature, the primary emphasis of ongoing research efforts has focused on the development of therapies to regenerate de novo muscle tissue in order to restore muscle strength [2, 5–14].

Animal models of VML are needed to determine the pathophysiology of the remaining traumatized musculature and to establish functional efficacy of developing therapies for clinical translation. A critical aspect of modeling VML is the recapitulation of persistent strength deficits [3], which distinguishes VML from the majority of established rodent injury models of different etiology

Michael Kyba (ed.), *Skeletal Muscle Regeneration in the Mouse: Methods and Protocols*, Methods in Molecular Biology, vol. 1460, DOI 10.1007/978-1-4939-3810-0_2, © Springer Science+Business Media New York 2016

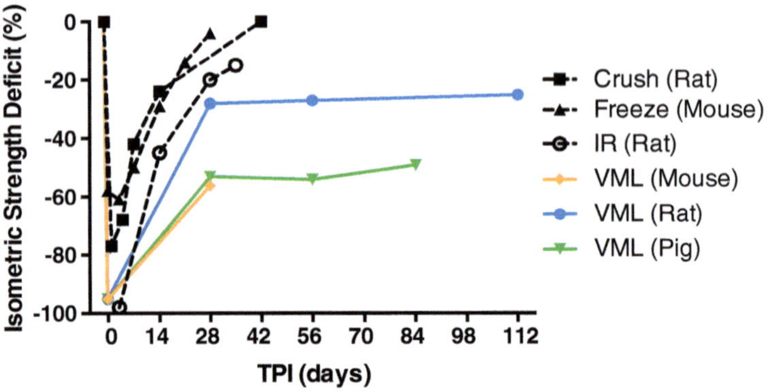

Fig. 1 Muscle function response to injuries of different etiology. Estimates of anterior crural muscle isometric strength deficits were plotted as a function of time post-injury (TPI). Strength deficit estimates were derived from the literature (Crush [15]; Freeze [17]; IR [16]; VML Rat [2, 7]) and unpublished data collected in our laboratory (VML mouse and pig). Unlike the majority of injury models, VML presents chronic strength deficits across species

(Fig. 1). For example, freeze, crush, and ischemia reperfusion injuries present a significant loss of strength acutely in rodents that can recover over the ensuing 4–6 weeks [15–17]. Additionally, while mechanistic understanding of the pathophysiology of VML may be elucidated in mouse and rat models, the translation of regenerative therapies for many clinical indications of VML requires progression from rodent to more clinically relevant large animal VML models. To facilitate the various efforts ongoing in the field, we have developed an interspecies (mouse, rat, and pig) sterile VML model in the anterior compartment of the lower hindlimb that is permissible to repeated in vivo neuromuscular strength assessments and current repair strategies.

2 Materials

All instruments should be sterile before use.

2.1 Rodent Procedures (Mouse and Rat)

1. Nose cone on a Bain circuit hooked to the rodent gas anesthesia machine.
2. Pre-surgical analgesic (*see* **Note 1**).
3. Electric hair clipper.
4. 10% Povidone iodine topical solution (Betadine solution).
5. 70% isopropyl alcohol.
6. Sterile instrument tray.
7. Surgical drapes (*see* **Note 2**).

8. Two towel clamps (*see* **Note 2**).
9. Cotton swabs (*see* **Note 3**).
10. 2 × 2 gauze (*see* **Note 3**).
11. 10 scalpel blade.
12. 6-0 Vicryl.
13. 6-0 Prolene.
14. Disposable biopsy punch (2 mm for mouse; 6 mm for rat).
15. Skin stapler.
16. Skin marker and ruler.
17. Cautery (*see* **Note 3**).
18. Adson tissue forceps, 1 × 2 teeth and suture platform heavier, 1.33 mm tip, 4.75 in.
19. Regular forceps, full curved, serrated 1 mm tip, 4 in.
20. Halstead mosquito hemostat forceps, curved regular, 5 in.
21. Halstead mosquito hemostat forceps, curved delicate, 5 in.
22. Miltex Stevens tenotomy scissors, straight sharp, 4.5 in.
23. Miltex Stevens tenotomy scissors, straight sharp, 4.125 in.
24. Olsen-Webster need holder and scissors, serrated delicate jaw, 4.75 in.
25. VWR spatula spoon, 6.5 in.

2.2 Pig Procedures

1. Presurgical medications as per veterinary guidance (*see* **Note 4**).
2. 10% Povidone iodine topical solution (Betadine solution).
3. 70% isopropyl alcohol.
4. Two Army-Navy retractors.
5. Two Schnidt tonsil clamps, slight curve, 7.75 in.
6. Four Mayo-Hegar needle drivers, straight, 6 in.
7. Two Mayo scissors, straight, 2 in.
8. Four Metzenbaum scissors, straight, 7 in.
9. Two Adson tissue forceps, 1 × 2 teeth and suture platform heavier, 1.33 mm tip, 4.75 in.
10. Two Debakey tissue forceps, 6 in.
11. 3-0 Vicryl.
12. 3-0 Prolene.
13. 3-0 Polydioxanone suture (PDS).
14. 10 scalpel blade.
15. 15 scalpel blade.

2.3 In Vivo Muscle Strength Assessments

1. Needle electrodes (*see* **Note 5**).
2. Electrical stimulator (*see* **Note 6**).
3. Torque measurement system (*see* **Note 7**).
4. Low noise coaxial cables.
5. USB multifunction data acquisition module (*see* **Note 8**).
6. Computer.
7. Data acquisition software (*see* **Note 9**).

3 Methods

Carry out all procedures in a sterile environment.

3.1 Rodent Surgical Creation of VML

1. Animals have food and water provided ad libitum, and unrestricted cage activity, prior to and after all procedures.
2. Once it is time for the procedure, the animal is anesthetized and the desired pain medication administered (*see* **Note 10**).
3. Remove hair from desired surgical limbs in the area of interest (*see* **Note 11**).
4. Clean the hairless, surgical area with isopropyl alcohol, Betadine and then isopropyl alcohol again to help reduce the potential of skin infection.
5. Place the animal in a surgical, sterile environment to begin the procedure.
6. Check the depth of anesthesia through a toe pinch just prior to starting the procedure.
7. Make a longitudinal incision along the lateral aspect of anterior compartment of the lower hind leg using a 10 blade scalpel (*see* **Note 12**).
8. Using blunt dissection, carefully separate the skin from the fascia overlying the tibialis anterior (TA) muscle (*see* **Note 13**).
9. Cut into the fascia by using the forceps to carefully lift a small portion in the middle third of the tibia and snipping into it with blunt tip scissors (*see* **Note 14**).
10. Separate the fascia from the muscle using blunt dissection to completely uncover the anterior surface of the TA.
11. Isolate the TA muscle from the adjacent extensor digitorum longus muscle (EDL) through blunt dissection by locating the demarcation on the lateral side of the leg and dissecting medially toward the tibia (*see* **Note 15**).
12. Insert a metal spatula between the TA and the underlying tibia and muscles to create a plane for biopsy punch (*see* **Note 16**).

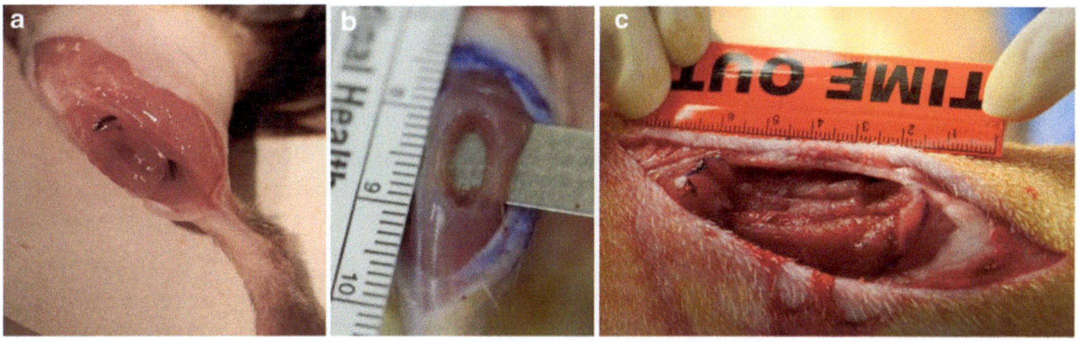

Fig. 2 Illustration of an interspecies lower hindlimb anterior compartment VML model. Pictures are presented of the surgical creation of VML in (**a**) mouse and (**b**) rat TA muscle, and the (**c**) pig PT muscle. Each injury model invokes a chronic loss of muscle tissue and strength and permits in vivo neuromuscular functional testing

13. Using a disposable biopsy punch in the middle third of the TA muscle, create a full-thickness defect by inserting the punch until it hits the spatula (*see* **Note 17**; Fig. 2).
14. Weigh the biopsied tissue.
15. Mark the defect with simple interrupted sutures of Prolene at the proximal, distal, medial, and lateral sides of the defect (*see* **Note 18**).
16. Close the injury by suturing the fascia with simple interrupted sutures (Vicryl) (*see* **Note 19**).
17. Close the skin incision with skin staples or simple interrupted sutures (Prolene) (*see* **Note 20**).
18. Allow animal to recover from anesthesia on a heated pad and then place back into housing environment with fresh bedding.
19. Following the procedure, check the animal at least twice daily for weight loss and their ability to ambulate for the first 72 h.

3.2 Pig Surgical Creation of VML

1. Fast pig overnight prior to surgery beginning the following morning (*see* **Note 21**).
2. Coral the pig into a transport cage and record body weight.
3. Inject glycopyrrolate (0.01 mg/kg, IM) as needed to reduce saliva secretion and block vagally mediated bradycardia during the surgical procedure.
4. Induce the pig with an IM injection of tiletamine–zolazepam (Telazol, 4–6 mg/kg) and initially maintain anesthesia with 5 % isoflurane in oxygen via a facemask.
5. Intubate the pig with an endotracheal tube and place on an automatic ventilator with the initial tidal volume at 10 ml/kg, peak pressure at 20 cm H_2O and respiration rate at 8–12

breaths per minute. The ventilator setting will be adjusted to maintain an end tidal PCO_2 of 40 ± 5 mmHg. Maintain anesthesia throughout the remainder of the procedure with 1–3% isoflurane in 30–37% oxygen.

6. Provide pre-surgical medications Buprenorphine sustained release (SR) (0.1–0.2 mg/kg; IM), Rimadyl (2.0–4.4 mg/kg; IM), and Excede (5.0 mg/kg; IM) for pain, swelling, and potential infection.
7. Insert ear vein and Foley catheters for fluid delivery and urine collection.
8. Shave and wash both hindlimbs in the pre-surgical area.
9. Transport intubated pig to surgical suite and place on operating table in a supine position.
10. Prepare the lower limbs for surgery using Betadine solution and sterile drapes. Wrap the hooves in sterile flexible cohesive bandage to allow for manipulation (*see* **Note 22**).
11. Create a longitudinal incision along lateral aspect of the anterior compartment using a scalpel (10 blade scalpel; *see* **Note 23**).
12. Separate the skin from the fascia both medially and laterally using blunt dissection with tonsil clamps and Metzenbaum scissors.
13. Separate and bluntly dissect the fascia layers from the underlying anterior muscle unit using Metzenbaum scissors.
14. Reflect the skin and fascia from the anterior surface of the anterior crural muscles using Army-Navy retractors.
15. Visualize the three primary muscles of the anterior compartment. From the landmark where the lateral and medial synergists converge over the peroneus tertius (primary muscle analogous to tibialis anterior muscle, PT), measure 1.5 cm distally and mark this point on the PT using a skin pen. From this mark, measure 3 cm distally and again mark the PT—these two marks will comprise proximal and distal boundaries of the VML defect. Within the proximal and distal boundaries, the lateral and medial boundaries are then marked 3 cm apart and centered to the long axis of the PT (*see* **Note 24**).
16. Using a scalpel (15 blade), surgically excise muscle tissue from the demarcated region within the middle of the PT muscle (Fig. 2) and weigh the excised tissue (*see* **Note 25**).
17. Place Prolene markers around the edges of excised area to indicate the original wound (*see* **Note 26**).
18. Close the fascia with simple interrupted sutures of Vicryl.
19. Close the skin incision with a monofilament subcuticular continuous suture (PDS).
20. If desired, repeat steps 10–20 for contralateral limb.

21. Place Bacitracin zinc ointment on the incision site and cover with a non-adherent pad (6 × 3 in.) followed by Ioban wrapping and then Elastikon elastic tape.
22. Following the procedure, check the animal at least twice daily for the first 72 h to assess pain, ambulation, and wound coverage (*see* **Note 27**).

3.3 In Vivo Muscle Strength Testing (Mice, Rats, and Pigs)

1. While under anesthesia, place the foot of the animal on the foot-plate of the force transducer that is appropriate for each species (*see* **Note 28**).
2. Stabilize the animal's knee and ankle at right angles (*see* **Note 29**).
3. Place percutaneous needle electrodes on the lateral side of the hind limb so they are stimulating the peroneal nerve (*see* **Note 30**).
4. Determine the maximal voltage and stimulation frequency that elicits peak isometric torque (*see* **Note 31**; Figs. 3 and 4).
5. Measure maximal isometric tetanic torque as a function of stimulation frequency and/or joint angle (*see* **Note 32**).

4 Notes

1. We use Buprenorphine SR (0.15–1.0 mg/kg) delivered subcutaneously.
2. These are to help isolate the hind limb of the animal and keep sterility.
3. Gauze and cotton swabs can both be used to apply light pressure to help control bleeding. If the fascia is severely bleeding, then cautery can be used to control bleeding. Do not use cautery on muscles.
4. We recommend conferring with your veterinarian as per experience. Our current medications are listed within the pig VML methods for guidance.
5. For rodents, our lab group prefers platinum subdermal needle electrodes from Grass Technologies, product number F-E2. For pigs, our lab group prefers disposable monopolar EMG needle electrodes that are 50 mm × 26 G from The Electrode Store.
6. For rodents, the electric stimulator that our lab uses is the A-M Systems Isolated Pulse Stimulator, Model 2100. For pigs, the Grass Technologies S88 Dual Output Square Pulse Stimulator is used.
7. We use Aurora Scientific torque measurement systems. Specifications for different systems are denoted within the methods.

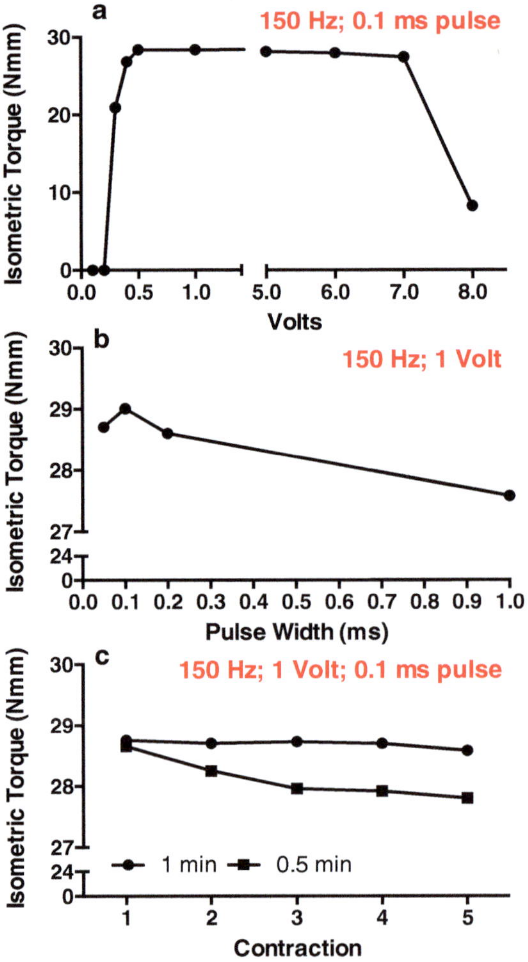

Fig. 3 Neural stimulation parameters for in vivo anterior crural muscle maximal isometric tetanic torque. Stimulation parameters and data were collected using rats, but are equivalent in mice. The common peroneal nerve was stimulated using percutaneous needle electrodes and all contractions were elicited by delivering electrical pulses at 150 Hz for a train duration of 400 ms. Core body temperature was maintained between 36 and 37 °C. Isometric torque was measured as a function of (**a**) stimulation voltage, (**b**) pulse width, and (**c**) contraction number using 0.5 and 1 min rest-intervals

8. The data acquisition module our lab group uses for both the rodents and the pigs is the National Instruments NI USB-6221.
9. We use MI-RAT, which is a custom-made data acquisition and analysis software program developed at the U.S. Army Institute of Surgical Research for the purpose of in vivo force measurement testing. It was created under the guidance of Thomas J. Walters, Ph.D. using LabVIEW software and allows for automated control of all measurement aspects. Other software packages are available from commercial vendors.

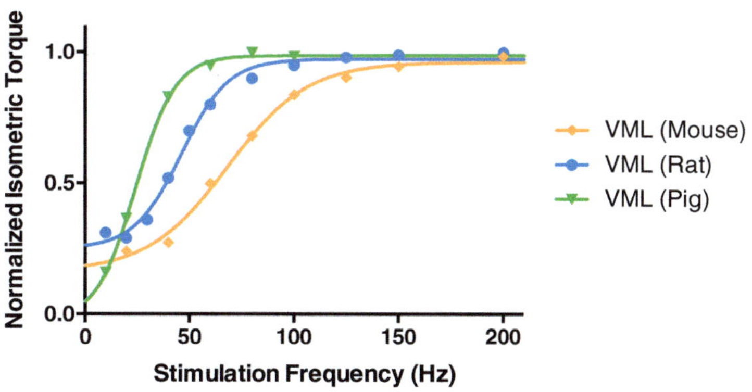

Fig. 4 Anterior crural muscle in vivo isometric torque as function of stimulation frequency across species. Isometric torque was elicited in anesthetized animals (isoflurane) by stimulating the peroneal nerve (0.1 ms pulse width) using needle electrodes in all species. Stimulation voltage was optimized for each experiment and train durations of 200, 400, and 800 ms were used for mouse, rat, and pigs, respectively. The core body temperature for each experiment was approximately 37 °C

10. The depth of anesthesia will be monitored throughout the surgery by noting a lack of response to a toe pinch test and slow breathing rate and depth. Foot withdrawal when pinched or blinking when the eyelashes are touched may be indicative of an inadequate depth of anesthesia for surgery.
11. For rodents, remove hair from the hip to ankle around the entire circumference of the leg.
12. The incision should extend from the knee to the ankle. A slightly curved incision helps prevent reopenings of the wound after surgery.
13. It would be helpful to then use a delicate hemostat to hold the skin separated on the medial size as to better expose the TA throughout the procedure. Curved scissors are useful for performing the blunt dissection, though it can also be done with straight edged scissors.
14. Typically 90° curved forceps are used for this procedure. Curved scissors may also be used to cut into the fascia.
15. Dissect straight under the TA until the tip of the scissors can feel the tibia. Then bluntly dissect slowly upward along with tibia until the TA is completely separated on the medial side of the TA. The TA should be dissociated from the tibia and underlying muscles when completed.
16. The spatula should be at least 4 mm in width for the mice and 8 mm in width for the rats. A spatula with the smallest depth possible is ideal as to provide a complete surface for the biopsy punch while minimizing damage to the underlying muscles.

To position the spatula in place, insert a regular hemostat from the lateral side between the TA and underlying muscles and pull the spatula through from the medial side to the lateral side.

17. A 2 and 6 mm biopsy punch will remove approximately 20 % of the TA muscle in mice and rats, respectively, which leads to an ~50 % and 30 % functional deficit in each respective species 4–8 weeks post-injury (Fig. 1). Any bleeding can typically be stopped with light pressure.
18. This will help indicate the original wound after euthanasia. Optionally, at this time you may treat the VML injury as per experimental design (e.g., [7, 18, 19]).
19. We did not perform this step in the mouse.
20. Be careful around the ankle. If the staples or sutures are too tight, the animal will open the wound to relieve pressure after surgery. A compressive wrap may be applied to the lower limb for 5–10 min after surgery.
21. Our lab currently uses female Yorkshire Cross pigs with a body weight at the beginning of the study of 40 kg.
22. Our lab currently employs a bilateral injury model, from which the pigs recover and exhibit normal cage activity.
23. The incision extends from the knee to the ankle. Mayo scissors are useful for extending the incision distally or proximally as needed to expose the anterior compartment.
24. A 3 × 3 cm region in the middle third of the PT muscle is targeted for VML.
25. Our lab currently removes 5 g of tissue from this region (~20 % of the muscle) which will result in a wound depth of ~1.5 cm and result in an ~40 % functional deficit out to 3 months post-injury compared to sham-operated controls (Fig. 1). Bleeding can be stopped with light pressure.
26. Tissue engineering and regenerative medicine therapies may be implanted to the defect site at this time.
27. Typically sustained release buprenorphine is administered every 72 h and Rimadyl is administered once daily through the first 7 days post-surgery. Typically bandage changes are required 3–4 days post-injury and incision closure occurs by 14 days post-injury.
28. Our lab uses dual-mode muscle levers 300C and 305C-LR with foot plate modifications purchased from Aurora Scientific for in vivo functional assessments in mice and rats respectively. For pig in vivo functional assessments, our lab uses a large animal testing system purchased from Aurora Scientific (890A). The servomotor input and force and displacement lever arm outputs are controlled and acquired, respectively, using a PC

equipped with a data acquisition board and custom designed Lab View based software program. The foot is secured to the foot plate using either tape (mouse and rat) or flexible cohesive bandage (pig). For similar in vivo methodological references, see [7, 20, 21]. It is important to monitor and maintain the animal's temperature at ~37 °C throughout the testing. Rodents should be placed on a heating pad or waterbed before testing begins to help maintain internal temperature throughout testing.

29. It is crucial that the knee is at a right angle for functional testing and the heel of the foot is firmly position to the bottom of the foot plate.

30. In particular for mouse assessments, the needle electrodes should be positioned and stabilized with alligator clips attached to a weighted stand. The percutaneous needle electrodes impart minimal damage to the surrounding musculature and do not damage the peroneal nerve unless pierced, allowing for repeated in vivo functional assessments as per study design. Alternatively, a nerve cuff may be implanted for repeated or terminal procedures [22].

31. Peak isometric torque occurs at different stimulation frequencies among species (Fig. 4). For voltage optimization, we use a stimulation frequency of 300, 150, and 100 Hz and train duration of 200, 400, and 800 ms for mice, rats, and pigs, respectively. For neural stimulation across species we use a pulse width of 0.1 ms. During this assessment, the torque waveform should only reflect dorsiflexion and have a smooth, rounded appearance. Perturbations of the waveform likely indicate either inadequate voltage, stimulation of the plantarflexor muscles, or anesthesia-related contractile depression [23] (Fig. 5).

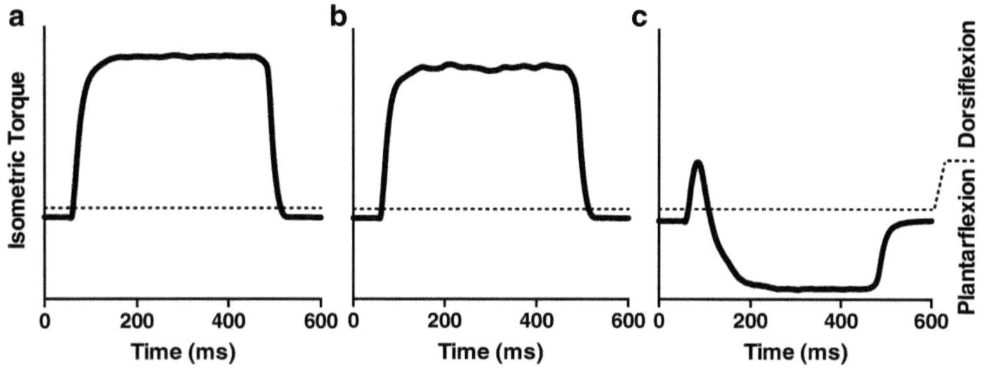

Fig. 5 Isometric torque waveforms measured with stimulating needle electrodes. (**a**) Optimal maximal tetanic isometric waveforms present a fluid square-wave appearance. (**b**) Inadequate electrical stimulation parameters can present jagged waveforms and (**c**) improper electrode placement can result in the recruitment of dorsiflexors (e.g., gastrocnemius), which will result in negative inflection of the waveform

32. Standard assessments of in vivo functional capacity following VML injury include measuring isometric torque as a function of stimulation frequency and/or joint angle. Stimulation frequency ranges will vary per species (Fig. 4) and should be carefully selected. The lever arm systems we use for mouse and rat testing is limited to 40° range of motion and does not capture the full anatomical range of motion in the rodent. If desired, for terminal experiments the TA and PT muscle may be isolated for functional experiments by tenotomizing synergist muscles [7].

Acknowledgements

All unpublished animal experiments were conducted in compliance with the Animal Welfare Act, the implementing Animal Welfare Regulations, and the principles of the Guide for the Care and Use of Laboratory Animals. This work was supported by Combat Casualty Care and Clinical and Rehabilitative Medicine Research Programs.

Disclaimer

The opinions or assertions contained herein are the private views of the author and are not be construed as official or as reflecting the views of the Department of the Army or the Department of Defense.

References

1. Corona BT, Rivera JC, Owens JC, Wenke JC, Rathbone CR (2015) Volumetric muscle loss leads to permanent disability following extremity trauma. J Rehabil Res Dev 52(7):785–792
2. Garg K et al (2015) Volumetric muscle loss: persistent functional deficits beyond frank loss of tissue. J Orthop Res 33(1):40–46
3. Grogan BF, Hsu JR (2011) Volumetric muscle loss. J Am Acad Orthop Surg 19(Suppl 1):S35–S37
4. Owens BD, Kragh JF Jr, Macaitis J, Svoboda SJ, Wenke JC (2007) Characterization of extremity wounds in Operation Iraqi Freedom and Operation Enduring Freedom. J Orthop Trauma 21(4):254–257
5. Aurora A, Roe JL, Corona BT, Walters TJ (2015) An acellular biologic scaffold does not regenerate appreciable de novo muscle tissue in rat models of volumetric muscle loss injury. Biomaterials 67:393–407
6. Chen XK, Walters TJ (2013) Muscle-derived decellularised extracellular matrix improves functional recovery in a rat latissimus dorsi muscle defect model. J Plast Reconstr Aesthet Surg 66(12):1750–1758
7. Corona BT et al (2013) Autologous minced muscle grafts: a tissue engineering therapy for the volumetric loss of skeletal muscle. Am J Physiol Cell Physiol 305(7):C761–C775
8. Corona BT et al (2012) Further development of a tissue engineered muscle repair construct in vitro for enhanced functional recovery following implantation in vivo in a murine model of volumetric muscle loss injury. Tissue Eng Part A 18(11–12):1213–1228
9. De Coppi P et al (2006) Myoblast-acellular skeletal muscle matrix constructs guarantee a long-term repair of experimental full-thickness abdominal wall defects. Tissue Eng 12(7):1929–1936

10. Garg K, Corona BT, Walters TJ (2014) Losartan administration reduces fibrosis but hinders functional recovery after volumetric muscle loss injury. J Appl Physiol (1985) 117(10):1120–1131
11. Li MT, Willett NJ, Uhrig BA, Guldberg RE, Warren GL (2014) Functional analysis of limb recovery following autograft treatment of volumetric muscle loss in the quadriceps femoris. J Biomech 47(9):2013–2021
12. Machingal MA et al (2011) A tissue-engineered muscle repair construct for functional restoration of an irrecoverable muscle injury in a murine model. Tissue Eng Part A 17(17–18): 2291–2303
13. Merritt EK et al (2010) Functional assessment of skeletal muscle regeneration utilizing homologous extracellular matrix as scaffolding. Tissue Eng Part A 16(4):1395–1405
14. Willett NJ et al (2013) Attenuated human bone morphogenetic protein-2-mediated bone regeneration in a rat model of composite bone and muscle injury. Tissue Eng Part C Methods 19(4):316–325
15. Stratos I, Graff J, Rotter R, Mittlmeier T, Vollmar B (2010) Open blunt crush injury of different severity determines nature and extent of local tissue regeneration and repair. J Orthop Res 28(7):950–957
16. Walters TJ, Garg K, Corona BT (2015) Activity attenuates skeletal muscle fiber damage after ischemia and reperfusion. Muscle Nerve 52(4):640–648
17. Warren GL et al (2005) Chemokine receptor CCR2 involvement in skeletal muscle regeneration. FASEB J 19(3):413–415
18. Corona BT et al (2013) The promotion of a functional fibrosis in skeletal muscle with volumetric muscle loss injury following the transplantation of muscle-ECM. Biomaterials 34(13):3324–3335
19. Ward CL, Ji L, Corona BT (2015) An autologous muscle tissue expansion approach for the treatment of volumetric muscle loss. Biores Open Access 4(1)
20. Corona BT et al (2010) Junctophilin damage contributes to early strength deficits and EC coupling failure after eccentric contractions. Am J Physiol Cell Physiol 298(2): C365–C376
21. Kheirabadi BS et al (2014) Long-term effects of Combat Ready Clamp application to control junctional hemorrhage in swine. J Trauma Acute Care Surg 77(3 Suppl 2):S101–S108
22. Call JA, Eckhoff MD, Baltgalvis KA, Warren GL, Lowe DA (2011) Adaptive strength gains in dystrophic muscle exposed to repeated bouts of eccentric contraction. J Appl Physiol (1985) 111(6):1768–1777
23. Ingalls CP, Warren GL, Lowe DA, Boorstein DB, Armstrong RB (1996) Differential effects of anesthetics on in vivo skeletal muscle contractile function in the mouse. J Appl Physiol (1985) 80(1):332–340

Chapter 3

Freeze Injury of the Tibialis Anterior Muscle

Gengyun Le, Dawn A. Lowe, and Michael Kyba

Abstract

Freeze injury is physically induced by exposing skeletal muscle to an extremely cold probe, and results in a robust degenerative and inflammatory response. One unique aspect of freeze injury is that it destroys not only the muscle fiber cells, but also all of the mononuclear cells in the zone of injury. Repair of the muscle is accomplished by satellite cells from outside of the zone of injury, which must migrate in and which may interact with inflammatory cells, hence the length of time before apparent histological recovery of the most damaged zone is typically somewhat longer with freeze injury than with other physical or chemical methods of injury. In this chapter, we present a detailed protocol for the freeze injury of the tibialis anterior (TA) muscle in mouse.

Key words Freeze injury, Tibialis anterior muscle, Regeneration

1 Introduction

Traumatic skeletal muscle injuries, such as those induced by extreme cold, crushing, or toxins, result in profound pathological changes and loss of muscle function, which require greater than 2 weeks to recover [1–3]. The regenerative potential of skeletal muscle under such experimental conditions is classically measured by inducing a defined amount of damage to muscle, waiting for a defined period of time, then monitoring muscle repair either histologically or with functional measurements. Many cell types contribute to the regeneration and repair of skeletal muscle, including the initial inflammatory infiltrate [4] the resident mesenchymal cell population [5–7] and of course satellite cells, the stem cells for skeletal muscle [8–12]. Inflammatory cells may directly influence differentiation, as coculture of myogenic cells with macrophages enhances formation of differentiated myotubes [13, 14]. Pro-inflammatory cytokines such as TNF-α and the chemokine receptor CCR2 are certainly necessary for complete and rapid regeneration of injured skeletal muscle [15–18]. Recent work, however, suggests that the inflammatory response orchestrates a

Michael Kyba (ed.), *Skeletal Muscle Regeneration in the Mouse: Methods and Protocols*, Methods in Molecular Biology, vol. 1460, DOI 10.1007/978-1-4939-3810-0_3,

flux in the activity of the mesenchymal cell population, which regulates the proliferation and differentiation of the activated satellite cells [19, 20].

Many commonly used injury models focus destruction on the contractile functional unit of skeletal muscle, the myofiber. These include agents such as BaCl and cardiotoxin that cause hypercontraction and death of fibers, physiological "overuse" injuries driven by electrically forced contraction, and physical crush injury, which disrupts the integrity of extremely large cells such as fibers, and disrupts the integrity of tissue components such as the vascular bed, but leaves mononuclear cells, for the most part, alive. An interesting counterpoint to these types of injury is the freeze injury, in which the rapid freeze–thaw kills all cells in the zone of injury, thus the repair process requires migration in of both the nonmyogenic support population as well as the myogenic progenitor cell population.

Studies in which an entire rat extensor digitorum longus (EDL) muscle was frozen in situ have suggested that myogenic progenitors cannot migrate in from adjacent muscles, nor can they transit through the circulation [21]. Grafts of freeze-killed EDL muscle sutured into the space left by removal of the TA and EDL can be repopulated by third party myogenic progenitor cells (injected into the frozen donor muscle at the time of grafting), which then form new muscle distinguished from the host by glucose-6-phosphate isomerase polymorphism [22]. Interestingly, this experiment provided evidence in favor of the possibility of immigration of myogenic progenitor cells, in the form of hybrid isoenzymes detected in some recipients.

Profiling of freeze injury compared to contraction-induced injury demonstrated a much greater inflammatory response [23]. In the protocol described below, a large zone in the belly of the TA muscle is killed, but the ends of the muscle are spared. Thus myogenic progenitors from the spared zones of the same TA muscle being frozen are largely responsible for its regeneration. As an inflammation-predominant muscle injury, freeze injury is an ideal model for research related to inflammatory response or inflammatory cell-driven regeneration.

The freeze injury protocol described below is ideally performed by a team of 3: an individual to prepare the animal for the procedure, a freeze-injury surgeon, and a suturing surgeon. This is especially convenient when a large number of animals are to be injured.

2 Materials

1. Isoflurane units (3× if possible).
2. Anesthesia tubing and nose cone.
3. Polycarbonate square.

4. O_2 tanks (3×).
5. Recirculating heating pad.
6. Dissecting microscope.
7. Hot bead sterilizer.
8. Isoflurane.
9. Buprenorphine.
10. Kimwipes.
11. Saline.
12. Artificial tears.
13. Nair hair remover.
14. Cotton-tipped swab.
15. Betadine solution and scrub.
16. Cotton swabs.
17. 70% alcohol.
18. Fine surgical scissors.
19. Fine tip forceps (×2).
20. Suture tool.
21. Dry ice.

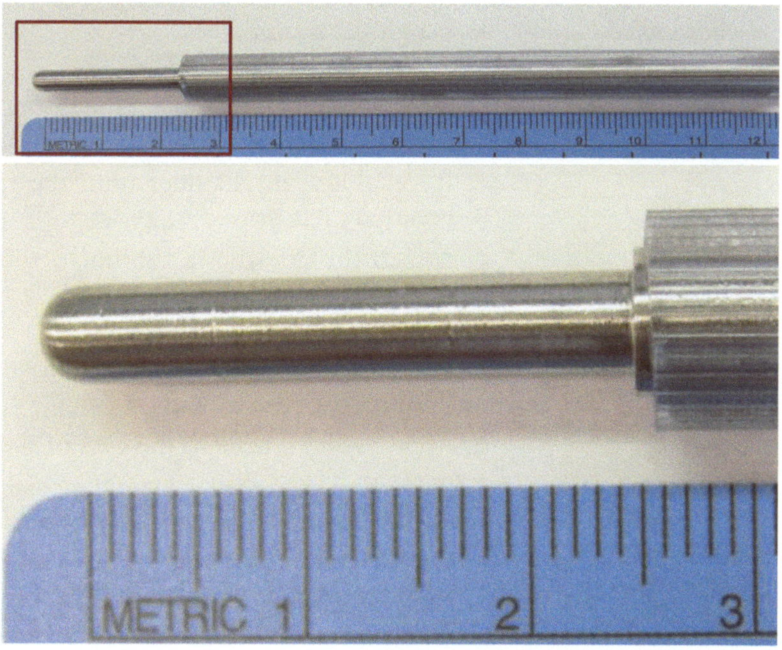

Fig. 1 Steel probe used to induce freeze injury. *Top*: A probe of length 25 mm is fixed to a 6-mm steel rod covered by autoclavable PVC (~14 mm outer diameter cylinder to the right). *Bottom*: Enlarged end of the freeze probe showing the shape of the tip

22. Steel freeze probe; for mouse TA muscle a 6-mm-diameter cylindrical rod works well (Fig. 1).
23. 6-0 silk sutures with needles.
24. Food pellets.
25. Small petri dish.
26. Hydrogen peroxide (H_2O_2).
27. Vetbond or superglue.
28. Distilled water.
29. Square steel platform for foot support.
30. Duct tape.
31. 0.5 mL or 1.0 mL syringe.
32. New Skin liquid bandage.
33. Metronidazole.
34. Surgical gloves.
35. Timer.

3 Methods

The procedures described below are for the mouse TA muscle, but can be adapted for other muscles, or for the rat.

3.1 Preparation

1. Weigh the mouse.
2. Attach tubing from isoflurane vaporizer to the anesthesia induction chamber and set the isoflurane concentration to 5 % and the O_2 gas flow rate to 1000 mL/min.
3. Place mouse in chamber until the mouse stops moving and breathing has slowed to 1–2 breaths per second.
4. Attach the anesthesia tubing to the nose cone and change isoflurane concentration to 2 % and the gas flow rate to 200 mL/min.
5. Place mouse in the supine position on the clean Kimwipes with its nose just inside the nose cone.
6. Apply artificial tears to the eyes of the mouse.
7. Shave hair from the lateral and medial aspects of the left/right/ both hindlimb(s). Shave up to the quadriceps about 3 mm above the knee and all the way down to the ankle.
8. Apply Nair to shaved areas with a cotton-tipped swab and leave on for ~2 min.
9. Remove Nair with Kimwipes and clean off the remaining Nair with distilled water.

10. Prepare the aseptic skin by applying Betadine in a circular pattern moving outward from the center of each shaved area three times, using a new, clean Q-tip with each new application.
11. Wipe off Betadine by applying cotton swab with 70% alcohol. Make sure each stroke of the cotton swab is with a new, clean swab. Try not to get the mouse's coat too wet with alcohol when cleaning.
12. If 3 stations are available, move the mouse to the surgical station. Make sure to clean nose cone and replace with clean Kimwipes between each mouse in this preparation station.

3.2 Injury

1. In surgical area, turn the gas flow rate to 100–200 mL/min and set the isoflurane concentration to 1–2%. Do this 2–3 min ahead of time; this is especially important for the first mouse of the day.
2. To injure the left TA muscle, place mouse in right lateral recumbency, secure the left hind foot, ideally by placing it on a 30 mm high block (*see* **Note 1**) and secure the proximal end of the tail with tape. If both hindlimbs are required, perform one side and then turn to the other side.
3. The surgeon should clean hands and lower arms with Betadine scrub thoroughly and put on surgical gloves to avoid contamination. During the injury, the surgeon is required to maintain an aseptic environment until end of process.
4. Make incision through the skin on the left hind limb to expose the TA muscle (Fig. 2a). This first incision is small (i.e., ~2 mm) and over the patellar tendon but perpendicular to the tendon. Use fine forceps and scissor to make the incision.

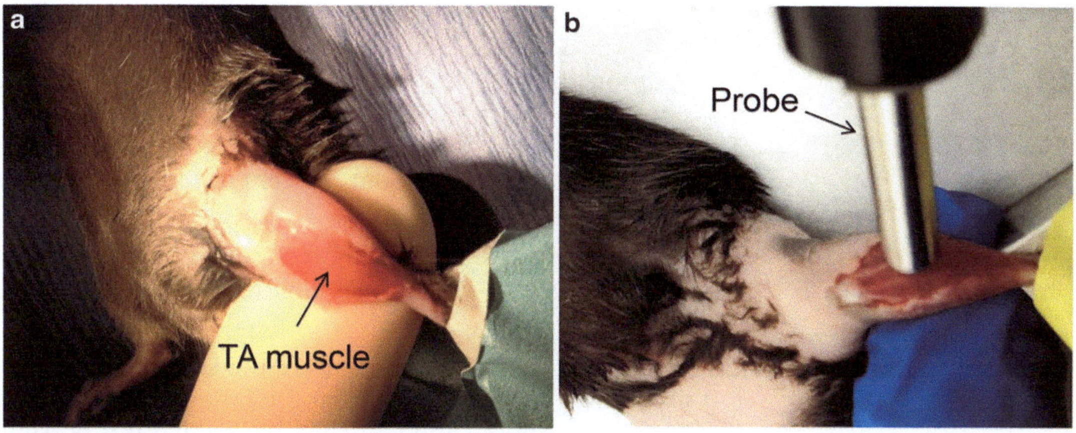

Fig. 2 Exposure of the TA muscle and positioning of the leg. *Left*: TA muscle is exposed by incision of the skin. *Right*: The probe, previously buried in dry ice for at least 30 min, is applied to the TA muscle without touching the skin. Note that the leg is positioned such that the surface of the TA muscle is horizontal so that when the probe is applied vertically the pressure is uniform

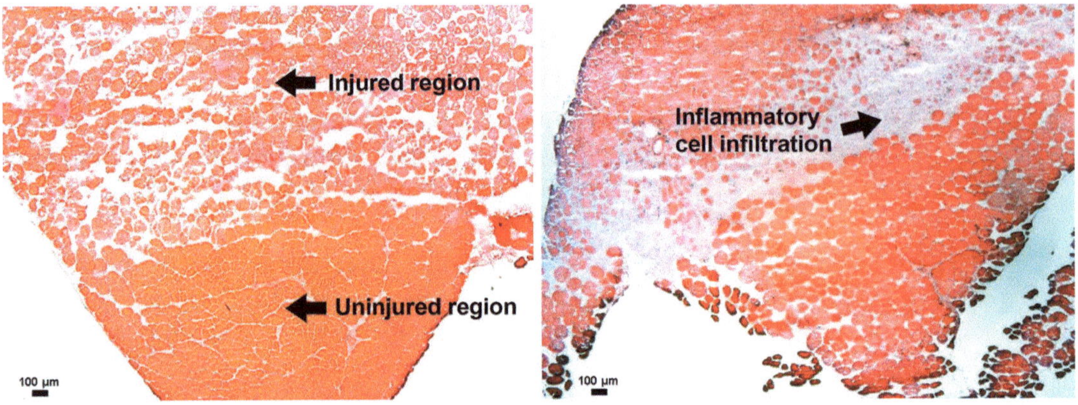

Fig. 3 Representative images of hematoxylin and eosin-stained cross sections of frozen mouse TA muscles. In both images the anterior surface of the muscle, that is, the freeze-injured portion, is at the top. *Left*: TA muscle 1 day post freeze injury. About 75 % of the muscle cross section is damaged. Compared to fibers in the uninjured region, fibers in the damaged region have lost their normal polygonal shape. There is a modest influx of inflammatory cells 1 day post-injury, mainly mononuclear cells, which is restricted to the damaged region adjacent to the deep uninjured region. *Right*: TA muscle 3 days post injury. At this time point, the level of inflammatory cell infiltration (indicated) has peaked. For additional images of freeze injured muscle, please see Ref. 23

5. Make a second incision continuing from the first incision but running perpendicular, i.e., over the lateral edge of the TA down to the ankle. Use fine scissors to make the incision while using fine tip forceps to hold the skin during the incision.
6. Use two pairs of fine tip forceps to loosen the skin from the underlying muscle.
7. Grab behind the knee with your left hand to support the left leg of the mouse.
8. Take the steel probe that has been precooled in dry ice (at least 30 min) and hold as if you were holding a pencil. Place the probe tip on the mid-belly of the TA (the fattest part of the muscle, Fig. 2b) and press firmly for exactly 10 s (*see* **Note 2**). The degree of injury from this freezing protocol is very large (Fig. 3), and muscle takes about 3 weeks to return to normal strength (Fig. 4).
9. Wait for ~1 min for the freeze-injured area to change from its frosted color back to the normal pink color, then apply 1–2 drops of saline.
10. Under a dissecting microscope: suture the skin incision using 6-0 silk suture. To start, ensure that the foot is taped down so the TA is fully exposed and stretched out so that it is possible to suture up the entire incision. Start suturing at the ankle. Make 4–5 throws per knot. Cut the suture as close to the knot as possible. Make sure there are no visible gaps in the skin.

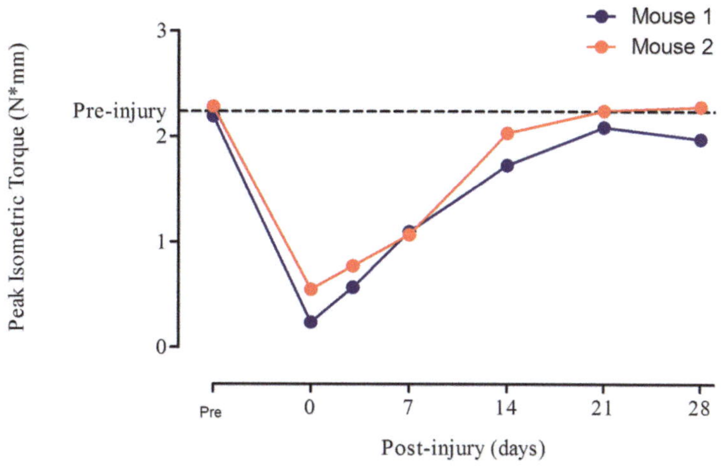

Fig. 4 Loss and recovery of muscle strength in two female mice after freeze injury to the left TA muscles. Maximal isometric torque of the left anterior crural muscles was measured immediately before and after injury was induced and again at 3, 7, 14, 21, and 28 days post-injury in the same two mice. Both mice had recovered to 90 % or more of the pre-injury level by day 21

11. Use a cotton tipped swab to clean the sutured incision with H_2O_2. Dry the area and place Vetbond or superglue on the sutured incision (*see* **Note 3**). Clean off excess glue with a cotton-tipped swab.
12. Wait until the Vetbond or superglue dries completely and then apply a mixture of New Skin and metronidazole by drawing a fine line to cover the sutures.
13. Sanitize tools in hot bead sterilizer and clean area with 70 % alcohol in between each mouse.
14. Give each mouse 0.1333 mg/kg subcutaneous buprenorphine injection. Place mouse in cage and place cage on a heating pad set to low (*see* **Note 4**).
15. Wait until the mouse is alert and moving around cage before returning the mouse to its housing room (*see* **Note 5**).
16. Soak food pellets in water for ~1 min and place four softened pellets in a small petri dish with additional water inside cage with the mouse. Make sure water bottle in the cage is positioned correctly (*see* **Note 6**).

3.3 Cleanup

1. Turn off isoflurane, heating pad, and O_2 tanks.
2. Soak all surgical instruments in warm, soapy water for 4–5 min.
3. Clean instruments with steel brush. Pat dry, place in autoclave bags and seal. Place autoclave indicator tape on bag and label with name and date. Autoclave for 45 min at 121 °C followed by a 10-min exhaust.

4. Soak anesthesia induction chamber and weighing container in warm soapy water for 4–5 min, rinse, dry and return to counter.
5. Clean recirculating heating pad, underlying surface, nose cone, and counter with 70 % alcohol.

4 Notes

1. This is important! The goal is for the lower leg to be parallel with the table so that when the probe is applied it is done so "flat" to produce an equal-force injury.
2. Set a timer. Keeping the probe in contact for too long can lead to freezing of the adjacent skin. This must be carefully avoided, to avoid infection and necrosis.
3. To avoid skin stiffness post-surgery, use as little glue as possible—control the flow of glue to minimize the amount of Vetbond or superglue on the incision site.
4. Take care not to set the heating pad too high as this is a common way to harm the mouse.
5. Mice can be group-housed after surgery.
6. Watch mice closely for the first 3-days post-injury (at least twice per day). In case signs of infection appear or the incision site reopens, take action immediately.

Acknowledgements

This work was supported by grants from the NIH (R01 AR055685 to MK and R01 AG031743 to DL) and Muscular Dystrophy Association (MDA351022). We would like to express our heartfelt gratitude to Dr. Gordon Warren for his significant contribution on development and validation of our freeze injury model.

References

1. Beiner JM, Jokl P, Cholewicki J, Panjabi MM (1999) The effect of anabolic steroids and corticosteroids on healing of muscle contusion injury. Am J Sports Med 27(1):2–9
2. Crisco JJ, Jokl P, Heinen GT, Connell MD, Panjabi MM (1994) A muscle contusion injury model. Biomechanics, physiology, and histology. Am J Sports Med 22(5):702–710
3. Pavlath GK et al (1998) Heterogeneity among muscle precursor cells in adult skeletal muscles with differing regenerative capacities. Dev Dyn 212(4):495–508
4. Tidball JG (1995) Inflammatory cell response to acute muscle injury. Med Sci Sports Exerc 27(7):1022–1032
5. Joe AW et al (2010) Muscle injury activates resident fibro/adipogenic progenitors that facilitate myogenesis. Nat Cell Biol 12(2): 153–163
6. Uezumi A, Fukada S, Yamamoto N, Takeda S, Tsuchida K (2010) Mesenchymal progenitors distinct from satellite cells contribute to ectopic fat cell formation in skeletal muscle. Nat Cell Biol 12(2):143–152

7. Murphy MM, Lawson JA, Mathew SJ, Hutcheson DA, Kardon G (2011) Satellite cells, connective tissue fibroblasts and their interactions are crucial for muscle regeneration. Development 138(17):3625–3637
8. Mauro A (1961) Satellite cell of skeletal muscle fibers. J Biophys Biochem Cytol 9: 493–495
9. Collins CA et al (2005) Stem cell function, self-renewal, and behavioral heterogeneity of cells from the adult muscle satellite cell niche. Cell 122(2):289–301
10. Schultz E, Gibson MC, Champion T (1978) Satellite cells are mitotically quiescent in mature mouse muscle: an EM and radioautographic study. J Exp Zool 206(3):451–456
11. Montarras D et al (2005) Direct isolation of satellite cells for skeletal muscle regeneration. Science 309(5743):2064–2067
12. Bosnakovski D et al (2008) Prospective isolation of skeletal muscle stem cells with a Pax7 reporter. Stem Cells 26(12):3194–3204
13. Cantini M et al (1994) Macrophages regulate proliferation and differentiation of satellite cells. Biochem Biophys Res Commun 202(3): 1688–1696
14. Cantini M, Carraro U (1995) Macrophage-released factor stimulates selectively myogenic cells in primary muscle culture. J Neuropathol Exp Neurol 54(1):121–128
15. Warren GL et al (2002) Physiological role of tumor necrosis factor alpha in traumatic muscle injury. FASEB J 16(12):1630–1632
16. Warren GL et al (2005) Chemokine receptor CCR2 involvement in skeletal muscle regeneration. FASEB J 19(3):413–415
17. Summan M et al (2003) Inflammatory mediators and skeletal muscle injury: a DNA microarray analysis. J Interferon Cytokine Res 23(5):237–245
18. Summan M et al (2006) Macrophages and skeletal muscle regeneration: a clodronate-containing liposome depletion study. Am J Physiol Regul Integr Comp Physiol 290(6):R1488–R1495
19. Heredia JE et al (2013) Type 2 innate signals stimulate fibro/adipogenic progenitors to facilitate muscle regeneration. Cell 153(2): 376–388
20. Lemos DR et al (2015) Nilotinib reduces muscle fibrosis in chronic muscle injury by promoting TNF-mediated apoptosis of fibro/adipogenic progenitors. Nat Med 21(7):786–794
21. Schultz E, Jaryszak DL, Gibson MC, Albright DJ (1986) Absence of exogenous satellite cell contribution to regeneration of frozen skeletal muscle. J Muscle Res Cell Motil 7(4):361–367
22. Morgan JE, Coulton GR, Partridge TA (1987) Muscle precursor cells invade and repopulate freeze-killed muscles. J Muscle Res Cell Motil 8(5):386–396
23. Warren GL et al (2007) Mechanisms of skeletal muscle injury and repair revealed by gene expression studies in mouse models. J Physiol 582(Pt 2):825–841

Chapter 4

Synergist Ablation as a Rodent Model to Study Satellite Cell Dynamics in Adult Skeletal Muscle

Tyler J. Kirby, John J. McCarthy, Charlotte A. Peterson, and Christopher S. Fry

Abstract

In adult skeletal muscles, satellite cells are the primary myogenic stem cells involved in myogenesis. Normally, they remain in a quiescent state until activated by a stimulus, after which they proliferate, differentiate, and fuse into an existing myofiber or form a de novo myofiber. To study satellite cell dynamics in adult murine models, most studies utilize regeneration models in which the muscle is severely damaged and requires the participation from satellite cells in order to repair. Here, we describe a model to study satellite cell behavior in muscle hypertrophy that is independent of muscle regeneration.

Synergist ablation surgery involves the surgical removal of the gastrocnemius and soleus muscles resulting in functional overload of the remaining plantaris muscle. This functional overload results in myofiber hypertrophy, as well as the activation, proliferation, and fusion of satellite cells into the myofibers. Within 2 weeks of functional overload, satellite cell content increases approximately 275 %, an increase that is accompanied with a ~60 % increase in the number of myonuclei. Therefore, this can be used as an alternative model to study satellite cell behavior in adulthood that is different from regeneration, and capable of revealing new satellite cell functions in regulating muscle adaptation.

Key words Satellite cells, Hypertrophy, Skeletal muscle, Functional overload, Synergist ablation

1 Introduction

Adult skeletal muscles have the remarkable ability to fully regenerate following extensive injury through the participation of resident stem cells, making skeletal muscles an ideal model system to study stem cell behavior in an adult tissue. Of the multiple stem cells in skeletal muscle, satellite cells are the primary myogenic stem cell that gives rise to mature myofibers. In order to study satellite cell dynamics in adult animals, the majority of studies employ a regenerative model, typically induced by injection of a myotoxin or through a damaging injury, such as a crush or freeze [1]. While these models clearly result in the activation, proliferation and differentiation of satellite cells, they also produce extensive damage to

Michael Kyba (ed.), *Skeletal Muscle Regeneration in the Mouse: Methods and Protocols*, Methods in Molecular Biology, vol. 1460, DOI 10.1007/978-1-4939-3810-0_4, © Springer Science+Business Media New York 2016

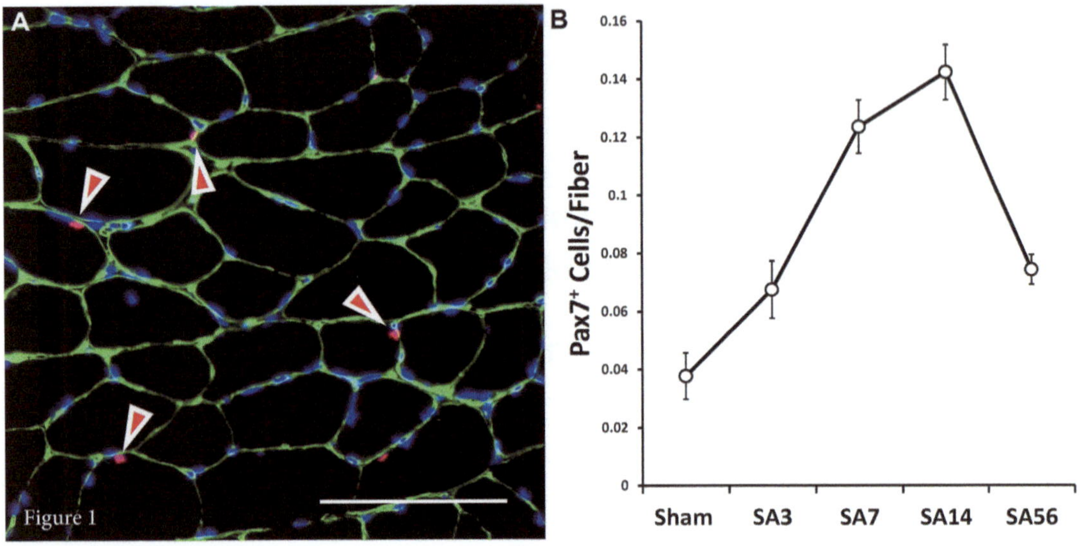

Fig. 1 Synergist ablation results in a rapid expansion of the satellite cell pool. (**a**) Satellite cells were identified immunohistochemically on mouse skeletal muscle cross sections using an antibody against the satellite cell-specific transcription factor, paired box 7 (*Pax7*). *White/red arrows* indicate Pax7+ cells. Myofibers can be visualized using an antibody against laminin. Scale bar—100 μm. (**b**) Satellite cells proliferate within 7 days of functional overload, further increasing in number by 14 days of overload, and then gradually returning to baseline by 56 days of overload

many cellular structures, in addition to the myofibers. Thus, the relevance to human muscle adaptation, for example in response to exercise training, may be limited.

Here, we present a surgical model which results in the induction of the same myogenic program in satellite cells, without requiring extensive damage to the myofibers [2, 3]. Synergist ablation involves the surgical removal of synergist muscles thereby placing functional overload on the remaining plantaris muscle. Using this model, we see rapid expansion of the satellite cell pool within 7 days of functional overload, which further increases by 14 days and starts to decline by 56 days (Fig. 1). Furthermore, this model induces a 60% increase in myonuclear number within 14 days of overload, which appears to be solely the result of satellite cells differentiating and then fusing to existing myofibers [3]. Therefore, this model can be used as an alternative method to study satellite cell behavior in adulthood that is distinct from regeneration, and thus, may involve different signaling events and/or molecular regulation.

2 Materials

2.1 Surgical Suite Setup (Fig. 2)

1. Fume hood.
2. Oxygen take.

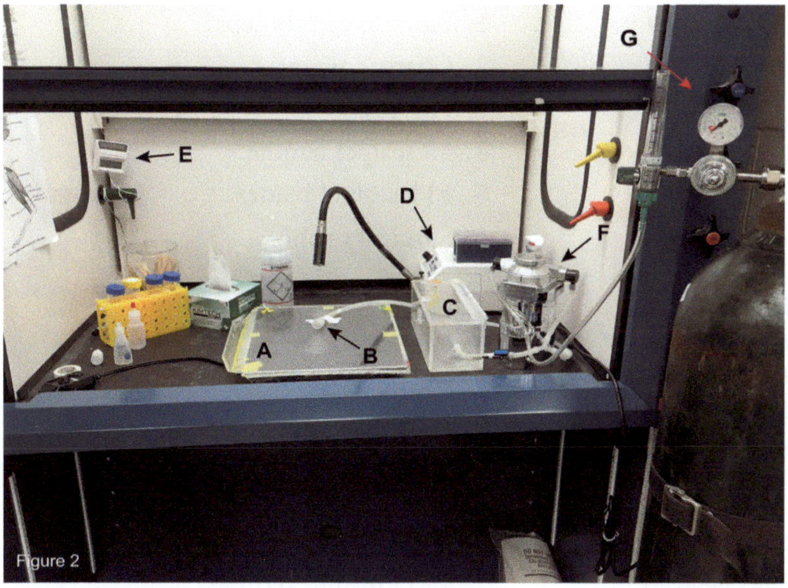

Fig. 2 Example of a surgery suite for performing rodent surgeries. (**a**) Heated Plexiglas surgical table; (**b**) Nose cone attached to isoflurane via tubing; (**c**) Closed induction chamber attached to isoflurane; (**d**) Light source; (**e**) Digital temperature gauge with probe located on surgical table; (**f**) Isoflurane vaporizer; (**g**) Oxygen tank with associated regulator

3. Isoflurane.
4. Isoflurane vaporizer machine.
5. Heated Plexiglas surgery pad.
6. Dissection microscope (optional).
7. Light source.
8. Tubing and plastic nose cone.
9. Directional control valves.
10. Digital temperature gauge with probe.

2.2 Animal Anesthetization, Analgesic and Restraint

1. Artificial tears.
2. 70 % ethanol.
3. Carprofen.
4. 27 G insulin syringe.
5. Labeling tape.

2.3 Surgical Site Preparation

1. Betadine surgical scrub.
2. #3 scalpel handle.
3. #11 scalpel blades.
4. 70 % ethanol.

2.4 Incision and Fascia Separation

1. Fine curved forceps.
2. Coarse forceps.
3. Fine scissors.
4. #3 scalpel handle.
5. #11 scalpel blades.
6. Surgical probe.

2.5 Separating and Cutting Away the Gastrocnemius/Soleus from the Plantaris

1. Fine curved forceps.
2. Surgical probe.
3. Fine scissors.
4. Cotton tip swabs.
5. 5-0 polypropylene surgical suture.

2.6 Contralateral Leg and Analgesic Administration

1. Carprofen.
2. 27 G insulin syringe.

2.7 Tissue Collection and Processing

1. Piece of scratch-resistant glass (10 in. by 10 in.).
2. RNAse removal solution (optional).

3 Methods

3.1 Surgical Suite Setup

1. Attach temperature probe to the top of the Plexiglas surgical table.
2. Turn on heating pad located on the underside of the Plexiglas table. We found that pads designed for animal aquarium use work well, since they are designed to be mounted underneath a glass surface. Temperature should be between 25 and 27 °C at the time of surgery.
3. Fill isoflurane vaporizer machine with isoflurane.
4. Turn oxygen tank on and set to a flow rate of 0.5–2 L of O_2 per minute.
5. Set flow to the closed chamber. Turn on isoflurane vaporizer machine to deliver 3–4 % isoflurane at a flow rate of 0.8–1.0 L/min.

3.2 Animal Anesthetization, Analgesic and Restraint

1. Using an insulin syringe, administer Carprofen (10 mg/kg) through an intraperitoneal (IP) injection 30 min prior to surgery.
2. Immediately prior to surgery, wipe down surgical table with 70 % ethanol.
3. Place animal in closed chamber connected to the isoflurane vaporizer machine.

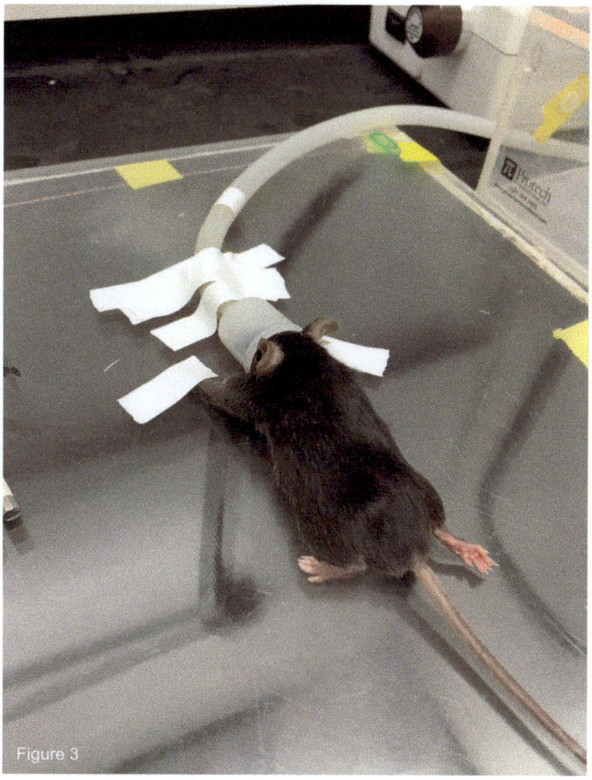

Fig. 3 Restraining the animal during surgery. Using standard laboratory labeling tape, tape the forepaws to make sure that the animals head remains in the nose cone for uninterrupted isoflurane administration

4. Once animal is sufficiently under anesthetic, quickly remove from box, apply artificial tears to the animals eyes to prevent their eyes from drying out, and immediately place face inside the nose cone.
5. Reduce isoflurane to 1–3 % and set flow to the nose cone.
6. Gently apply tape to the forepaws to avoid the animals head from dislodging from the nose cone during surgery (Fig. 3).

3.3 Surgical Site Preparation

1. Apply 70 % ethanol to the posterior side of both hindlimbs.
2. Using a #11 scalpel blade, gently shave hair away from the posterior side of both hindlimbs.
3. Secure the first hindlimb by placing tape across the underside of the hindpaw, placing the leg in full plantar flexion.
4. Apply Betadine in a circular motion from the inside to the outside of the site of incision followed by alcohol in a similar fashion. This should be repeated two times.

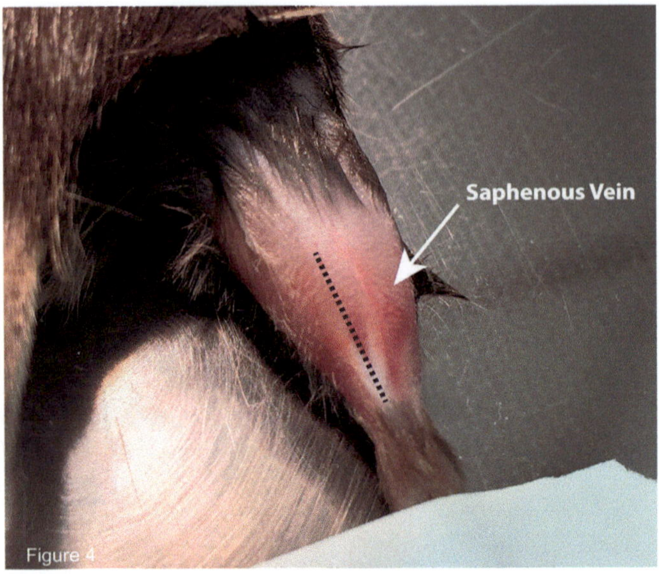

Fig. 4 Locating incision line. The incision (*dotted line*) should be made starting at the mid-belly of the gastrocnemius and run directly between the medial and lateral tendons. Careful consideration should be taken to avoid to superficial vasculature, particularly the saphenous vein

3.4 Incision and Fascia Separation

1. Place the scalpel just medial to the great saphenous vein at the mid-belly of the gastrocnemius. Incision should run in between the medial and lateral tendons of the gastrocnemius (Fig. 4) (*see* **Note 1**).
2. With constant pressure, make an incision down the length of the hindlimb stopping just proximal to the heel (*see* **Notes 2** and **3**).
3. Use forceps to grab the skin and fascia on the medial side, while simultaneously using the surgical probe to gently separate the skin and fascia from the gastrocnemius.
4. For the lateral side, place the probe on the lateral tendon of the gastrocnemius and gently position the probe between the lateral head of the gastrocnemius and the lateral hamstring (*see* **Note 4**).

3.5 Separating and Cutting Away the Gastrocnemius/Soleus from the Plantaris

1. Use forceps to grab the gastrocnemius tendon and gently pull upwards.
2. Pulling the gastrocnemius upwards should reveal the bifurcation between the gastrocnemius tendon and the plantaris tendon (Fig. 5a).
3. From the medial side, insert the surgical probe between this bifurcation and *gently* run it up and down the distal 1/2 of the muscle to fully separate the gastrocnemius/soleus from the plantaris (Fig. 5b) (*see* **Note 5**).

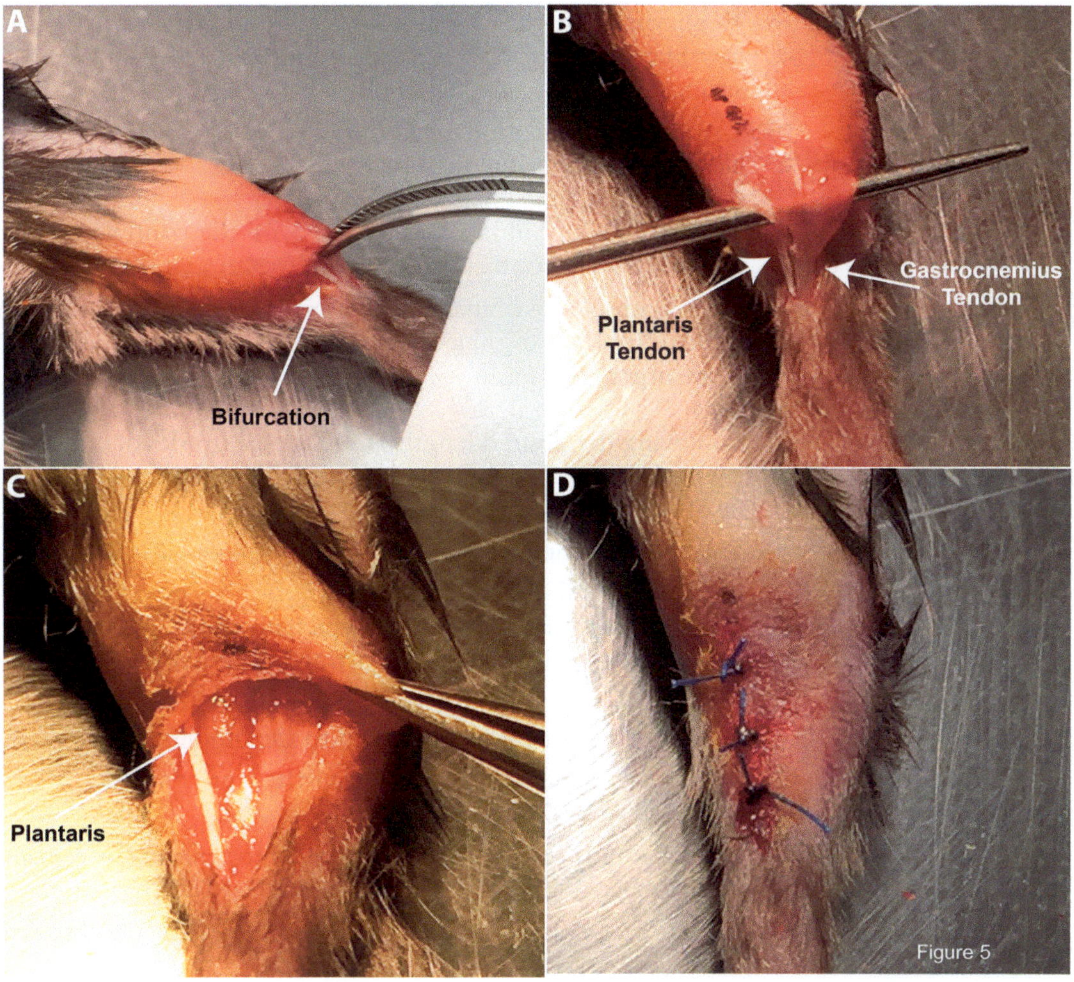

Fig. 5 Isolating the plantaris tendon and removing the gastrocnemius and soleus. (**a**) Gently pull up on the gastrocnemius tendons to identify the bifurcation from the plantaris tendon; (**b**) Place surgical probe in between plantaris and gastrocnemius/soleus tendon (note medial gastrocnemius tendon has been cut to expose plantaris tendon located underneath); (**c**) Surgical area after gastrocnemius and soleus have been removed; (**d**) Sutures closing the incision site

4. Once the probe is between the two sets of tendons, leave the probe in place and use a fine pair of scissors to cut the gastrocnemius/soleus tendon which is sitting above the probe.
5. Take toothed-forceps and grab the severed gastrocnemius/soleus tendon and slowly and gently lift up and away from the plantaris.
6. Cut the gastrocnemius/soleus complex at approximately mid-belly and remove from the hindlimb (Fig. 5c) (*see* **Note 6**).
7. Bleeding should be minimal, however use cotton swabs to absorb any blood that may accumulate at the surgical site

8. Depending on the amount of satellite cell activation and myofiber hypertrophy desired, additional variations to the surgical procedure can be employed (*see* **Note 7**).
9. Use standard suturing procedures to place 3–4 sutures and close the incision (Fig. 5d) (*see* **Note 8**).

3.6 Contralateral Leg and Analgesic Administration

1. Repeat steps of Subheadings 3.2–3.5 on the contralateral leg (*see* **Note 9** for control surgery).
2. As a postoperative analgesic Carprofen (10 mg/kg IP) or Meloxicam (2–5 mg/kg IP) should be given at 24 h and 48 h post-surgery.

3.7 Tissue Collection and Processing

1. Following the desired length of functional overload, euthanize the animal according to the guidelines approved by the institution.
2. Use 70% ethanol to wet the fur and remove all fur and skin from the hindlimb.
3. Dissect away both medial and lateral hamstrings from the plantaris and remaining proximal portion of the gastrocnemius.
4. Remove as much of the residual gastrocnemius as possible while the plantaris is still attached to the hindlimb.
5. Dissect away the plantaris at the furthest proximal origin and distal insertion points.
6. Place plantaris on glass and use a scalpel to remove any remaining gastrocnemius muscle (*see* **Notes 10** and **11**).
7. Finish processing tissue in accordance with downstream analyses.

4 Notes

1. A dissection scope may be used, particularly during the learning phase, in order to aid in identification of anatomical landmarks. Modifications will be needed to be made to the surgical suite setup in order to accommodate the microscope, while still maintaining proper care of the animal.
2. The larger the incision, the easier it will be to locate, cut, and remove the gastrocnemius and soleus. However, this will inevitability result in a greater healing response. As the surgeon becomes more proficient, a smaller incision can be made in order to minimize trauma and recovery.
3. Careful consideration should be taken to make sure the incision reaches at least mid-way down the Achilles tendon. This will make identification and separation of the gastrocnemius/soleus tendon from the plantaris tendon much easier.

4. If the incision was done correctly, the saphenous vein should be just lateral to the incision. It is important to make sure these vessels stay intact during the surgery. Using the probe, it is important to separate the lower hamstring which has the saphenous vein and nerves from the gastrocnemius muscle.
5. The more distal along the tendon bifurcation that you insert your probe, the greater the likelihood that you will successfully remove the soleus in addition to the gastrocnemius. This is because the soleus tendon attaches into the gastrocnemius tendon more proximal on the Achilles tendon. Make sure to not probe underneath the saphenous vein, which should have been adequately separated away in the previous step.
6. After removing the complex, closely examine the surgical site to make sure that no part of the soleus remains. Despite being careful, sometimes the probe will not get all the way under the soleus tendon when probing between the gastrocnemius/soleus tendon and plantaris tendon. If this occurs, you may not end up removing all of the soleus muscle. However, the presence of the remaining soleus should be easy to identify based on the dark red color of the soleus relative to the plantaris. To remove the rest of the soleus, gently guide the probe between the plantaris and soleus and cut the soleus away. At this point, removal of the soleus may be challenging; however, as long as it has been severed it should not affect the functional overload of the plantaris.
7. Based on empirical evidence, the more gastrocnemius that is removed, the greater the mechanical load placed on the plantaris. Removing a large portion (i.e., >3/4) of the gastrocnemius may result in a regenerative response, in addition to hypertrophy of existing myofibers, characterized by the appearance of small, centrally nucleated myofibers. Removing only the lower 1/3 of the gastrocnemius should minimize the regenerative response. It is recognized that this surgical model results in a robust growth response, which may be undesirable for some studies. As an alternative, researchers may wish to leave the soleus intact resulting in a dual soleus/plantaris overload.
8. Since the plantaris is now exposed, careful consideration should be taken to not accidentally damage it while suturing. The distal aspect of the incision can be the most challenging, since there is little excess skin to suture through. In this case, the surgeon should be careful to not accidentally suture through the plantaris tendon.
9. We believe that in order to get maximal functional overload of the plantaris, surgical removal of the gastrocnemius/soleus should be performed on both legs. If one leg is left as a contralateral control, the animal with undoubtedly favor this leg and

not put as much weight on the leg with which the surgery was performed. To address this, we always delegate animals to act as "sham" operated controls. For these animals, perform all of the procedures up until Subheading 3.4, **step 4** (i.e., cutting the tendon) and then skip to Subheading 3.5 (suturing).

10. Typically if the plantaris is overloaded for longer than 14 days, the small amount of gastrocnemius that remains will fuse to the proximal portion of the plantaris. It is important to remove this excess tissue as best as possible, particularly if the muscle will be homogenized for RNA or protein analysis. Since this excess gastrocnemius has not been subjected to mechanical overload, it will undoubtedly possess a different molecular signature from that of the plantaris.
11. If tissue is to be used for RNA analyses, apply RNAse removal solution liberally to the glass surface to minimize potential RNA degradation.

References

1. Charge SB, Rudnicki MA (2004) Cellular and molecular regulation of muscle regeneration. Physiol Rev 84:209–238
2. Fry CS, Lee JD, Jackson JR, Kirby TJ, Stasko SA, Liu H, Dupont-Versteegden EE, McCarthy JJ, Peterson CA (2014) Regulation of the muscle fiber microenvironment by activated satellite cells during hypertrophy. FASEB J 28:1654–1665
3. McCarthy JJ, Mula J, Miyazaki M, Erfani R, Garrison K, Farooqui AB, Srikuea R, Lawson BA, Grimes B, Keller C, Van Zant G, Campbell KS, Esser KA, Dupont-Versteegden EE, Peterson CA (2011) Effective fiber hypertrophy in satellite cell-depleted skeletal muscle. Development 138: 3657–3666

Chapter 5

Inducing and Evaluating Skeletal Muscle Injury by Notexin and Barium Chloride

Matthew T. Tierney and Alessandra Sacco

Abstract

Models of skeletal muscle injury in animal models are invaluable tools to assess muscle stem cell (MuSC)-mediated tissue repair. The optimization and comprehensive evaluation of these approaches have greatly improved our ability to assess MuSC regenerative potential. Here we describe the procedures for skeletal muscle injury with notexin and $BaCl_2$ and assessment of the dynamics of tissue regeneration.

Key words Skeletal muscle injury, Muscle regeneration

1 Introduction

Muscle stem cells (MuSC), known to be essential for skeletal muscle development and repair (reviewed in ref. [1]), have been studied in a number of experimental settings. While ex vivo experiments have yielded useful information regarding the molecular characterization of isolated MuSC, the rapid loss of self-renewal potential in conventional culture systems precludes their functional assessment [2–4]. Transplantation models overcome this limitation to allow for evaluation of their regenerative potential in vivo but do not avoid modifications or cellular selection that may result from the isolation process. Thus, the gold standard to functionally assess stem cell activity is within their native environment in situ. In high-turnover tissues, such as blood and epithelia, stem cell function can be readily assessed in steady-state conditions [5]. Conversely, MuSC are not sufficiently active to monitor their behavior during normal skeletal muscle homeostasis. Instead, they largely reside in quiescent state, rarely entering the cell cycle except upon stress or injury [6, 7]. Thus, an intervention is required in order to assess their function, i.e., inducing an injury and assessing the ability of endogenous MuSC to face this challenge and efficiently repair the damaged tissue. Over the years, different modes of skeletal muscle

Michael Kyba (ed.), *Skeletal Muscle Regeneration in the Mouse: Methods and Protocols*, Methods in Molecular Biology, vol. 1460, DOI 10.1007/978-1-4939-3810-0_5, © Springer Science+Business Media New York 2016

injury in animal models have been utilized and range from natural toxins (cardiotoxin, notexin), medications (bupivacaine), chemical agents (barium chloride), and physical stressors, i.e., freezing or crush injury, tissue ablation and contraction-induced injury. While this wide spectrum of interventions differs in both the type and extent of myofiber damage and consequently provoke slightly different dynamics of tissue regeneration, they are useful experimental settings to evaluate MuSC-mediated tissue repair. In addition, they have been invaluable tools to investigate the complex interplay between MuSC and their immediate microenvironment, including how different tissue resident cell types temporally coordinate tissue repair. Here we provide details on the procedures utilized in our laboratory to induce skeletal muscle injury in mice and assess tissue repair in vivo [3, 4, 8, 9].

2 Materials

Prepare and store all reagents and antibodies at 4 °C (unless indicated otherwise). Follow all waste disposal regulations when disposing waste materials.

2.1 Skeletal Muscle Injury and Tissue Collection

1. Isoflurane vaporizer, supply gas (oxygen), flowmeter, and induction chamber.
2. Small animal hair clippers.
3. Insulin syringes, 0.3 mL with 29G × 0.5″ ultra-fine needles.
4. PBS, pH = 7.4 (1×).
5. Biopsy cryomolds, 10 × 10 × 15.
6. O.C.T. compound.
7. Notexin, 10 μg/ml in PBS.
8. Barium chloride, 1.2 % in normal saline.

2.2 Tissue Cryosectioning and Immunostaining

1. Research cryostat, CM3050 S (Leica).
2. Microscope slides, Superfrost Plus, 25 × 75 × 1.0 mm.
3. Hydrophobic PAP pen.
4. Slide staining humidity box, black cover.
5. Fixative solution: PBS, pH = 7.4 (1×) containing 1.5 % paraformaldehyde.
6. Blocking buffer: PBS, pH = 7.4 (1×) containing 20 % normal goat serum, 0.1 % Triton X-100.
7. Antigen retrieval solution: deionized water containing 100× citrate-based antigen unmasking solution, pH = 6.0. Add 0.5 mL antigen unmasking solution to 49.5 mL deionized water and store at 4 °C (*see* **Note 1**).

8. Antibodies
   (a) Rat anti-laminin B2, clone A5, 0.5 mg/mL (catalog number 05-206) (Millipore).
   (b) Mouse anti-Pax7 concentrate (catalog number Pax7) (Developmental Studies Hybridoma Bank).
   (c) Mouse anti-embryonic myosin heavy chain (catalog number F1.652) (Developmental Studies Hybridoma Bank).
   (d) Alexa Fluor secondary antibodies.
9. Hoechst 33258, 10 mg/mL.
10. Fluoromount-G mounting medium.
11. Microscope cover glass, 24 × 50 mm.

3 Methods

Carry out all procedures at room temperature unless otherwise specified.

3.1 Skeletal Muscle Injury

1. The researcher must choose to perform either (1) a focal injury to a portion of the tibialis anterior muscle or (2) to completely injure the entire muscle (*see* **Note 2**). Prepare the desired volume of barium chloride or notexin to be injected (10 μl for focal injury, 50 μl for complete injury) in 0.3 mL insulin syringes with a 29 G × 0.5″ ultra-fine needle.
2. Turn on the flowmeter to 200 mL/min oxygen and the isoflurane vaporizer to 2 %. Anesthetize the mouse by placing inside the induction chamber and wait 2–3 min until breathing has slowed and the mouse appears asleep. Remove the mouse from the induction chamber, divert flow from the flowmeter and vaporizer to an anesthesia mask on a surgical bench and quickly place this mask on the mouse. Wait 2–3 min and confirm it is nonresponsive to a foot or tail pinch.
3. Remove hair from the injection site at the anterior hind limb with small animal hair clippers. Use an alcohol wipe to disinfect.
4. To perform a focal injury, insert the needle at the center of the tibialis anterior muscle, maintaining an angle of 45° between the needle and the length of the tibialis. Slightly withdraw, maintaining the needle tip still in the center of the tibialis anterior muscle, to allow space for the solution volume. Inject 10 μl barium chloride or notexin solution, holding the needle in place for 2–3 s to ensure the volume will not leak out from the muscle once the needle has been removed.
5. To perform a full injury, first perform 10–20 jabs over the entire area of the tibialis anterior muscle to allow for complete

diffusion of the injected solution. Inject 50 μl barium chloride or notexin solution in a total of five injection sites, again such that the entire muscle has been injured. Some leakage of the volume is expected and normal, but the majority of solution should remain within the damaged muscle.

6. Remove the anesthesia mask and return the recipient mouse to its cage. Allow for up to 5 min for the mouse to awaken. Reflect on the animal's health status to confirm normal activity and proper ambulation after an adjustment period of up to 15 min (*see* **Note 3**).

3.2 Tissue Harvest and Cryosectioning

1. At the desired time of recovery post-injury, collect the recipient mouse for sacrifice and prepare the fixative solution (*see* **Note 4**).
2. Turn on the flowmeter to 200 mL/min oxygen and the isoflurane vaporizer to 2 %. Anesthetize the mouse by placing inside the induction chamber and wait 2–3 min until breathing has slowed and the mouse appears asleep. Remove the mouse from the induction chamber, confirm it is nonresponsive to a foot or tail pinch and sacrifice by cervical dislocation.
3. Place the mouse on a surgical bench and soak the hind limbs in 70 % ethanol. Remove the skin covering the hind limb muscles by making an incision near the hip and slowly pull the skin down the limb and over the ankle. Use a razor blade to sever the distal tendon of the tibialis anterior muscle. Hold this tendon with forceps, gently pulling towards the knee while sliding a razor blade between the muscle and the tibialis bone until fully separated. Temporarily place the muscle in PBS.
4. Prepare for tissue freezing by gathering dry ice and liquid nitrogen. Chill a volume of 2-methylbutane sufficient to submerge the biopsy cryomold in liquid nitrogen until it begins to solidify. Transfer the recipient muscle to biopsy cryomolds containing O.C.T. compound, positioned longitudinally so cross sections can be achieved by sectioning from top to bottom. Plate the bottom of the cryomold on top of the chilled 2-methylbutane until it has completely frozen over and then submerge for 1 min. Remove the cryomold and place on dry ice. Frozen samples can be stored at −80 °C until ready to be cryosectioned.
5. Set the chamber temperature to −20 °C, place samples into the cryostat and allow them to come to temperature for 30 min. Cut and place two 10 μm sections serially on 4–8 slides, subject to the number of immunostainings that will be performed. Once two sections have been placed on each slide, eliminate 200 μm of tissue and resume normal cryosectioning. Repeat this process until the core 80 % of the muscle has been

sectioned so as to represent the entire muscle and allow for analysis of the injured area. Once completed, label and store slides at −20 °C.

3.3 Immunostaining

1. Remove slides from storage at −20 °C and let come to room temperature. Encircle sections with a hydrophobic PAP pen to retain solution on the mounted tissue. Place slides in a humid incubation box (dark, UV safe), where all subsequent steps should be performed.
2. Fix sections in chilled acetone for 10 min at −20 °C (embryonic myosin heavy chain/laminin stain) or in PBS, pH = 7.4 containing 1.5 % paraformaldehyde for 15 min (Pax7/laminin stain).
3. Aspirate fixative and wash in PBS, pH = 7.4 for 5 min. Block and permeabilize sections in blocking buffer for 1 h.
4. Incubate sections with the following primary antibodies, diluted in blocking buffer for 2 h (or overnight): mouse anti-embryonic myosin heavy chain (1:500) and/or rat anti-laminin B2 (1:200 dilution). Skip to **step 9** for embryonic myosin heavy chain/laminin stain; continue to **step 5** for Pax7/laminin stain.
5. Aspirate primary antibodies and perform three washes in PBS, pH = 7.4 for 5 min each. Incubate sections with the appropriate secondary antibodies (*see* **Note 5**), diluted in blocking buffer for 1 h.
6. Aspirate secondary antibodies and perform three washes in PBS, pH = 7.4 for 5 min each. Post-fix sections in PBS, pH = 7.4 containing 1.5 % paraformaldehyde for 15 min to preserve the signal specific to all antibodies used thus far. Antigen retrieval procedures are somewhat harsh and can greatly reduce signal without this step.
7. Aspirate fixative and wash in PBS, pH = 7.4 for 5 min. Immerse slides in pre-warmed antigen retrieval solution contained either a plastic or ceramic Coplin jar and incubate in a heated water bath maintained at >80 °C for 15 min.
8. Remove Coplin jar containing slides from the heated water bath and let cool for 5 min. Wash in PBS, pH = 7.4 for 5 min. Re-block and permeabilize sections in blocking buffer for 30 min. Incubate sections with mouse anti-Pax7 (1:100 dilution) in blocking buffer overnight.
9. Aspirate primary antibodies and perform three washes in PBS, pH = 7.4 for 5 min each. Incubate sections with the appropriate secondary antibodies and Hoechst 33258 (1:1000 dilution) in blocking buffer for 1 h.
10. Aspirate secondary antibodies and wash in a Coplin jar with PBS, pH = 7.4 for 30 min. Mount slides with Fluoromount-G water-based mounting medium and microscope cover glasses. Once

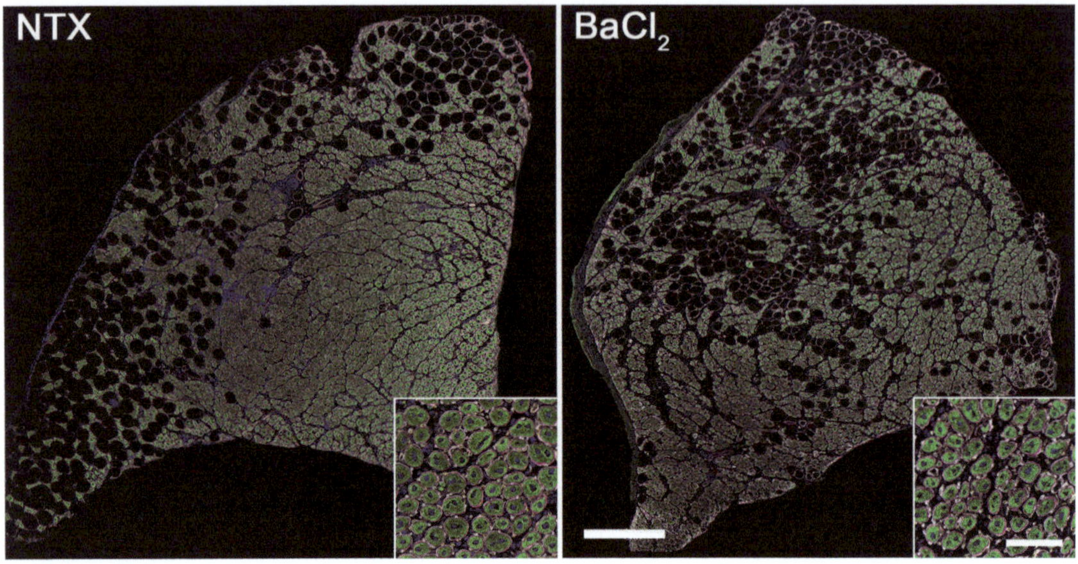

Fig. 1 Representative images of tibialis anterior 5 days after tissue injury induced by 50 μl injection of either notexin (*left*) or barium chloride (*right*). *Green* = embryonic myosin heavy chain, *White* = Laminin, *Blue* = Hoechst. Scale bar, 500 μm (100 μm, *insert*)

the cover glass has been placed on the slide, utilize forceps to gently push visible air bubbles to the side of the slide as they can dry the sample and generate significant autofluorescence. Store slides at 4 °C until ready for imaging (*see* **Note 6**) (Fig. 1).

4 Notes

1. Antigen retrieval solution at working concentration can be reused up to five times if stored at 4 °C in between uses.
2. Focal injuries are sufficient for single injury experiments, while complete injuries are required for serial injury studies to ensure that the portion of the muscle ultimately analyzed has gone through the regeneration process at each step.
3. Injury to the tibialis anterior muscle may temporarily result in an impaired ability to dorsiflex from the ankle; any gait abnormalities should resolve within 1 day post-injury.
4. Time points at early stages (3–5 days post-injury), late stages (8–14 days) and following the completion of muscle repair (30–45 days) should be chosen to comprehensively assess the dynamics of skeletal muscle regeneration in context of the experimental intervention.
5. Preference for any secondary antibody is dependent upon the available or preferred spectra given the researcher's imaging microscope.

6. To quantify Pax7+ cell density or the cross-sectional area of regenerating myofibers, the entire slide should be scanned by eye and a single section closest to the core of the injured area should be selected for quantification. If the researcher is unsure, several sections can be chosen for analysis and the section containing the largest number of regenerating myofibers should be chosen. As mononucleated cells are not expected to span the >40 μm length given >4 serial slides and there is little risk of scoring the same cell twice, several sections may be analyzed to measure Pax7+ cell density if desired. For myofibers quantification, analysis of several sections should be avoided, as myofibers are likely to span a significant portion of the muscle but cannot be distinguished. Regenerating myofibers can be identified by embryonic myosin heavy chain up to ~7 days post-injury; following its downregulation during myofiber maturation, central nucleation can alternatively be used. A minimum of 500 myofibers per samples should be analyzed, to accurately represent the distribution of myofiber cross-sectional area throughout the muscle. Several means of automated image analysis are available and may be used to identify, outline and calculate myofiber cross-sectional area [10–12].

Acknowledgements

This work was supported by the US National Institutes of Health (NIH) grants R01 AR064873, R03 AR063328 and P30 AR06130303 to AS, and US National Institutes of Health (NIH) grant F31 AR065923-01 to MT. We thank Kenny Venegas and Buddy Charbono for technical support.

References

1. Dumont NA, Wang YX, Rudnicki MA (2015) Intrinsic and extrinsic mechanisms regulating satellite cell function. Development 142(9):1572–1581
2. Montarras D, Morgan J, Collins C, Relaix F, Zaffran S, Cumano A, Partridge T, Buckingham M (2005) Direct isolation of satellite cells for skeletal muscle regeneration. Science 309(5743):2064–2067
3. Sacco A, Doyonnas R, Kraft P, Vitorovic S, Blau HM (2008) Self-renewal and expansion of single transplanted muscle stem cells. Nature 456(7221):502–506
4. Gilbert PM, Havenstrite KL, Magnusson KE, Sacco A, Leonardi NA, Kraft P, Nguyen NK, Thrun S, Lutolf MP, Blau HM (2010) Substrate elasticity regulates skeletal muscle stem cell self-renewal in culture. Science 329(5995):1078–1081
5. Rando TA (2006) Stem cells, ageing and the quest for immortality. Nature 441(7097): 1080–1086
6. Dhawan J, Rando TA (2005) Stem cells in postnatal myogenesis: molecular mechanisms of satellite cell quiescence, activation and replenishment. Trends Cell Biol 15(12): 666–673
7. Abou-Khalil R, Brack AS (2010) Muscle stem cells and reversible quiescence: the role of sprouty. Cell Cycle 9(13):2575–2580
8. Mozzetta C, Consalvi S, Saccone V, Tierney M, Diamantini A, Mitchell KJ, Marazzi G, Borsellino G, Battistini L, Sassoon D et al (2013) Fibroadipogenic progenitors mediate

the ability of HDAC inhibitors to promote regeneration in dystrophic muscles of young, but not old Mdx mice. EMBO Mol Med 5(4):626–639
9. Tierney MT, Aydogdu T, Sala D, Malecova B, Gatto S, Puri PL, Latella L, Sacco A (2014) STAT3 signaling controls satellite cell expansion and skeletal muscle repair. Nat Med 20(10):1182–1186
10. Miller GR, Stauber WT (1994) Use of computer-assisted analysis for myofiber size measurements of rat soleus muscles from photographed images. J Histochem Cytochem 42(3):377–382
11. Mula J, Lee JD, Liu F, Yang L, Peterson CA (2013) Automated image analysis of skeletal muscle fiber cross-sectional area. J Appl Physiol (1985) 114(1):148–155
12. Kim YJ, Brox T, Feiden W, Weickert J (2007) Fully automated segmentation and morphometrical analysis of muscle fiber images. Cytometry A 71(1):8–15

Chapter 6

Cardiotoxin Induced Injury and Skeletal Muscle Regeneration

Glynnis A. Garry, Marie Lue Antony, and Daniel J. Garry

Abstract

Skeletal muscles have a tremendous capacity for repair and regeneration in response to injury. This capacity for regeneration is largely due to a myogenic stem cell population, termed satellite cells, which are resident in adult skeletal muscles. In order to decipher the mechanisms that govern myogenic stem cell quiescence, activation, differentiation, and self-renewal, a reproducible injury model is required. Therefore, we have utilized the delivery of the myonecrotic agent, cardiotoxin, to examine the molecular mechanisms of myogenic stem cells in response to injury. Here, we describe our experience using cardiotoxin as a potent myonecrotic agent to study skeletal muscle regeneration. We provide a detailed protocol to examine skeletal muscle injury and regeneration using morphological analyses.

Key words Cardiotoxin, Skeletal muscle regeneration

1 Introduction

Skeletal muscles have a remarkable capacity for regeneration due to a myogenic stem cell population that is resident in adult skeletal muscles (Fig. 1) [1, 2]. This capacity for regeneration can be limited by muscle diseases such as muscular dystrophy or by aging (i.e., sarcopenia). For example, studies using genetic mouse models of muscular dystrophy (i.e., mdx or mdx:utrophin double knockout mice) demonstrate impaired muscle regeneration and premature lethality [3]. Similarly, parabiosis experiments where young mice are conjoined to aged mice demonstrate improved skeletal muscle regeneration in the parabiosed aged mouse compared to the age matched control. These studies emphasize the importance of deciphering the mechanisms that govern the response of myogenic stem cells to injury and the corresponding regenerative response that completely restores the cellular architecture of muscle that is indistinguishable from its unperturbed experimental control (Fig. 1).

Michael Kyba (ed.), *Skeletal Muscle Regeneration in the Mouse: Methods and Protocols*, Methods in Molecular Biology, vol. 1460, DOI 10.1007/978-1-4939-3810-0_6, © Springer Science+Business Media New York 2016

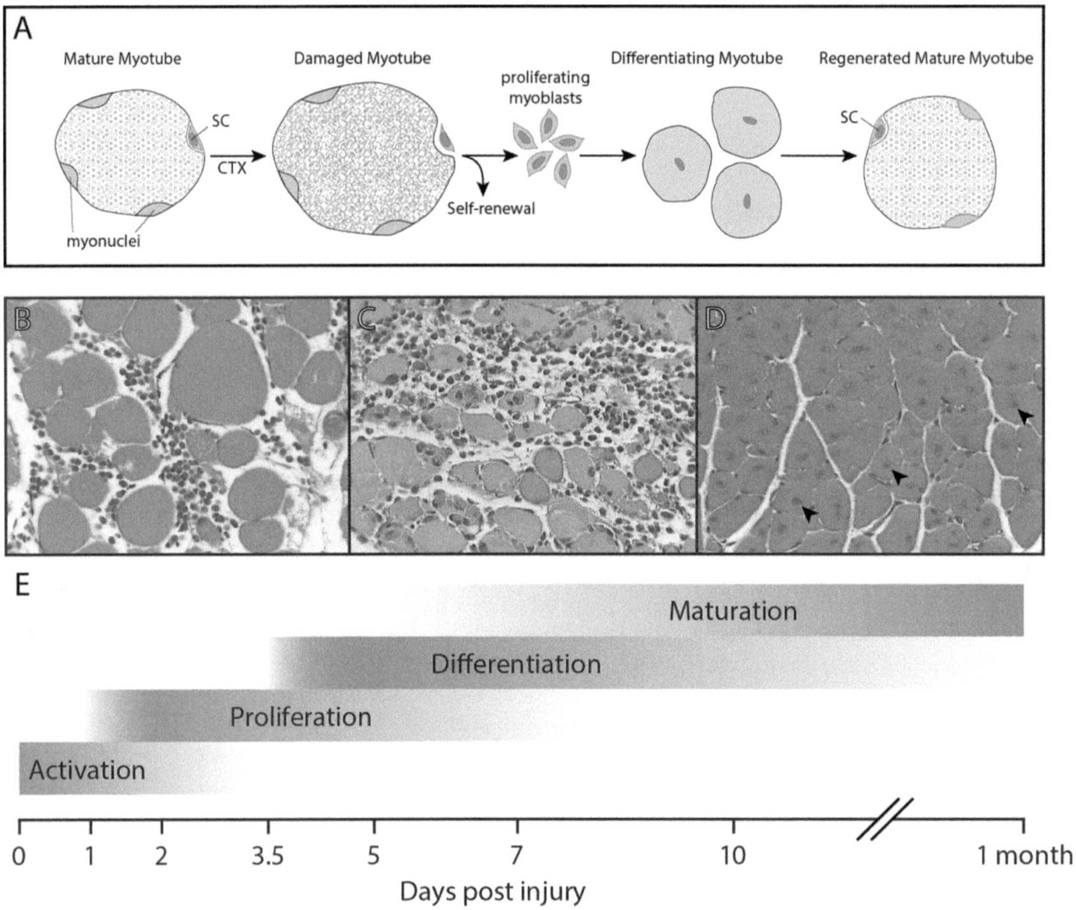

Fig. 1 Cardiotoxin induced skeletal muscle regeneration. (**A**) Schematic highlighting the unpertubed multinucleated myotube that contains quiescent satellite cells occupying a sublaminar position. Following intramuscular delivery of cardiotoxin, satellite cells are activated and proliferate or undergo self-renewal. Satellite cells exit the cell cycle and form centronucleated myotubes (the hallmark of skeletal muscle regeneration) and reestablish their cellular architecture. (**B–D**) Histological analysis of skeletal muscle following cardiotoxin injury at 24 h (**B**), 2 days (**C**) and 10 days (**D**). Note the centronucleated myotubes at 2 and 10 days following cardiotoxin injury. (**E**) Schematic highlighting the time course for the activation, proliferation, and differentiation of satellite cells/myoblasts and the maturation of the regenerated myofibers

Previous studies have utilized a number of different injury models to examine the mechanisms that govern skeletal muscle regeneration. These strategies include freeze injury, crush injury, surgical muscle removal, and others. While there may be clear advantages associated with these muscle injury protocols, in our experience, cardiotoxin induced skeletal muscle injury is easy to deliver, does not require a surgical procedure and produces a comparable and reproducible response in age and gender-matched adult mice (Fig. 1) [3–5].

2 Materials

2.1 Cardiotoxin Induced Injury

1. Cardiotoxin stock solution: cardiotoxin (from *Naja mossambica mossambica*, Sigma Chemical) is resuspended in sterile PBS at a concentration of 1 mg/ml and stored at −20 °C.
2. Cardiotoxin working solution (10 μM) dispensed in 1.5 ml Eppendorf tubes and stored at 4 °C.
3. 1 ml disposable syringe.
4. 70% ethanol in a 250 ml plastic wash bottle.
5. Gloves (Nitrile) × 2 pairs.

2.2 Delivery of BrdU to Mark Proliferative Cellular Events

1. BrdU dissolved in 100 ml water (1 mg/ml).
2. Foil (cut in 6 × 6 in. sheets).

2.3 Skeletal Muscle Fixation and Processing

1. Avertin (2,2,2-tribromoethanol) stock solution (20 mg/ml diluted in PBS) freshly prepared and stored on ice or at 4 °C.
2. 40 ml 4% paraformaldehyde dissolved in PBS (pH 7.2) on ice.
3. Surgical instruments [Mouse kit which includes: micro bulldog clamp, serrated; Dissecting scissors, straight 10 cm; Iris forceps, curved, serrated; needle holder; Dumont #5].
4. 50 ml disposable falcon conical tube.
5. 10% sucrose in PBS.
6. 50 ml 2-methylbutane (isopentane).
7. Ice bucket (4 L capacity).
8. Liquid nitrogen.
9. Peel-A-Way disposable embedding molds (Electron Microscopy Sciences).
10. O.C.T. embedding compound.
11. Disposable gloves.
12. Permanent marker (black).

2.4 Frozen Sections and Morphological Analyses

1. Leica or Bright Hacker Cryostat cooled to −14 to −18 °C.
2. High Profile Leica 818 Microtome disposable blades (Leica Microsystems).
3. Colorfrost Plus Precleaned microscope slides (25 × 75 × 1.0 mm).
4. Mini Super PAP liquid-repellent slide marker pen.
5. Hematoxylin 560 stain.
6. Alcoholic eosin stain Y 515.
7. Blue Buffer 8 (Leica Biosystems).
8. Define (Leica Biosystems).

9. 2 N HCl (16.6 ml concentrated in 83.4 ml dH2) (prepared fresh each day).
10. Borate buffer [0.1 M boric acid, pH 8.5 (0.62 g boric acid in 100 ml dH_2O)].
11. Anti-bromodeoxyuridine (BrdU) mouse monoclonal primary antibody (1:25 dilution, Roche).
12. Secondary antibody (fluorescein isothiocyanate or FITC-conjugated goat anti-mouse antibody, diluted 1:50 in PBC).
13. Vectashield (Vector Laboratories).
14. Precleaned cover glass (25 × 50, #1.5).

3 Methods

3.1 Preparation and Delivery of Cardiotoxin

The investigators typically wear protective eyeware, a lab coat and are double gloved. A 1 mg/ml stock solution of cardiotoxin is prepared and a working solution (10 μM) is aliquoted into disposable Eppendorf vials and placed on ice.

1. Mice are anesthetized with Avertin (250 mg/kg body wt) and placed in a prone position.
2. Approximately 150 μL of cardiotoxin (10 μM solution) is withdrawn in a 1.0 ml disposable syringe.
3. The posterior aspect of the adult mouse hindlimb is rinsed with 70 % ethanol and the calcaneal tendon is exposed (Fig. 2).
4. A direct intramuscular injection is made into the mid-belly of the gastrocnemius muscle, approximately 1.0 cm superior to the calcaneal tendon insertion into the belly of the muscle (Fig. 2A). The needle is initially aspirated to assure the delivery is outside the vascular system and then slowly delivered intramuscularly (**Note 1**).
5. The needle is withdrawn and the animal is replaced in the cage.
6. All syringes are disposed in a chemical hazard sharps container. The gloves and mask are disposed in the chemical hazard waste container.

Similarly, injury can be achieved in the tibialis anterior (TA) muscle by sterilizing the region with 70% ethanol. Approximately 50 μL of cardiotoxin (10 μM) should be injected into the belly of the TA muscle, with needle placement 1 cm inferior to the proximal insertion of the TA. The needle should be inserted in a shallow angle along the belly of the TA, with the syringe at a 20° angle. Initially aspirate, then inject slowly along the belly of the TA while simultaneously removing the needle to ensure diffuse muscular injury. Great care should be taken to minimize variability in injection pattern from mouse to mouse. In all instances of cardiotoxin injury, the contralateral leg should be maintained as an uninjured control.

3.2 BrdU Pulsing

BrdU (1 mg/ml) is administered ad libitum in the drinking water of mice immediately following cardiotoxin induced injury (or at a specified time period prior to sacrifice).

1. Place BrdU-drinking water in a conventional water delivery system, which is completely covered with foil as BrdU is light sensitive (Fig. 3) [6].
2. Change the BrdU-drinking water solution every 3–5 days. Prepare the BrdU-water solution fresh. The remaining BrdU is discarded in the hazardous waste container.

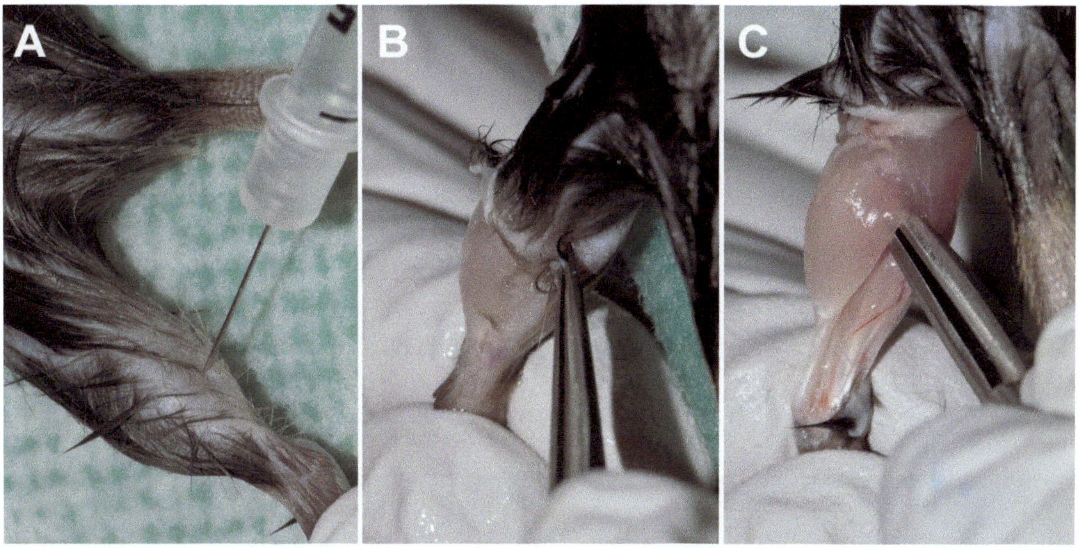

Fig. 2 Cardiotoxin induced muscle injury. (**A**) Cardiotoxin is introduced into the gastrocnemius muscle. (**B**, **C**) At specified time periods following cardiotoxin injury, the animal is sacrificed, the overlying skin of the posterior crural musculature is removed (**B**) and the gastrocnemius muscle is isolated and removed (**C**)

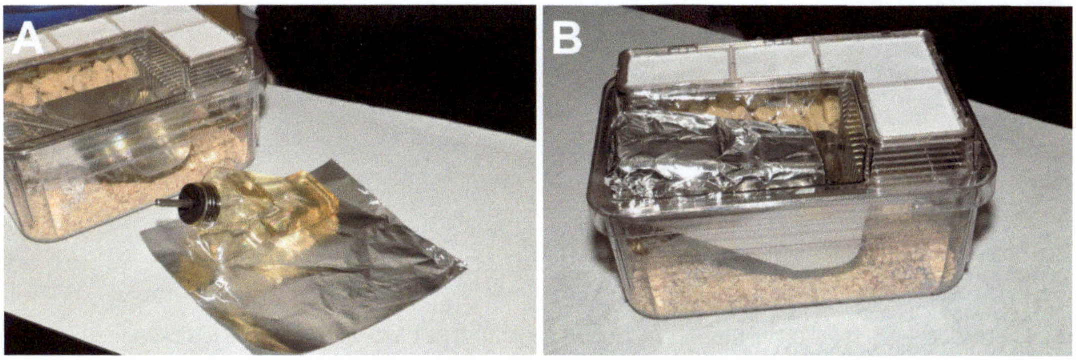

Fig. 3 BrdU delivery to adult mice. (**A**) BrdU solution is prepared and placed in a drinking water bottle. (**B**) The BrdU solution in the drinking water is covered by foil as it is light sensitive

3.3 Animal Perfusion and Fixation of Tissues

At specified time periods following injury, selected animals are anesthetized with Avertin (250 mg/kg body wt) for perfusion and fixation of tissues.

1. Determine whether the animal is adequately sedated using a toe-pinch strategy to examine for paw withdrawal. Once adequate sedation is achieved, the mouse is placed in the supine position.
2. The thorax and subxiphoid regions are rinsed with 70 % ethanol (in a water bottle), the skin is removed, and a midline thoracotomy is undertaken to expose the mediastinum and the rib cage is retracted. The heart is exposed and the course of the aorta is defined.
3. A small probe is placed posterior to the aortic arch, a small incision (using the scissors) is made in the left ventricular apex.
4. The cannula is introduced into the left ventricle and advanced to the ascending aorta. The end of the cannula is placed in the ascending aorta and stabilized using a mini bulldog clamp.
5. Approximately 10 ml of ice-cold calcium free Tyrode's solution is introduced at a rate of 1.5 ml/min using a conventional minipump (Fig. 4).
6. A small incision is introduced into the right atrium to allow the effluent to exit the animal. Note that the liver will begin to blanch and the lower extremities will hyperextend.

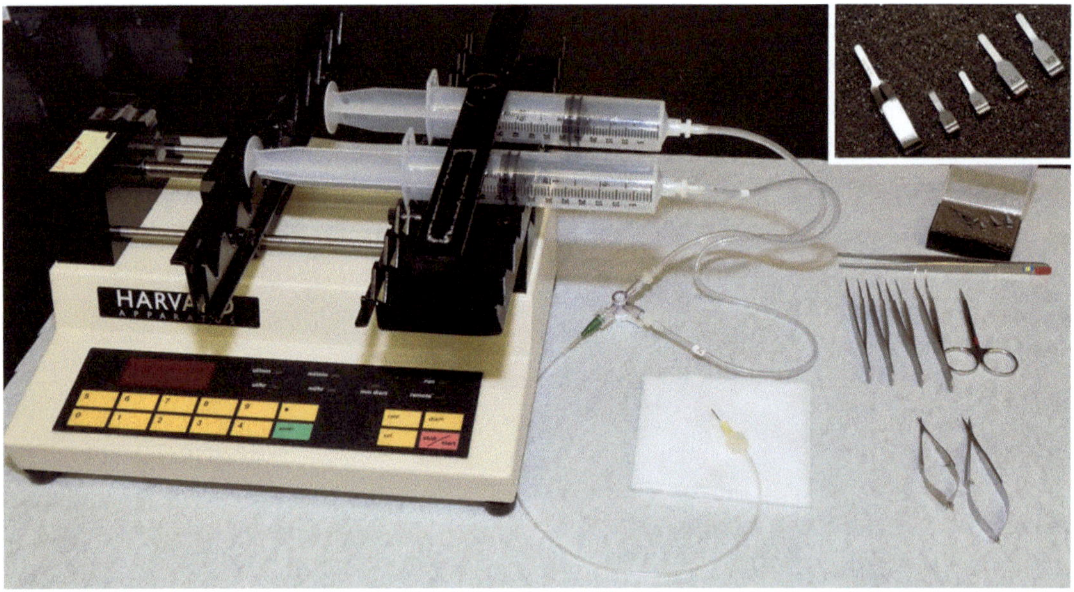

Fig. 4 Perfusion setup for skeletal muscle analysis. A minipump setup to deliver a rinse (Calcium Free Tyrode's Solution) followed by 4 % paraformaldehyde via intraventricular-aortic cannulation, which is secured using a microbulldog clamp (*inset*). Surgical instruments for thoracotomy, cannulation and skeletal muscle harvesting are shown

7. The animal is then perfused with 15 ml of ice-cold 4% paraformaldehyde dissolved in PBS, which was prepared within a seven period (but no older than 10 days) and perfused at a rate of 1.5 ml/min (Fig. 4).
8. Following paraformaldehyde perfusion, the animal is placed in the prone position. The skin of the lower extremity is rinsed with 70% ethanol. Using scissors, the skin is removed beginning at the ankle and continuing in a superior fashion to the knee (Fig. 2B).
9. The calcaneal tendon is then identified and cut with the scissors. The calcaneal tendon is then held with the forceps and the posterior crural musculature (including the gastrocnemius and underlying soleus muscles) are freed from the fascia and removed at the tibial plateau (Fig. 2C).
10. The musculature is then rinsed in ice-cold PBS and post-fixed at 4 °C for an additional 12–16 h (Fig. 5). Following this post-perfusion fixation, the muscle is placed in a 50 ml disposable conical tube containing 10% sucrose dissolved in PBS and stored at 4 °C overnight.

3.4 Tissue Freezing and Cryosectioning

The fixed and cryopreserved muscle specimen is then ready to be embedded. An embedding mold is cut to allow for one side to be held by a long forceps and the sides cut to allow for the frozen block to be easily mounted (Fig. 6A).

1. The mold is filled with O.C.T. and the tissue specimen is then oriented for either longitudinal or cross-sectional analysis (Fig. 6A–C).
2. Using the long forceps, the embedding mold containing the muscle and OCT are frozen by immersing the bottom 1/3 of the mold into the isopentane that is supercooled by liquid nitrogen (Fig. 6C). The embedding mold is removed once it is completely frozen with the exception of the very center of the block (2 mm diameter) and placed on dry ice (**Note 2**).
3. The O.C.T. block is removed from the plastic mold and affixed to the chuck using O.C.T. and allowed to equilibrate in the cryostat for approximately 1 h. A new disposable microtome blade is then placed in the knife holder to equilibrate to the temperature (Fig. 6D).
4. The equilibrated chuck containing the O.C.T./muscle block is mounted to the specimen head of the cryostat and 4–7 micron cryosections are obtained (at −14 to −18 °C) and placed on plus coated slides and further affixed to the slide by placing the section/slide at room temperature for 15 min.

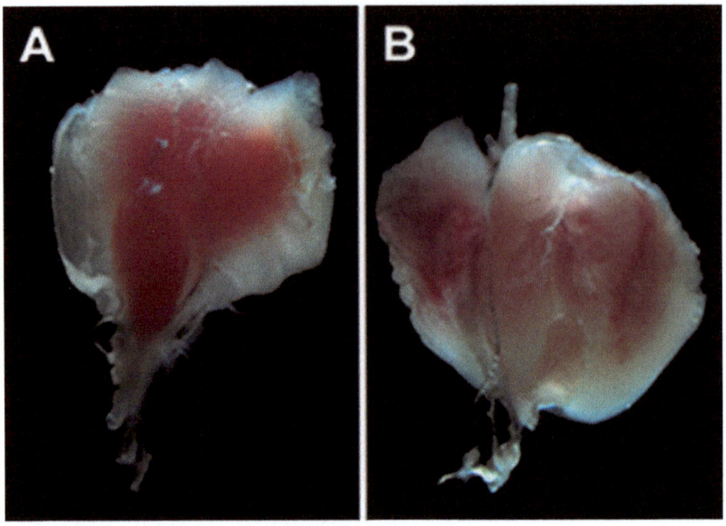

Fig. 5 Skeletal muscle injury secondary to cardiotoxin. (**A**, **B**) Whole-mount image of uninjured (**A**) and perturbed (2 days) gastrocnemius muscles following cardiotoxin injury (**B**)

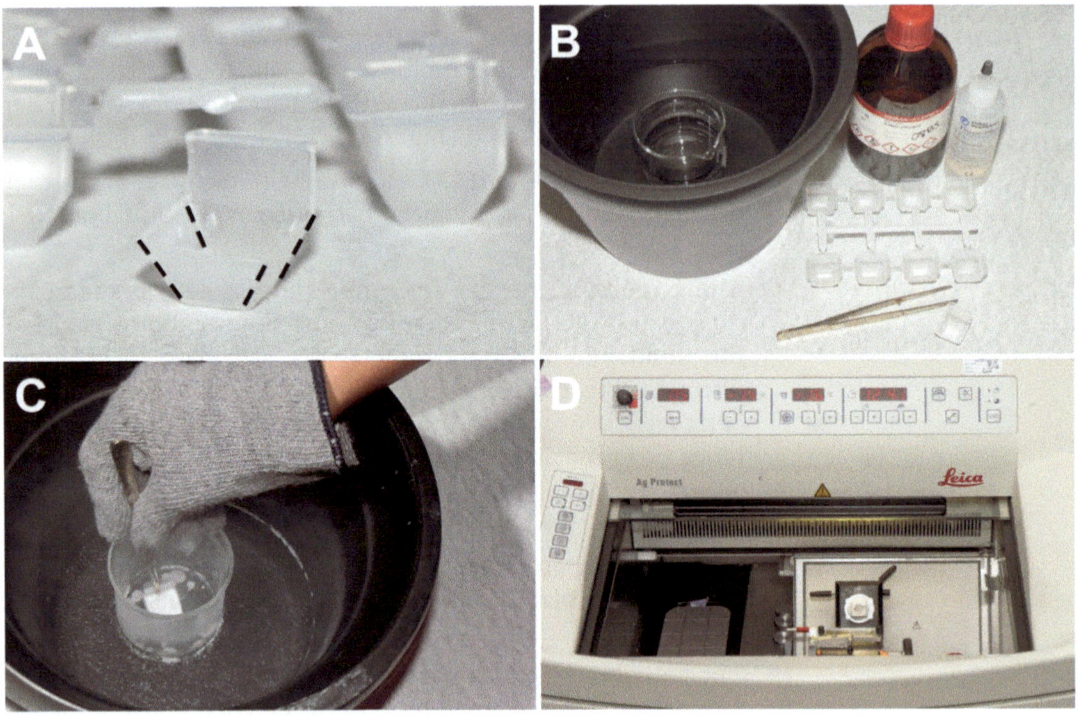

Fig. 6 Skeletal muscle cryoanalysis. (**A**) A paraffin mold is cut as shown in order to handle the specimen and easily remove the frozen block (the mold is cut along the *black dashed lines*). (**B**) The O.C.T. and skeletal muscle are placed in the mold. A plastic ice bucket is filled (2 in.) with liquid nitrogen and a beaker containing the isopentane is placed in the bottom of the ice bucket. (**C**) The O.C.T.-skeletal muscle containing mold is frozen by contacting the bottom of the mold to the liquid nitrogen super cooled isopentane until a central 2 mm region remains and then the block is placed in the precooled microtome (**D**) for cryosectioning

5. The slides are then placed in a labeled plastic slide box containing a desiccant capsule, sealed with Parafilm and stored in the −80 °C freezer.

3.5 Morphological and Immunohistochemical Analyses

Cryosections on plus coated slides stored in a slide box at −80 °C are removed from the freezer and equilibrated at room temperature (approximately 30–45 min). The Parafilm seal is removed from the slide box, the slides are removed and hydrated in PBS (in either a Coplin jar or staining dish).

The hematoxylin and eosin staining protocol is as follows:

1. Hydrate in distilled water for 2 min.
2. Stain in hematoxylin 560 for 2 min.
3. Rinse in running water for 1 min.
4. Counterstain with eosin for 1.5 min.
5. Rinse in tap water for 1 min.
6. Rinse in blue buffer for 1.5 min.
7. Rinse in tap water for 1 min.
8. Wash in 80 % ethanol for 30 s.
9. Stain in eosin stain for approximately 10–15 s.
10. Wash in 80 % ethanol for 0.5 min, followed by two changes of 100 % ethanol (each 30 s).
11. Three changes of Formula 83 for 1 min each and then mount with Permount and #1 coverslip.

3.6 Protocol for Immunohisto chemical Analysis

1. The PAP pen is used to outline the section on the slide to limit the use of primary and secondary antisera.
2. Sections are hydrated in PBS for 30 min.
3. Sections are permeabilized with 0.3 % Triton X-100/PBS for 20 min.
4. Wash with PBS 5 min × 3; PBS is then removed by wicking with a Kimwipes.
5. Sections are immediately incubated with 150 μL of 10 % normal goat serum in 0.01 % Triton X-100 (or the species specific serum from which the secondary antibody was produced) for 60 min in a humid chamber. The blocking solution is removed by wicking.
6. Incubate sections with 150 μl of the primary antibody diluted in 1 % normal goat serum/0.01 % Triton X-100 overnight at 4 °C in a humid chamber.
7. The following day the sections are washed with PBS (in a Coplin jar for 5 min × 3).

8. Incubate sections with 150 μL of the fluorophore-conjugated secondary antibody (diluted in 1 % normal goat serum/0.01 % Triton X-100) for 45 min (**Note 3**).
9. Wash sections in PBS in Coplin jar (3 × 5 min each); remove PBS by wicking.
10. Coverslip using Vectashield and a #1.5 coverslip. Seal slides using clear nail polish. The slides are placed in flat slide holder and stored at 4 °C until imaged.

3.7 BrdU Immunohistochemical Protocol Utilized to Identify Cells That Have Undergone a Proliferative Event

1. Hydrate sections/slides in PBS contained in a Coplin jar (3 × 5 min PBS).
2. Denature the DNA by placing the slides in a Coplin jar containing 2 N HCl and 0.5 % Triton X-100 at 37 °C for 60 min [6, 7].
3. Neutralize with Borate buffer (3 × 5 min changes).
4. Wash with PBS (3 × 5 min) and apply PAP pen circles around the sections on the slide.
5. Block with nonspecific binding solution (i.e., the normal serum from the species used to generate the fluorophore conjugated secondary antibody) such as 10% normal goat serum for 30 min at room temperature.
6. Apply the monoclonal BrdU primary antibody diluted 1:25 in PBS with 0.5 % Triton X-100 and incubate overnight at 4 °C in a humidified chamber [6, 7].
7. The next day, rinse the slides in PBS (3 × 5 min) and incubate with 150 μL of the fluorophore conjugated secondary antibody (1:50 dilution of FITC-goat anti-mouse in PBS, 0.5 % Triton X-100) at room temperature in the humid chamber for 1 h.
8. Rinse in PBS (3 × 5 min) and then coverslip with Vectashield anti-fade mounting medium or proceed to immunostain with another species of primary antibody.

4 Notes

1. Note that great care should be taken when injecting the cardiotoxin to prevent inadvertent self-injection. Double gloving is strongly encouraged.
2. The freezing of the skeletal muscle in OCT should not involve complete insertion of the block into the liquid nitrogen super cooled isopentane as it will result in the cracking of the block. Equilibration of the frozen block once placed in the cryostat should be established after approximately 30–45 min.
3. To separate the free fluorophore from the conjugated secondary antibody, quick spin the antibody in a tabletop centrifuge for 15 s and then remove and use the supernatant.

Acknowledgements

This work was supported by grants from the National Institutes of Health (U01 HL100407 and 1R01 HL122576) and the Department of Defense (11763537). GAG is a research fellow funded by the Sarnoff Cardiovascular Research Foundation. The authors acknowledge Cynthia Dekay and Stefan Kren for their assistance with the figures.

References

1. Hawke TJ, Garry DJ (2001) Myogenic satellite cells: physiology to molecular biology. J Appl Physiol 91:534–551
2. Shi X, Garry DJ (2006) Muscle stem cells in development, regeneration and disease. Genes Dev 20:1692–1708
3. Chamberlain JS, Metzger J, Reyes M, Townsend D, Faulkner JA (2007) Dystrophin-deficient mdx mice display a reduced life span and are susceptible to spontaneous rhabdomyosarcoma. FASEB J 21(9):2195–2204
4. Goetsch SC, Hawke TJ, Gallardo TD, Richardson JA, Garry DJ (2003) Transcriptional profiling and regulation of the extracellular matrix during muscle regeneration. Physiol Genomics 14:261–271
5. Hawke TJ, Jiang N, Garry DJ (2003) Absence of p21 rescues myogenic progenitor cell proliferative and regenerative capacity in Foxk1 null mice. J Biol Chem 278:4015–4020
6. Naseem RH, Meeson AP, DiMaio JM, White MD, Kallhoff J, Humphries C, Goetsch SC, DeWindt LJ, Williams MA, Garry MG, Garry DJ (2007) Reparative myocardial mechanisms in adult C57BL/6 and MRL mice following injury. Physiol Genomics 30:44–52
7. Yan Z, Choi S, Liu X, Zhang M, Schageman JJ, Lee SY, Hart R, Lin L, Thurmond FA, Williams RS (2003) Highly coordinated gene regulation in mouse skeletal muscle regeneration. J Biol Chem 278(10):8826–8836

Chapter 7

Fibrosis-Inducing Strategies in Regenerating Dystrophic and Normal Skeletal Muscle

Patrizia Pessina and Pura Muñoz-Cánoves

Abstract

The excessive accumulation of collagens (fibrosis) impairs the function of vital tissues and organs. Fibrosis is a hallmark of severe muscular dystrophies, such as the incurable Duchenne Muscular Dystrophy (DMD), where skeletal muscle is substituted by scar (fibrotic) tissue as disease advances. One of the major obstacles in increasing our ability to combat fibrosis-driven muscular dystrophy progression is that no optimal in vivo models of muscle fibrosis are currently available, limiting fibrosis research and the development of novel therapies. In this chapter we describe different experimental strategies to accelerate and enhance muscle fibrosis in vivo in the widely used animal model for DMD, the mdx mouse. Since excessive tissue scarring also hampers the normal regeneration process after muscle injury, we have extended these fibrogenic strategies to the muscle of normal (non-diseased) mice. These strategies will allow fibrosis induction and assessment in a wide array of genetically modified mouse lines in physiological and pathological conditions of muscle regeneration. They should eventually improve our ability to combat fibrosis and foster muscle regeneration in DMD.

Key words Fibrosis, Skeletal muscle regeneration, Duchenne muscular dystrophy, mdx mice, Chronic injury

1 Introduction

Aberrant deposition of collagens (fibrosis) during the abnormal repair process of damaged tissues leads to impaired function and chronifies disease in a large variety of vital organs. In skeletal muscle, fibrosis is often associated with the muscular dystrophies, including the lethal Duchenne muscular dystrophy (DMD). DMD is caused by the lack of the dystrophin protein due to genetic mutations, which leads to fragility of the muscle fiber membrane that becomes susceptible to contraction-induced damage (reviewed in [1, 2]). The chronicity of tissue damage in DMD leads to vicious cycles of degeneration, regeneration and inflammation, that eventually cause muscle wasting and replacement by fibrotic (and fat) tissue, resulting in lethal respiratory and cardiac failure [3].

Michael Kyba (ed.), *Skeletal Muscle Regeneration in the Mouse: Methods and Protocols*, Methods in Molecular Biology, vol. 1460, DOI 10.1007/978-1-4939-3810-0_7,

Furthermore, the destruction of the muscle architecture and function by the fibrotic tissue progressively reduces the amount of target tissue available for gene- and cell-based therapies for DMD, especially in patients at advanced age [4, 5].

In the widely used mdx mouse model for human DMD, the clinical manifestations are less severe and fibrosis only becomes apparent at old age, except for the poorly accessible diaphragm muscle, that is difficult to handle and offers limited material for experimentation [6, 7]. Indeed, the easily accessible limb muscles only develop substantial fibrosis in mdx mice after 18 months of age. In order to hasten and exacerbate the DMD phenotype, additional transgenic mouse models have been generated through intercrossing of mdx with other mouse lines, but manipulation of these double- or triple-mutant mice is time- and cost-consuming, and, furthermore, these mice do not always recapitulate adequately the dystrophic process observed in humans [6, 8–10]. Therefore, there is an increasing need to develop mouse models that present fibrosis at early stages in life and that more closely mimic human DMD, and which can be easily used for research purposes.

We have recently reported optimized strategies to both anticipate the appearance and prolong the duration of fibrosis in hindlimb muscles of young mdx mice, and which can also induce de novo fibrosis in muscle of WT mice [11, 12]. So far, studies on muscle damage in non-dystrophic animal models have been performed classically with a single injection of myotoxins (for example, cardiotoxin [13, 14]). However, this standard single-injury procedure is not an appropriate fibrosis-inducing strategy, as the damage resolves rapidly and collagen deposition is very mild and only transient [1].

Here we describe in detail distinct fibrosis-inducing protocols, ranging from increasing muscle damage on limb muscles of young mdx mice, by myotoxin injection, to surgically induced trauma (laceration or denervation) or injection of the profibrotic growth factor TGFβ. We also describe the application of these fibrogenic protocols to muscle of normal non-dystrophic mice. These methods accelerated and increased muscle fibrosis in young mdx mice, which persisted over time, and correlated with reduced muscle force, thus mimicking the severe DMD phenotype. Furthermore, enhanced and more stable fibrosis was also obtained by combining these procedures in muscles of normal mice, mimicking abnormal repair after severe damage.

These procedures will facilitate rapidly assessing fibrosis in the easily accessible limb muscles of young mdx mice, without the need of using aged mice. Transferring these fibrogenic strategies to the muscle of WT (non-diseased) mice will also permit inducing and studying fibrosis in several preexisting genetically modified mice, which in turn will contribute to shed light on the underlying mechanisms of fibrosis development. These fibrogenic regimes should enhance our capacity to eventually halt fibrosis during dystrophy progression and abnormal muscle regeneration.

2 Materials

2.1 Reagent and Chemicals

1. Cassettes 15 × 15 × 5 mm (Tissue-Tek Cryomold, Sakura).
2. OCT (Tissue-Tek Cryomold, Sakura).
3. Hematoxylin.
4. Eosin.
5. DPX mounting medium.
6. Primary antibodies: rabbit polyclonal collagen I (Coll 1) (Millipore), rabbit polyclonal fibronectin (FN) (Abcam), and rabbit polyclonal phosphorylated-Smad2/3 (P-Smad2/3) (BD Pharmingen).
7. DAPI: stock solution 1 mg/mL, final concentration 1 μg/mL.

2.2 Equipment

1. Sterile scissors.
2. Sterile tweezers.
3. Hamilton syringe.
4. Surgical scalpel.
5. Surgical clamp.
6. Surgical thread.

2.3 Reagents Setup

2.3.1 Fibrosis-Inducing Strategies

1. *Cardiotoxin solution*: Reconstitute cardiotoxin (CTX; Latoxan) at 10^{-5} M in distilled water. Store in 1 mL aliquots at −20 °C.
2. *TGFβ1 stock solution*: Reconstitute TGFβ1 (rhTGFβ1; R&D Systems) at 10 μg/mL in sterile 4 mM HCl containing 1 mg/mL human or bovine serum albumin. Store in 50 μl aliquots at −20 °C. 50 μl *of TGFβ solution*: Dilute ten times with PBS (to a final concentration of 10 ng/mL) and use 50 μl per muscle (for a total of 50 ng per muscle).

2.3.2 Histological Analysis

1. *Bouin solution*: 71 mL saturated picric acid, 24 mL formaldehyde 37 %, 5 mL acetic acid.
2. *Picric-Sirius solution*: 10 mL Direct Red80 and 90 mL saturated picric acid.
3. *0.1 % Fast green solution*: 100 mg Fast Green in 100 mL saturated picric acid.
4. *0.1 % Fast Green and 0.1 % Direct Red80*: 100 mg Fast Green and 100 mg Direct Red in 100 mL saturated picric acid.
5. *Permeabilization*: PBS 1×, 0.5 % Triton.
6. *Blocking solution*: 10 % goat serum/10 % BSA/PBS 1×.

3 Methods

3.1 Mice

Three-month-old normal C57Bl/6J mice (the classic standard laboratory mouse strain, hereafter referred to as wild type (WT)) and dystrophic C57Bl/10scsn-mdx (mdx) male mice were used for the experiments: the background strain for mdx mice is similar to, but not identical, the C57Bl/6J strain. Mice were housed in standard cages under 12-h light-dark cycles and fed ad libitum with a standard chow diet (*See* **Note 1**).

3.2 Single Fibrotic Treatments in Muscles of WT and mdx Mice

3.2.1 Myotoxin-Induced Injury by Cardiotoxin Injection

1. Sterilize the hind leg with 70% ethanol.
2. Shave the leg between the knee and the ankle, to avoid contaminations or infections.
3. Cut the skin above the tibialis anterior (TA) muscle with sterile scissors in order to identify exactly the muscle.
4. Carefully inject 50 μl of 10^{-5} M CTX (*see* **Notes 2** and **3**).
5. Suture the skin with surgical clamp and thread.

3.2.2 Laceration

1. Sterilize the hind leg with 70% ethanol.
2. Shave the leg between the knee and the ankle, to avoid contaminations or infections.
3. Carefully cut and separated the skin from the underlying tissue.
4. Cut horizontally the TA muscle of one leg at its middle length by making a lesion through 75% of its width and 50% of the muscle thickness with a scalpel (*see* **Notes 4–6**).
5. Suture the skin with surgical clamp and thread.

3.2.3 Denervation

1. Sterilize the hind leg with 70% ethanol.
2. Shave the leg between the hip and the knee, to avoid contaminations or infections.
3. Cut and separate carefully the skin from the underlying tissue.
4. Remove surgically a 5 mm segment of the sciatic nerve down to the gluteus from one leg (*see* **Note** 7).
5. Suture the skin with surgical clamp and thread.

3.2.4 Pro-fibrotic Treatment by TGFβ Injection

1. Sterilize the hind leg with 70% ethanol.
2. Shave the leg between the knee and the ankle, to avoid contaminations or infections.
3. Cut and separate carefully the skin from the underlying tissue.
4. Carefully inject 50 ng of TGFβ1 in the TA muscle in a volume of 50 μl of PBS (vehicle). Two injections (one per week) are made in TA muscle (*see* **Notes 8** and **9**).
5. Suture the skin with surgical clamp and thread.

3.3 Combined Fibrotic Treatments in Muscles of WT Mice

3.3.1 Cardiotoxin Injury Combined with Denervation

1. Sterilize the hind leg with 70% ethanol.
2. Shave the leg between the hip and the ankle, to avoid contaminations or infections.
3. Carefully cut and separated the skin from the underlying tissue.
4. TA muscles are injected with 50 μl of 10^{-5} M CTX immediately after denervation, as indicated above (*see* **Notes 2** and **3**).
5. Suture the skin with surgical clamp and thread.

3.3.2 Cardiotoxin Injury Combined with TGFβ Treatment

1. Sterilize the hind leg with 70% ethanol.
2. Shave the leg between the knee and the ankle, to avoid contaminations or infections.
3. Carefully cut and separated the skin from the underlying tissue.
4. TA muscles are injected with 50 μl of 10^{-5} M CTX (*see* **Notes 2** and **3**).
5. Suture the skin with surgical clamp and thread.
6. 50 μl of TGFβ1 solution are injected intramuscularly twice at day 7 and 10 after CTX injection, without cutting the skin (*see* **Notes 8** and **9**).

3.4 Histological Analysis of Fibrosis

3.4.1 Analysis of Muscle Tissue Architecture with Hematoxylin–Eosin Staining

1. Collect the damaged muscles and put them in cassettes with OCT.
2. Freeze the embedded muscles in isopentane cooled with liquid nitrogen and store at −80 °C until analysis.
3. Cut the muscle samples in 10 μm sections with a cryostat.
4. Let the sections dry at room temperature (RT).
5. Cover the section with hematoxylin for 4 min.
6. Wash with tap water for 5 min.
7. Rinse with acid EtOH for 10 s.
8. Rinse with water for 10 s.
9. Cover the section with eosin for 40 s.
10. Wash with EtOH 70% for 5 min.
11. Wash with EtOH 96% for 5 min.
12. Wash with absolute EtOH for 5 min.
13. Wash with xylol for 5 min.
14. Mount the slide with DPX.

3.4.2 Analysis of Collagen Deposition by Sirius Red Staining

Steps from 1 to 4 as in Subheading 3.4.1.

5. Leave the section overnight (O/N) in Bouin solution.
6. Wash with tap water for 5 min.
7. Cover with picric-Sirius solution (Sirius Red solution) for 1 h.

8. Rinse rapidly with 2 % acetic acid.
9. Wash with EtOH 70 % for 5 min.
10. Wash with EtOH 96 % for 5 min.
11. Wash with absolute EtOH for 5 min.
12. Wash with xylol for 5 min.
13. Mount the slide with DPX.
14. The relative area of the sections occupied by Sirius Red staining is obtained on digital pictures of muscle samples processed with ImageJ software after conversion of the original pictures on binary images. Collagen values are expressed as the percentage of the total area of the section occupied by Sirius Red staining (Fig. 1).

3.4.3 Biochemical Quantification of Collagen Content

Steps from 1 to 4 as in Subheading 3.4.1.

5. Collect 10 10 μm-cryosections in an Eppendorf tube and incubate them with a solution containing 0.1 % Fast Green in saturated picric acid 15 min on a rotary shaker.
6. Wash with distillated water three times.
7. Incubate with 0.1 % Fast Green and 0.1 % Direct Red80 in saturated picric acid for 30 min on a rotary shaker.
8. Wash with distillated water three times.
9. Gently resuspend in a solution of 0.1 M NaOH in absolute methanol (1 vol:1 vol).
10. Absorbance is measured in a spectrophotometer at 540 and 605 nm wavelengths and used to calculate total protein and collagen.

3.4.4 Immunofluorescence Analysis of Fibrosis Markers on Muscle Sections

Steps from 1 to 4 as in Subheading 3.4.1.

5. Fix the section by adding 4 % PFA for 10 min at RT.
6. Wash three times with PBS.
7. Permeabilize for 20 min at RT.
8. Wash three times with PBS.
9. Add blocking solution and incubate for 1 h at RT.
10. Incubate the cells with the primary antibody diluted in blocking solution O/N at 4 °C. For the detection of extracellular matrix proteins and activated fibroblasts (active TGFβ pathway) the following primary antibodies can be used: Collagen I (Coll 1), Fibronectin (FN) and phosphorylated-Smad2/3 (P-Smad2/3).
11. Wash three times with PBS.
12. Incubate the section with secondary antibody diluted in blocking buffer and DAPI at RT for 1.5 h.

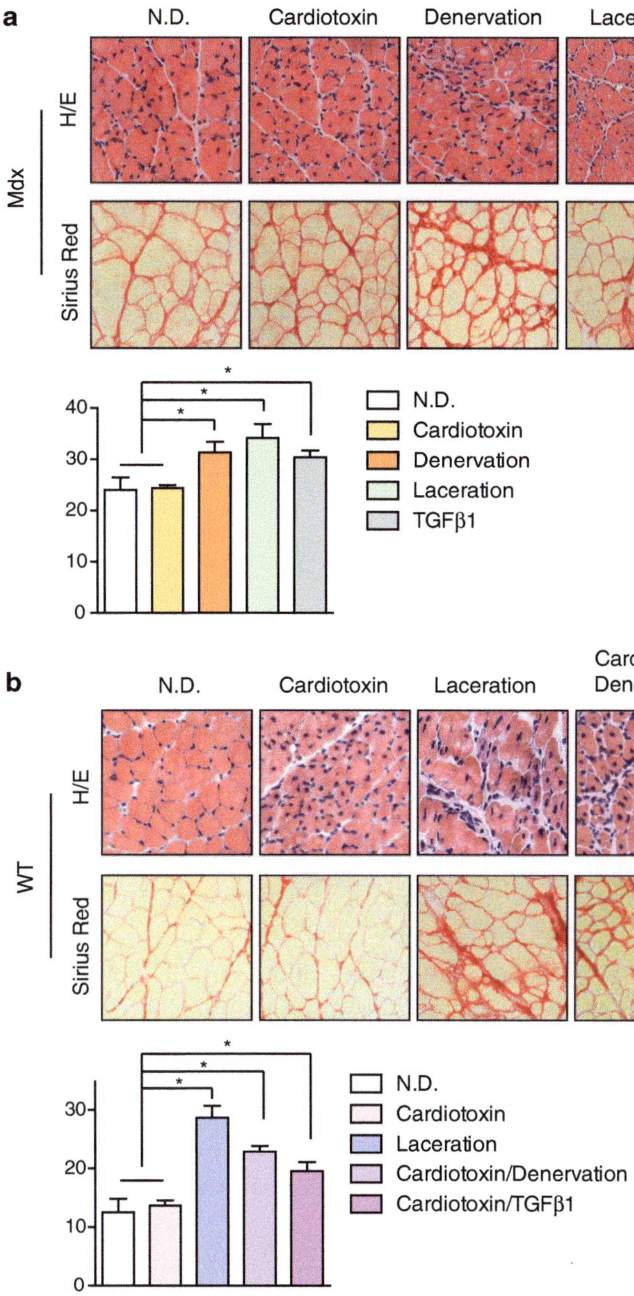

Fig. 1 (**a**) *Upper panel*: H/E and Sirius Red staining of mdx TA muscle 2 months after cardiotoxin injury, denervation, laceration, or injection of TGFβ in muscle of 3-month-old mdx mice compared to non-damaged (ND) mdx control muscle. *Lower panel*: quantification of collagen content in TA muscle of young mdx mice 2 months after the different treatments, as described above. Values represent mean ± SEM; $n = 4$ on each group. Nonparametric Mann–Whitney *U* test; $^*P < 0.05$ versus ND. (**b**) *Upper panel*: H/E and Sirius Red staining of cardiotoxin-injured, laceration, cardiotoxin/denervation, and cardiotoxin/TGFβ-injured muscles of wild type (WT) mice at 1 month after injury. *Lower panel*: quantification of collagen content 1 month after injury of WT muscle. Values represent mean ± SEM; $n = 4$ on each group. Nonparametric Mann–Whitney *U* test; $^*P < 0.05$ versus ND WT muscle

Table 1
Primer list. List of primers used to check fibrotic genes by qRT-PCR

Gene	Forward primer	Reverse primer
L7	GAAGCTCATCTATGAGAAGGC	AAGACGAAGGAGCTGCAGAAC
Collagen I	GGTATGCTTGATCTGTATCTGC	AGTCCAGTTCTTCATTGCATT
Fibronectin	GCGACTCTGACTGGCCTTAC	CCGTGTAAGGGTCAAAGCAT
CTGF	CAGGCTGGAGAAGCAGAGTCGT	CTGGTGCAGCCAGAAAGCTCAA
TIMP-1	TTCCAGTAAGGCCTGTAGC	TTATGACCAGGTCCGAGTT
TGFβ1	TATGACCAGGTCCGAGTT	CTGGTGCAGCCAGAAAGCTCAA

13. Wash three times with PBS.
14. Mount the slide with Mowiol mounting medium.

3.5 Molecular Analysis of Fibrosis Markers

1. Total RNA is isolated from muscle tissue using TRIzol.
2. cDNA is synthesized from 1 μg of total RNA using the First Strand cDNA Synthesis kit and random priming according to the manufacturer's instructions.
3. RT-PCR is performed on a LightCycler 480 System using LightCycler 480 SYBR Green I Master Mix with 10 μM each primer and normalized to L7 ribosomal RNA as a housekeeping gene.
4. Thermocycling conditions: initial step of 10 min at 95 °C, then 50 cycles of 15 s of denaturation at 94 °C, 10 s of annealing at 60 °C and 15 s of extension at 72 °C.
5. Primers of fibrosis-related genes (mouse) are listed in Table 1.

3.6 Functional Analysis: Muscle Force Measurement

1. After the indicated days of treatment, mice are sacrificed and the TA is rapidly excised into a dish containing oxygenated Krebs–Ringer solution.
2. The optimum muscle length (L_o) is determined from micromanipulations of muscle length to produce the maximum isometric twitch force.
3. Maximum isometric-specific tetanic force is determined from the plateau of the curve of the relationship between specific isometric force with a stimulation frequency (Hz) ranging from 1 to 200 Hz for 450 ms, with 2 min of rest between stimuli.
4. The force is normalized per total muscle fiber cross-sectional area (CSA), to calculate the specific net force (mN/mm^2).

4 Notes

1. All operations are performed after intraperitoneal injection (i.p.) of ketamine–medetomidine anesthesia (50 mg/kg and 1 mg/kg body weight). Atipamezol (1 mg/kg body weight) by subcutaneous injection is used to reverse the effects of anesthesia.
2. Use Hamilton syringes for muscle injections for better accuracy.
3. We suggest to perform 2/3 injections in the same muscle (for a total volume of 50 μl) to make sure that the whole muscle area is covered.
4. We advise to lacerate just one TA per mouse, due to the severity of the procedure.
5. Laceration has the disadvantage that the affected area is relatively small thereby limiting the amount of material available for downstream processing. Therefore, subsequent cellular analysis of fibrotic muscle by techniques such as fluorescence-activated cell sorting (FACS) may not be possible in this type of model, but it will be very useful for histological analysis.
6. Since laceration affects a limited area of the muscle, we recommend this procedure for histological analysis on the specific area damaged, as the extent of fibrosis might be underestimated in undamaged areas of the muscle.
7. Denervation of just one leg per mouse is recommended, due to the severity of the procedure.
8. In the cardiotoxin injury combined with TGFβ1 treatment, ketamine/metedomidine anesthesia is recommended just for the first injury (CTX): in the following rounds of injuries (TGFβ1 delivery), mice can be temporally anesthetized with isofluorane.
9. The volumes of injections described above are specific for TA muscle. If these methods were to be applied to Gastrocnemius (GC) or Quadriceps (QC) muscles, the volume should be raised up to 75 μl, both for CTX and TGFβ1.

Acknowledgment

Work in the authors' laboratory has been supported by MINECO-Spain (SAF2012-38547, PIE1400061), AFM, E-Rare, Fundació Marató TV3, MDA, EU-FP7 (Myoage, Optistem, and Endostem), and DuchennePP-NL. PP was supported by a postdoctoral fellowship from AFM. The help of Drs. Antonio L Serrano, E. Perdiguero, Y. Kharraz, and M. Jardí in the original studies where these protocols were first used is greatly appreciated.

References

1. Mann CJ, Perdiguero E, Kharraz Y, Aguilar S, Pessina P, Serrano AL, Munoz-Canoves P (2011) Aberrant repair and fibrosis development in skeletal muscle. Skelet Muscle 1(1):21. doi:10.1186/2044-5040-1-21
2. Serrano AL, Munoz-Canoves P (2010) Regulation and dysregulation of fibrosis in skeletal muscle. Exp Cell Res 316(18):3050–3058. doi:10.1016/j.yexcr.2010.05.035
3. Muntoni F (2003) Cardiac complications of childhood myopathies. J Child Neurol 18(3): 191–202
4. Benedetti S, Hoshiya H, Tedesco FS (2013) Repair or replace? Exploiting novel gene and cell therapy strategies for muscular dystrophies. FEBS J 280(17):4263–4280. doi:10.1111/febs.12178
5. Muir LA, Chamberlain JS (2009) Emerging strategies for cell and gene therapy of the muscular dystrophies. Expert Rev Mol Med 11:e18. doi:10.1017/S1462399409001100
6. Ardite E, Perdiguero E, Vidal B, Gutarra S, Serrano AL, Munoz-Canoves P (2012) PAI-1-regulated miR-21 defines a novel age-associated fibrogenic pathway in muscular dystrophy. J Cell Biol 196(1):163–175. doi:10.1083/jcb.201105013
7. Carnwath JW, Shotton DM (1987) Muscular dystrophy in the mdx mouse: histopathology of the soleus and extensor digitorum longus muscles. J Neurol Sci 80(1):39–54
8. Chandrasekharan K, Martin PT (2010) Genetic defects in muscular dystrophy. Methods Enzymol 479:291–322. doi:10.1016/S0076-6879(10)79017-0
9. Wehling-Henricks M, Jordan MC, Gotoh T, Grody WW, Roos KP, Tidball JG (2010) Arginine metabolism by macrophages promotes cardiac and muscle fibrosis in mdx muscular dystrophy. PLoS One 5(5):e10763. doi:10.1371/journal.pone.0010763
10. Zhou L, Rafael-Fortney JA, Huang P, Zhao XS, Cheng G, Zhou X, Kaminski HJ, Liu L, Ransohoff RM (2008) Haploinsufficiency of utrophin gene worsens skeletal muscle inflammation and fibrosis in mdx mice. J Neurol Sci 264(1–2):106–111. doi:10.1016/j.jns.2007.08.029
11. Pessina P, Cabrera D, Morales MG, Riquelme CA, Gutierrez J, Serrano AL, Brandan E, Munoz-Canoves P (2014) Novel and optimized strategies for inducing fibrosis in vivo: focus on Duchenne Muscular Dystrophy. Skelet Muscle 4:7. doi:10.1186/2044-5040-4-7
12. Pessina P, Kharraz Y, Jardi M, Fukada S, Serrano AL, Perdiguero E, Munoz-Canoves P (2015) Fibrogenic cell plasticity blunts tissue regeneration and aggravates muscular dystrophy. Stem Cell Reports 4(6):1046–1060. doi:10.1016/j.stemcr.2015.04.007
13. Perdiguero E, Sousa-Victor P, Ruiz-Bonilla V, Jardi M, Caelles C, Serrano AL, Munoz-Canoves P (2011) p38/MKP-1-regulated AKT coordinates macrophage transitions and resolution of inflammation during tissue repair. J Cell Biol 195(2):307–322. doi:10.1083/jcb.201104053
14. Suelves M, Vidal B, Serrano AL, Tjwa M, Roma J, Lopez-Alemany R, Luttun A, de Lagran MM, Diaz-Ramos A, Jardi M, Roig M, Dierssen M, Dewerchin M, Carmeliet P, Munoz-Canoves P (2007) uPA deficiency exacerbates muscular dystrophy in MDX mice. J Cell Biol 178(6):1039–1051. doi:10.1083/jcb.200705127

Part II

Evaluating Progenitor Cells

Chapter 8

Isolation, Cryosection and Immunostaining of Skeletal Muscle

Huascar P. Ortuste Quiroga, Katsumasa Goto, and Peter S. Zammit

Abstract

Adult skeletal muscle is maintained and repaired by resident stem cells called satellite cells, located between the plasmalemma of a muscle fiber, and the surrounding basal lamina. When needed, satellite cells are activated to form proliferative myoblasts, that then differentiate and fuse to existing muscle fibers, or fuse together to form replacement myofibers. In parallel, a proportion of satellite cells self-renew, to maintain the stem cell pool. To date, Pax7 is the marker of choice for identifying quiescent satellite cells. Co-immunostaining of skeletal muscle with Pax7 and laminin allows both identification of satellite cells, and the myofiber that they are associated with. Furthermore, satellite cells can be followed through the early stages of the myogenic program by co-immunostaining with myogenic regulatory factors such as MyoD. To test genetically modified mice for satellite cell expression, co-immunostaining can be performed for Pax7 and reporter genes such as eGFP. Here, we describe a method for identification of satellite cells in skeletal muscle sections, including muscle isolation, cryosectioning and co-immunostaining for Pax7 and laminin.

Key words Satellite cell, Stem cell, Skeletal muscle, Pax7, Muscle fiber, Cryosection, Immunostaining

1 Introduction

The primary function of skeletal muscle is conversion of chemical energy into mechanical energy, through interaction of actin and myosin in the sarcomeres. This enables voluntary and involuntary movement, and postural support. The main components of skeletal muscle are bundles of muscle fibers (myofibers), usually arranged in parallel. A myofiber is a syncytium, with the post-mitotic myonuclei located at the periphery. Skeletal muscle represents approximately 38 % of body mass in men, and 31 % in women [1].

As muscle precursor cells (myoblasts) enter terminal differentiation and fuse to form muscle fibers during development, growth and regeneration, they lose the ability to reenter the cell cycle [2]. Thus, to supply new myonuclei for maintenance and repair, skeletal muscle contains a population of resident stem cells, termed satellite cells due to their location on the plasmalemma of a muscle

Michael Kyba (ed.), *Skeletal Muscle Regeneration in the Mouse: Methods and Protocols*, Methods in Molecular Biology, vol. 1460, DOI 10.1007/978-1-4939-3810-0_8, © Springer Science+Business Media New York 2016

fiber, within the ensheathing basal lamina [2–4]. Mammalian adult skeletal muscle has little turnover of myonuclei [5], and hence satellite cells are usually mitotically quiescent [6]. However, in response to muscle damage, satellite cells are rapidly activated to generate myoblast progeny that proliferate, undergo myogenic differentiation, and fuse to repair damaged myofibers, resulting in regeneration of a functional muscle [7]. Non-satellite cell populations have been shown to have myogenic potential [8], but when the satellite cell population is genetically ablated, skeletal muscle does not regenerate [7].

Pax7, a member of the paired box containing family of transcription factors, is expressed in quiescent and activated satellite cells [9] where it plays an important role in satellite cell function [7, 9–11]. Pax7 is a widely used molecular marker for satellite cell detection in mammalian, avian, and amphibian species [12–14] due to its expression being limited to satellite cells in skeletal muscle, and availability of effective antibodies. Co-immunostaining for Pax7 and laminin allows both identification of satellite cells, and the myofiber that they are associated with [15]. Progression of satellite cells through the myogenic program can also be tracked by co-immunostaining for Pax7 and myogenic regulatory factors such as MyoD and myogenin: activated satellite cells co-express Pax7 and MyoD, while differentiating satellite cells express Myogenin, but not Pax7 [13, 16]. To test for satellite cell expression in genetically modified mice, co-immunostaining can be performed for reporters such as β-galactosidase or eGFP and Pax7 e.g., [17].

Nevertheless, immunostaining for Pax7 to detect satellite cells in muscle sections using conventional techniques can be problematic. Here, we describe an immunostaining protocol that optimizes detection of Pax7 in satellite cells, which is also compatible with co-immunostaining for multiple anitgens. While generally applicable to most skeletal muscles, we only describe isolation of the tibialis anterior (TA) and extensor digitorum longus (EDL) muscle, as these are well characterized, e.g., [18, 19].

1.1 Advantages of this Technique

Aldehyde-based fixatives are often used for tissue and cell fixation. However, a disadvantage of this class of fixatives is that they work by cross-linking proteins, resulting in the inability of some protein epitopes to bind antibodies [20, 21]. This protocol employs a heat induced epitope retrieval (HIER) procedure, which improves immunostaining efficiency by modifying the molecular configuration of target proteins through exposure to a heated buffer solution [22]. The high temperature causes cross-linked protein epitopes to unfold, while the buffer solution maintains the conformation of the unfolded protein. Post-fixation with cold methanol further improves antigenicity by removing lipids, which further increases antibody access [23].

This protocol complements others in *Methods in Molecular Biology*, where we describe isolation of skeletal muscles and preparation of single myofibers, for either grafting [24] or tissue culture and immunostaining [25].

2 Materials

2.1 Materials for Dissection and Cryogenic Freezing of Isolated Muscles

1. Stereo dissection microscope with transmission illumination.
2. Cork dissection board.
3. Vannas microscissors.
4. Dissection pins.
5. Fine forceps with grip teeth—1 pair.
6. Fine forceps—2 pairs.
7. Large (30.5 cm) forceps—1 pair.
8. Sterile disposable scalpels No. 11.
9. Cork disks—13 × 13 mm (or cryomolds).
10. Optimal Cutting Temperature (OCT) embedding medium for frozen tissue specimens.
11. Liquid nitrogen Dewar—1.8 L capacity.
12. Isopentane.
13. Clamp stand.
14. Metal beaker—70 × 45 mm.
15. Ethanol solution—70 %.
16. Disposable gloves.

2.2 Materials for Cryosection of Muscles

1. Cryostat (e.g., Leica CM1950).
2. Mounting slides (e.g., Polysene® slides, Thermo Scientific, Ref # 10143265).
3. Microtome blades.
4. Anti-roll microtome glass.
5. Indelible pen.
6. Fine forceps—2 pairs.
7. Mounted razor blade—13 × 38 mm.
8. Brush—width: 13 mm.
9. Light microscope.

2.3 Materials for Immunostaining Muscle Sections

1. Microwave oven—capacity between 500 and 800 W.
2. Epifluorescence microscope.
3. Phosphate Buffer Saline (PBS).

4. Tween 20.
5. Triton X-100 Surfact-Amps Detergent solution.
6. Paraformaldehyde—4 % in PBS.
7. Methanol.
8. Maleic acid.
9. Sodium chloride.
10. Sodium hydroxide solution (5 M).
11. Citric acid monohydrate.
12. Trisodium citrate dihydrate.
13. Roche® blocking reagent (Sigma-Aldrich, Ref# 11096176001).
14. Primary antibodies, including Mouse Anti-Chicken Pax7 Monoclonal Antibody (Ref # AB_528428) (Table 1).
15. Fluorescein-conjugated Alexa Flour® secondary antibodies (Table 2).
16. Coplin jars—60 ml capacity.
17. Humidified slide chamber.
18. Reagent bottles—0.5 L capacity.
19. Microwave-resistant container—0.5 L capacity.
20. Plastic microwave-resistant slide holder.
21. Filter paper.
22. Liquid-repellent slide marker pen (e.g. liquid blocker Super PAP pen).
23. Cover glass—22 × 25 × 0.5 mm.
24. Vectashield containing DAPI (Vector Laboratories, Ref # H–1200).

Table 1
Primary antibodies

Description	Source	Reference	Working dilution	Clonality
Mouse anti-chicken Pax7 (5ea) supernatant	Developmental Studies Hybridoma Bank (DSHB)	AB_528428	1:10	Monoclonal
Mouse anti-chicken Pax7 (5ea) concentrate	Developmental Studies Hybridoma Bank (DSHB)	AB_528428	1:200	Monoclonal
Mouse anti-Pax7	Abcam	AB_55494	1:200	Monoclonal
Rabbit anti-laminin	Abcam	AB_11575	1:200	Polyclonal
Rabbit anti-laminin	Dako	Z0097	1:200	Polyclonal

Table 2
Secondary antibodies

Description	Source	Reference	Working dilution	Clonality
Goat anti-mouse IgG (H+L) Alexa Fluor® 594	Thermo Fisher Scientific	A-11032	1:500	Polyclonal
Donkey anti-mouse IgG (H+L) Alexa Fluor® 594	Thermo Fisher Scientific	R37115	1:500	Polyclonal
Goat anti-rabbit IgG (H+L) Alexa Fluor® 488	Thermo Fisher Scientific	R37116	1:500	Polyclonal
Donkey anti-rabbit IgG (H+L) Alexa Fluor® 488	Thermo Fisher Scientific	A-21206	1:500	Polyclonal

IgG (H+L): gamma immunoglobulins (heavy and light chains)

25. Nail varnish.
26. Cling film.
27. Heat-resistant gloves.

3 Methods

3.1 Isolation and Freezing of Tibialis Anterior and Extensor Digitorum Longus Muscles from Mouse Hind Limb

1. Wear gloves throughout. Layout equipment for muscle dissection (Fig. 1a).
2. Fill a 1.8 L Dewar three quarters full with liquid nitrogen. Attach the metal beaker to the clamp stand and fill three-quarters with isopentane (Fig. 2a).
3. Euthanise mouse by cervical dislocation (*see* **Note 1**).
4. Saturate fur of the hind limbs and lower body with 70% ethanol. Use a scalpel to shave the hind limbs and remove shaved fur with towel paper.
5. Position a pair of scissors at the back of the hind limb at the level of the lower thigh, make an incision through the skin and cut skin down to the paw, stopping just proximal to the digits. Grasp skin using the toothed forceps and pull towards the paw, revealing the underlying musculature and tendons (Fig. 1b).
6. Pin mouse onto a cork dissecting board, with one forelimb pinned through the palm. Pin the contralateral hind limb through the dorsal side of the paw. Extend the tail to the side and the contralateral hind limb to the opposite side, in a cross configuration (Fig. 1b).
7. Using forceps, gently tear and remove the connective tissue overlying the tibialis anterior (TA) while not damaging the muscle.

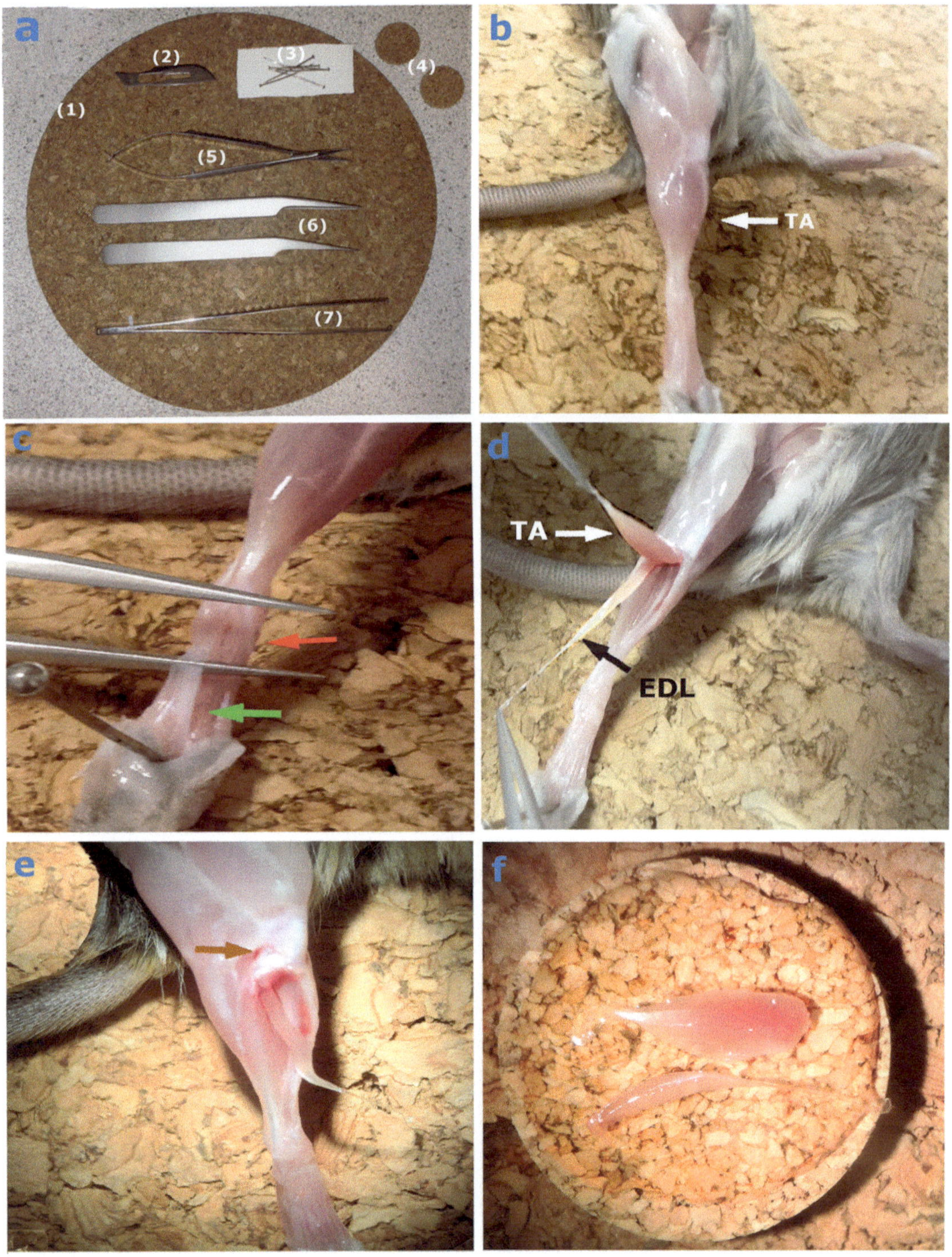

Fig. 1 Dissection of TA and EDL muscles. (**a**) Equipment required for dissection. (*1*) Dissection board, (*2*) scalpel blade, (*3*) dissection pins, (*4*) cork disks, (*5*) microscissors, (*6*) fine forceps × 2, and (*7*) fine forceps with grip teeth. (**b**) Mouse pinned in the cross configuration with skin removed from leg for dissection of the TA (indicated by a *white arrow*) and EDL (located beneath TA). (**c**) Distal tendons of the TA (indicated by *red arrow*). The distal EDL tendons (indicated by *green arrow*) have been lifted with forceps. The tendon is located lateral to

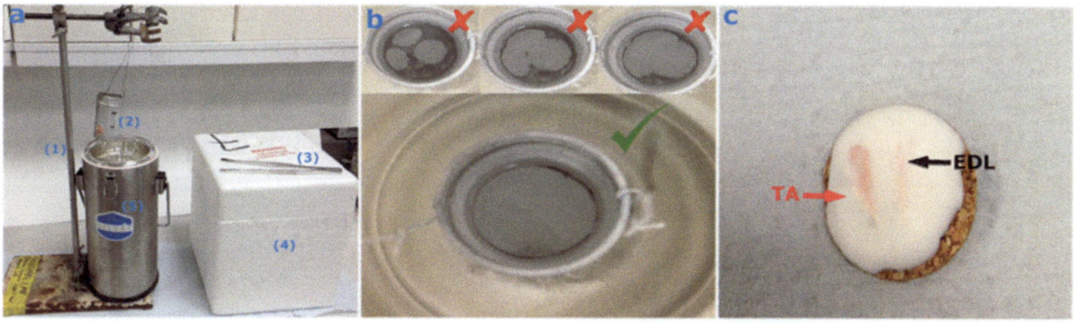

Fig. 2 Freezing of isolated TA and EDL muscles. (**a**) Equipment required for freezing. (*1*) Clamp stand, (*2*) metal beaker, (*3*) large (30.5 cm) forceps, (*4*) Styrofoam box containing dry ice, and (*5*) 1.8 L liquid nitrogen Dewar containing liquid nitrogen. (**b**) Freezing of isopentane at the base of the metal beaker. *Top panels* show progression of the freezing process (*red crosses*), with an acceptable degree of isopentane freezing indicated with a *green tick*. (**c**) Frozen TA (*red arrow*) and EDL (*black arrow*) muscles mounted in OCT, following 30–40 s submersion in liquid nitrogen-cooled isopentane

8. Locate the extensor digitorum longus (EDL) tendons proximal to the extensor retinaculum (annular ligament), laying lateral to the tendon of the TA, on the same side as the extended tail (Fig. 1c). Position forceps under the tendons and gently cut them just distal of the extensor retinaculum. Carefully loop out the tendons, pulling the cut ends through the extensor retinaculum, while ensuring that all four tendons are present.
9. Locate the distal tendon of the TA in the anteromedial dorsal aspect of the foot, close to the ankle, and cut (Fig. 1c).
10. Grip the distal TA tendon with forceps and carefully pull the TA muscle away from the underlying musculature (Fig. 1d), freeing with forceps/scissors as necessary.
11. Hold the TA perpendicular to the leg and using scissors, cut muscle close to the knee joint and ventral crest of the tibia, and remove.
12. Label a cork disk (or cryomold) with indelible pen, invert and pour a fine layer of OCT mounting medium onto it. Place the muscle into the OCT on the cork disk, ensuring that the muscle is fully submersed. Avoid creating bubbles in the OCT, especially near the muscle.
13. Grip the distal tendons of the EDL and gently ease the muscle away from the underlying musculature and bone.

Fig. 1 (continued) the tendon of the TA, on the same side as the extended tail. (**d**) Partially removed TA (*white arrow*) and EDL (*black arrow*) held by their respective distal tendons. (**e**) The EDL muscle is fully revealed when the TA is removed. *Brown arrow* indicates the approximate location of the proximal EDL tendons. (**f**) Isolated TA and EDL muscles in OCT, resting on a cork disk

14. Move the musculature from around the knee on the same side as the contralateral leg to expose the two tendons of the EDL, laying to the side of the knee (Fig. 1e). Carefully cut the two proximal tendons where the muscle arises from the lateral epicondyle of the femur. The EDL should slide free when pulled by the distal tendon. If not, ensure that the proximal tendons are cut.
15. Transfer the EDL onto a labeled, OCT covered, cork disk. Alternatively you can pair both TA and EDL on the same cork disk (Fig. 1f).
16. Lower the metal beaker so that it is half submerged in liquid nitrogen (Fig. 2a) and wait until a frozen (white) layer of isopentane forms at the base of the beaker (Fig. 2b) (*see* **Note 2**).
17. Use the large (30.5 cm) forceps to submerge the cork disk-mounted muscle(s) in the isopentane for 30–40 s. Remove the frozen muscle (Fig. 2c) and store in a covered foam cooler of dry ice, if freezing further samples. Alternatively, keep frozen mounted muscle(s) in liquid nitrogen (*see* **Note 3**).
18. Remove skin from contralateral hind limb (**step 5**) and re-pin the contralateral forelimb and hind limb into the cross configuration (**step 6**). Repeat **steps 7–17** to remove and freeze the contralateral TA and EDL.
19. Store OCT-mounted muscles at –80 °C.

3.2 Cryosection of Skeletal Muscle

1. Allow OCT-mounted muscles to equilibrate to the cryostat chamber temperature (–18 to –21 °C) for 60 min. Ensure tools (forceps, brushes, and mounted razor blade) are kept cold by also storing them in the cryostat chamber (Fig. 3) to prevent melting OCT during processing.
2. Use the mounted razor blade to carefully ease the muscle off the cork disk and trim excess OCT, handling the block using cooled forceps. Use OCT medium to attach the muscle to a cryostat stub. For transverse sections, adjust so that the muscle is perpendicular to the stub and wait for the OCT to freeze to secure in place. If the cryostat has a quick freeze bar (Fig. 3b) for mounted stubs, wait for the specified time (usually 10 min).
3. While the OCT freezes to attach the muscle to the stub, label/number clean, dust free glass slides (include a "test" slide). Place slides on storage shelf (Fig. 3a).
4. Attach stub to the object holder, so that the muscle is perpendicular to the cryostat blade. Fold out the anti-roll plate (Fig. 3c) and lower the muscle near the blade stage. Do not lower the muscle quickly, as misjudging the distance of the blade can dislodge the sample from the stub. If necessary, use the motorized drive controls (Fig. 3b) to position the muscle in line with the cryostat blade (*see* **Note 4**).

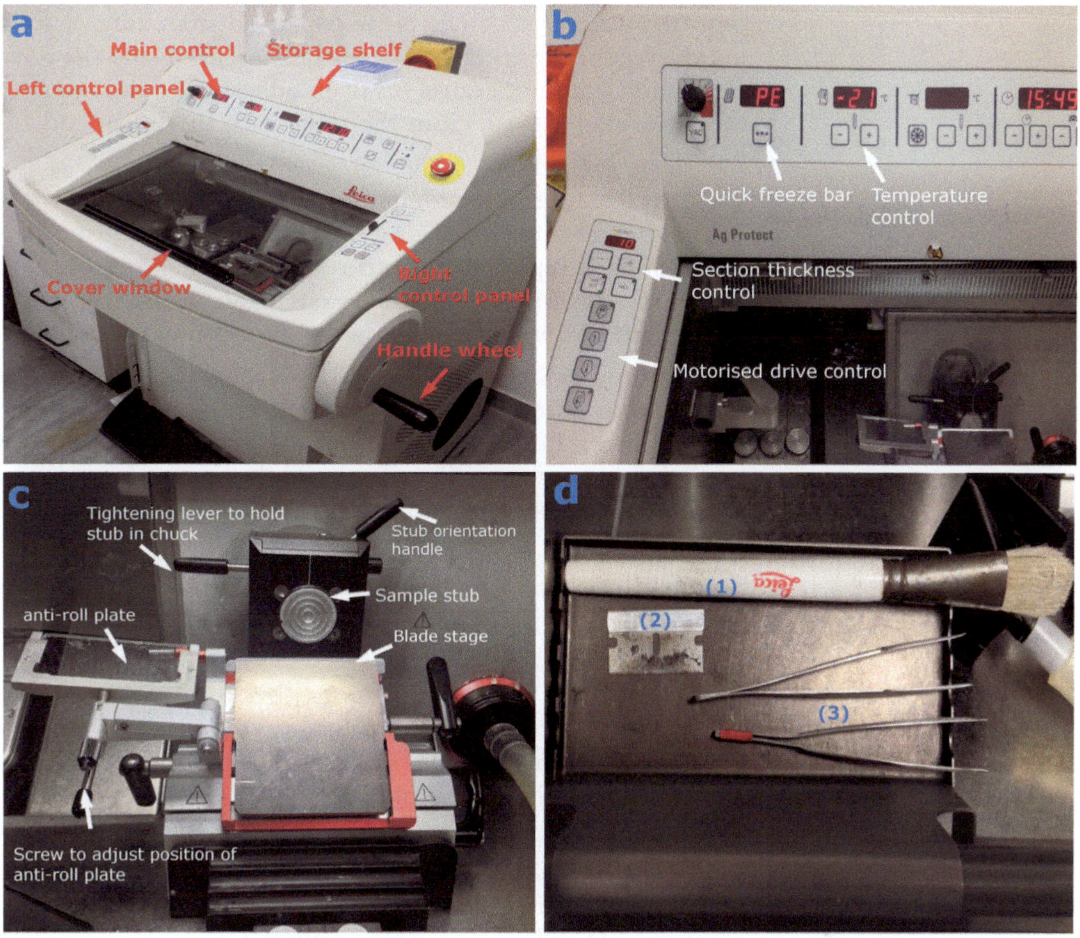

Fig. 3 Cryostat cross-sectioning of skeletal muscle. (**a**) A Leica CM1950 cryostat with parts labeled. (**b**) Detail of cryostat control panels. (**c**) Detail of stub sample holder and blade stage, with adjustment controls highlighted. (**d**) Equipment required for sectioning. (*1*) 13 mm brush, (*2*) mounted razor blade, and (*3*) fine forceps

5. Fold the anti-roll plate back and using the rotating handle, lower the sample onto the blade, ensuring a clean cut. At a thickness of 20–22 μm, section into the frozen OCT medium to reveal the muscle. Then switch section thickness to 7–10 μm (Fig. 3b) and continue sectioning until a smooth transverse cryosection is produced by each cutting cycle. Adjust the screw of the anti-roll plate to obtain a uniform flat section. Use the stub orientation handle to adjust the angle of the stub if needed (Fig. 3c). Regularly fold the anti-roll plate out and use the 13 mm brush to sweep debris from the blade stage.
6. When ready for collecting cryosections, fold the anti-roll plate over the blade stage and turn the rotating wheel in a smooth motion (*see* **Note 5**). When the muscle cryosection is cut, fold the anti-roll plate out and lower the test slide over the section. Static attraction will draw the cryosection to adhere and melt

onto the warmer slide (*see* **Note 6**). Avoid stretching or folding the section by keeping a steady pressure. Pressing the muscle section may sometimes result in adhesion of the tissue to the blade stage. If so, clean the stage with the brush and cut the next cryosection.

7. View the test muscle section under a light microscope to check orientation and quality. If unsatisfactory, further adjust orientation of the sample and/or anti-roll plate, cut another section and collect on the test slide for examination with the microscope. Repeat until a satisfactory test cryosection is obtained.
8. Fold the anti-roll plate back over the blade stage, cut the next muscle cryosection and begin to collect on the first labeled slide (number 1). Cut the next muscle section and collect on labeled slide number 2. Continue until the required number of slides have a muscle section from this initial sampling area of the muscle.
9. To obtain the next sampling area, fold out the anti-roll plate and section the muscle another 20–25 times without collecting on a slide, removing the sections from the blade stage with the brush.
10. Cut a muscle cryosection from this second sampling area and collect, placing next to the section from the initial sampling area, starting with slide number 1.
11. Repeat **steps 9** and **10** throughout the muscle, so that each slide contains a representative cryosection from the entire length of the muscle, and each sampling area has consecutive sections from slide 1 (Fig. 4a) (*see* **Note 7**).
12. Leave slides at room temperature to dry (approximately 40 min).
13. Place dried slides in a slide case and store at −80 °C.

3.3 Immunostaining

1. Prepare 500 ml of a 100 mM citric acid monohydrate solution, diluted in deionized water (*see* **Note 8**).
2. Prepare 500 ml of a 100 mM trisodium citrate dihydrate solution, diluted in deionized water (*see* **Note 8**).
3. Prepare 8.2 mM trisodium citrate dihydrate/1.8 mM citric acid monohydrate buffer solution in a reagent bottle, by adding 41 ml of 100 mM trisodium citrate dihydrate and 9 ml of 100 mM citric acid monohydrate solution to 450 ml of deionized water.
4. Prepare maleic acid buffer in deionized water, by dissolving maleic acid to a final concentration of 100 mM and sodium chloride to 150 mM. While on a magnetic stirrer, adjust pH to ~7.5 by careful addition of 5 M NaOH (*see* **Note 8**).
5. Prepare Tween Phosphate Buffer Saline (TPBS) solution by adding Tween 20 to a final concentration of 0.1 % in PBS (*see* **Note 8**).

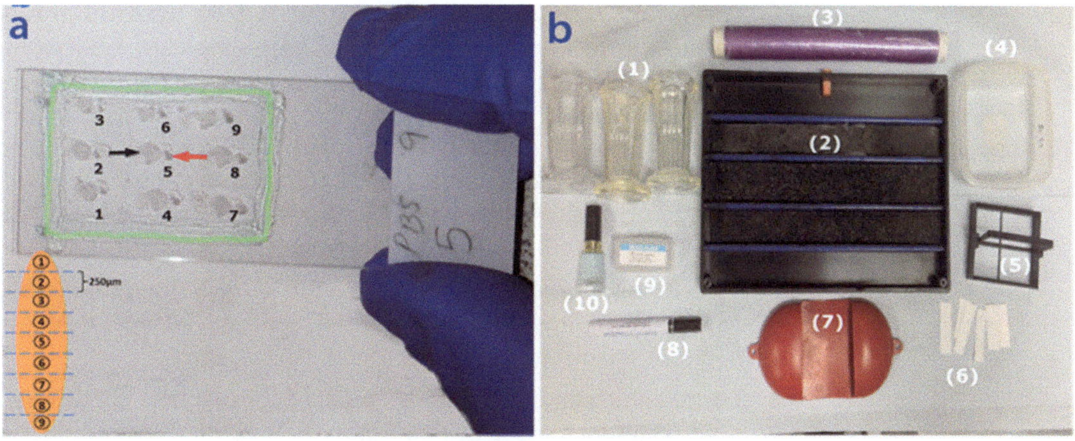

Fig. 4 Slide of muscle cryosection and equipment for immunostaining. (**a**) Glass slide containing cryosections of both TA (*black arrow*) and EDL (*red arrow*) muscles. Diagram represents a skeletal muscle divided into nine sampling regions ~250 μm apart (*dashed blue line*). Numbers on the slide correspond to sampling regions on the diagram, indicating the area of the muscle from which the cryosection was cut. Furthermore, each sampling area has consecutive cryosection from slide 1. Liquid-repellent perimeter drawn around sectioned samples (highlighted in *green*). (**b**) Equipment required for immunostaining. (*1*) Coplin jars, (*2*) humidity slide chamber, (*3*) cling-film, (*4*) 0.5 L microwave-resistant container, (*5*) microwave-resistant slide holder, (*6*) filter paper strips, (*7*) heat-resistant rubber glove, (*8*) Liquid-repellent slide marker pen (e.g. liquid blocker Super PAP pen), (*9*) glass coverslips, and (*10*) nail varnish

6. Prepare the blocking solution by diluting the 10% Roche blocking reagent to 1% in 100 mM maleic acid buffer (*see* **Note 9**).
7. Place methanol at 4 °C—this will be required for post-fixation (*see* **step 13**).
8. Prepare equipment for immunostaining (Fig. 4b).
9. Remove slides form the −80 °C freezer and equilibrate to room temperature.
10. Fill three-quarters of a 60 ml Coplin jar with TPBS and place slides in the jar at room temperature for 10 min. Decant and replace with fresh TPBS and again incubate for 10 min. Repeat once more (*see* **Note 10**).
11. To fix the tissue, fill a Coplin jar with 4% paraformaldehyde/PBS, immerse slides and incubate for 15 min.
12. Wash slides three times with TPBS, 10 min per wash.
13. Place slides in cooled methanol in a Coplin jar for 15 min.
14. Wash slides three times with TPBS, 10 min per wash.
15. Place slides in a plastic (microwave-resistant) slide holder filled two-thirds with trisodium citrate dihydrate/citric acid monohydrate buffer solution. Cover the container with cling film to minimize evaporation and prevent boil over of buffer solution, but leave a small (approximately 3 cm) opening (*see* **Note 11**).

16. Microwave at high power for 3 min. Repeat until the buffer solution boils. This usually takes 2 × 3 min cycles. After solution has boiled, perform the heating cycle for a further six times, 3 min per cycle. Wear heat-resistant gloves as the container becomes extremely hot. Buffer solution may evaporate slightly, so maintain volume. After the final heating cycle, carefully remove the container from the microwave and leave to cool at room temperature for 60 min.
17. Remove slides from cooled trisodium citrate dihydrate/citric acid monohydrate buffer solution and wash three times in TPBS in a Coplin jar, 10 min per wash.
18. Thoroughly remove TPBS by tilting slides so that the leading edge touches a sheet of towel paper. Wipe the reverse side of the slide to remove excess TPBS and remove any remaining TPBS by using fine strips of filter paper, running the strips along the perimeter of the muscle sections.
19. Use a liquid-repellent slide marker pen (e.g. liquid blocker Super PAP pen) to draw a perimeter around the muscle sections (Fig 4a).
20. Saturate the sponge strips of a humidity slide chamber and place slides across the rests of the chamber. Gently pipette 150–200 μl of blocking solution per slide, ensuring that the solution covers the muscle sections. Block the sections for 30 min at room temperature.
21. Prepare the antibody vehicle solution by adding Triton-X 100 to a final concentration of 0.5 % in blocking solution. Dilute primary antibodies to the working dilution, preparing 150–200 μl per slide (*see* **Note 12**). Satellite cells are visualized using monoclonal mouse anti-chicken Pax7 antibody. Co-immunostaining with a polyclonal rabbit anti-laminin antibody can be used to delimit muscle fibers (Table 1).
22. Remove the blocking solution and gently pipette 150–200 μl of the primary antibody solution per slide, ensuring that all muscle sections are covered. Close the lid and incubate the humidified chamber on a level surface at 4 °C overnight.
23. Remove primary antibody solution (can be stored at 4 °C for reuse). If needed, redraw the perimeter with the liquid-repellent slide marker pen.
24. Wash slides in a Coplin jar with TPBS 3 times, 10 min per wash.
25. Dilute fluorochrome-conjugated secondary antibodies to the working dilution in the antibody vehicle solution (Table 2) preparing 150–200 μl per slide.
26. Remove slides from TPBS, place back in the humidified chamber and gently pipette 150–200 μl of secondary antibody solution per slide (*see* **Note 12**). Incubate slides in the dark at room temperature for 60–120 min.

27. Remove secondary antibody solution and wash slides in TPBS in Coplin jar for three times, 10 min per wash. Ensure slides are not exposed to light during washing cycles, e.g., by wrapping Coplin jars in aluminum foil (*see* **Note 13**).
28. Remove TPBS, place slides on a flat surface and add a few (4–5) drops of Vectashield containing DAPI. Carefully lower a 50×22×0.5 mm glass coverslip over the immunostained sections, avoid trapping air bubbles. Nail varnish can be brushed along the edges of the cover slip to secure it. Leave for 10 min.
29. View slides under an epifluorescence microscope (Figs. 5 and 6).
30. Store slides at 4 °C in the dark.

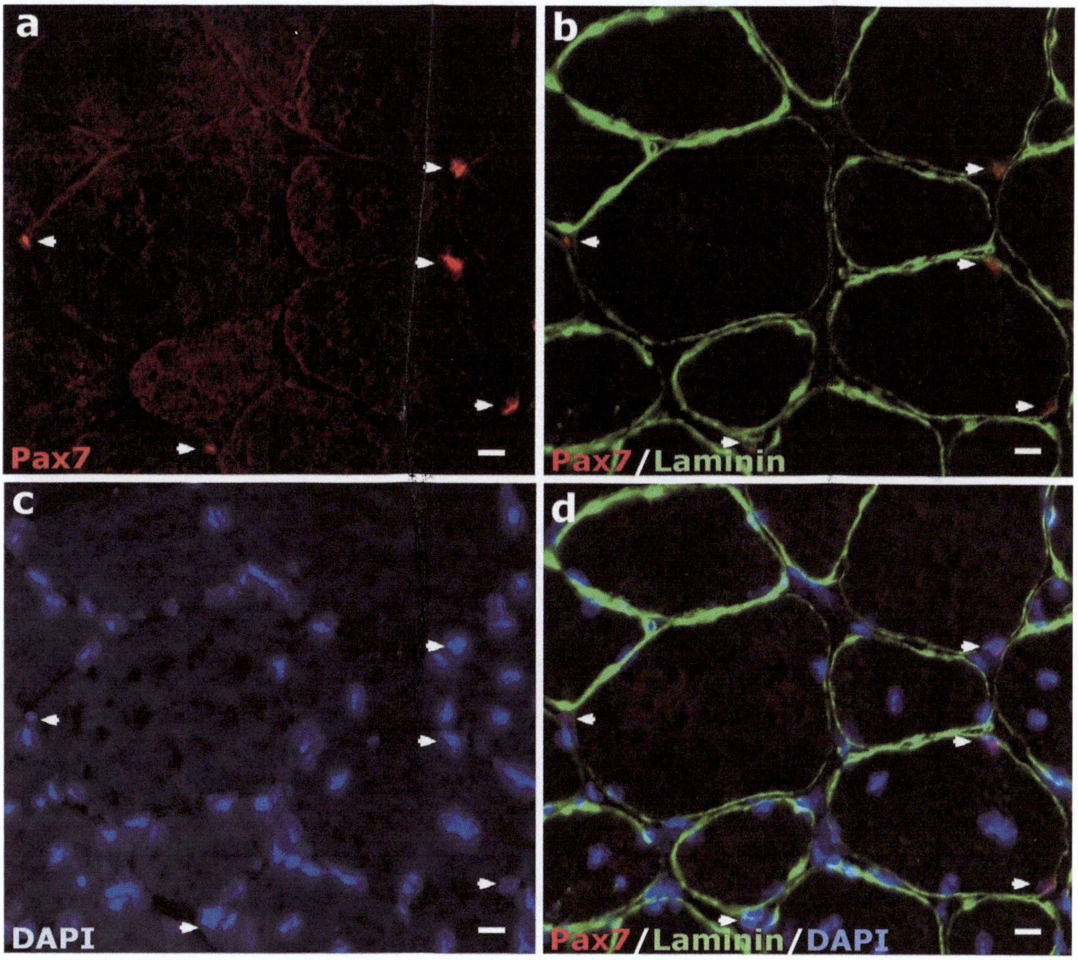

Fig. 5 Immunostaining of transverse cryosectioned tibialis anterior muscle. Satellite cells are indicated by *white arrows*. (**a**) Pax7 immunosignal (*red*) identifies quiescent satellite cells. (**b**) Co-immunostaining of Pax7 with the extracellular matrix protein laminin (*green*) to delimit individual myofibers. (**c**) All nuclei are marked by a DAPI counterstain. (**d**) Merged image of (**a**)–(**c**). Micrographs acquired using Axiovision software (version 4.8.1) on a Zeiss Axiovert 200 M Inverted Microscope. Scale bar represents 10 μm

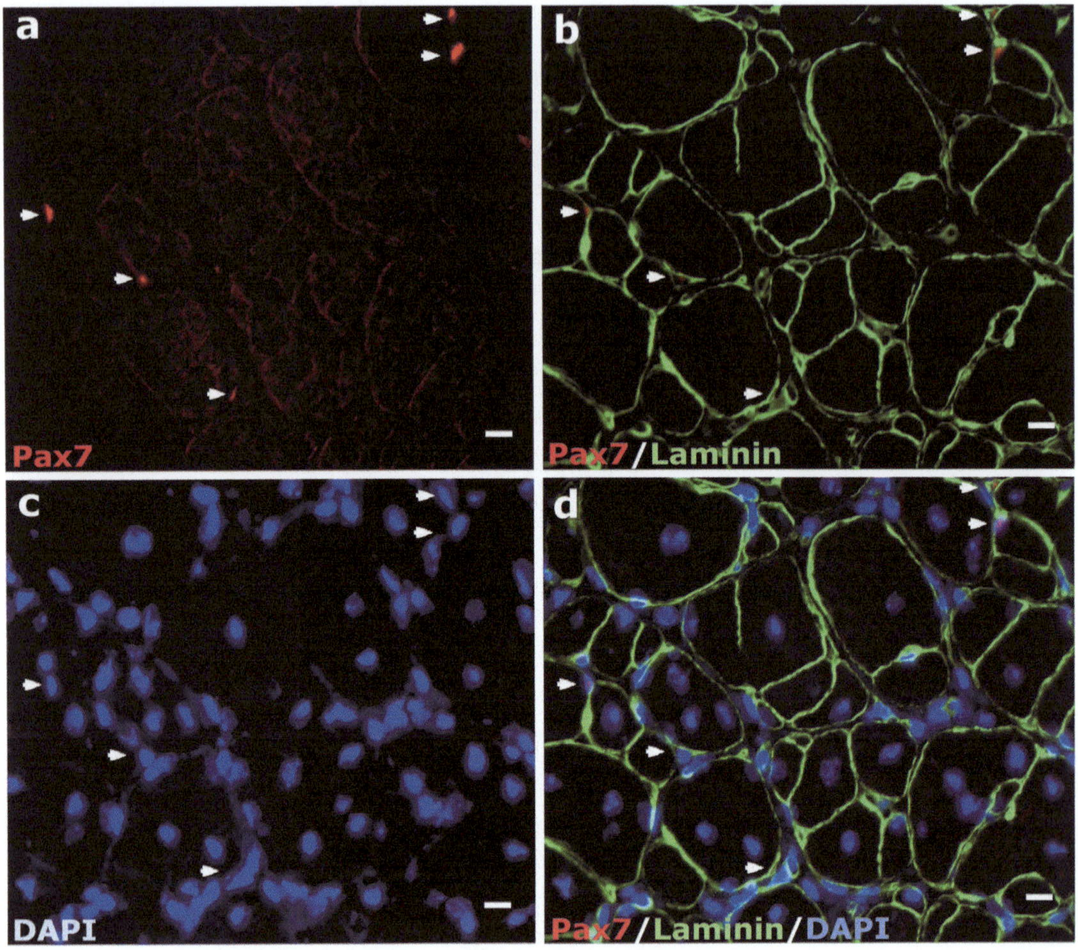

Fig. 6 Immunostaining of transverse cryosectioned regenerated tibialis anterior muscle. Injury was induced in the TA by intramuscular injection of cardiotoxin [26], and the muscle isolated 10 days later. Satellite cells are indicated by *white arrows*. (**a**) Pax7 immunosignal (*red*) identifies quiescent/self-renewed satellite cells. (**b**) Co-immunostaining of Pax7 with laminin (*green*) to delimit individual myofibers. (**c**) All nuclei are marked by a DAPI counterstain. (**d**) Merged image of (**a**)–(**c**). Note the centrally located myonuclei present within the myofibers: a hallmark of skeletal muscle regeneration. Micrographs acquired using Axiovision software (version 4.8.1) on a Zeiss Axiovert 200 M Inverted Microscope. Scale bar represents 10 μm

4 Notes

1. Cervical dislocation is favored over carbon dioxide inhalation as it prevents muscles from becoming hypoxic.
2. It may be useful to have a partner to freeze samples to facilitate rapid muscle isolation.
3. If holding frozen muscles in another liquid nitrogen filled Dewar, avoid direct contact of the samples with liquid nitrogen as this may damage the muscle tissue. Place the cork disks in a

sealable plastic bag and tape or clip the bag to the side of the Dewar. Do not entirely seal the sealable bag.

4. It is advisable to use the rotating handle to bring the sample to the blade stage to prevent dislodging the muscle from the stub.
5. Stopping or hesitating during the cutting cycle can cause varying sample thickness and dislodge the sample from the stub.
6. 2–3 mm of OCT embedding medium surrounding the tissue allows for flipping or curling of the cryosection, before making contact with the slide.
7. To obtain a representative set of muscle cryosections, alter the sectioning interval between sampling areas depending on the length of the muscle. For example, for smaller muscles, shorten the distance between sampling regions to 100–150 μm.
8. Can prepare these solutions in advance and store at room temperature in sealed reagent bottles. Due to viscosity, easiest to dilute Tween 20 and Triton X-100 to make 10% stock solutions, using a graduated 50 ml falcon tube.
9. Can prepare a stock of blocking solution and store at 4 °C for a few days, or at −20 °C longer term.
10. Washes can be performed directly on the slide using 150–200 μl of TPBS here and after antibody incubations.
11. Excessive evaporation causes the salt concentration of buffers to fluctuate. Similarly, boil-over of buffer solution can lead to exposure of the tissue to the atmosphere.
12. Gently spin primary and secondary antibody vials to reduce aggregates, which will minimize immunofluorescent background.
13. If carrying out washes on slides directly, ensure that the slide chamber lid is closed to avoid evaporation.

Acknowledgements

We would like to thank members of the Zammit and Goto labs for help and advice. We also gratefully acknowledge colleagues who shared antibodies through the Developmental Studies Hybridoma Bank developed under the Auspices of the NICHD and maintained by the University of Iowa. The laboratory of Professor Peter Zammit is supported by the Medical Research Council, Muscular Dystrophy UK, Association Française contre les Myopathies, FSHSoc and Rational Bioactive Materials Design for Tissue Regeneration (BIODESIGN) (262948 from the European Commission Seventh Framework Program.

References

1. Janssen I, Heymsfield SB, Wang ZM, Ross R (2000) Skeletal muscle mass and distribution in 468 men and women aged 18–88 yr. J Appl Physiol 89:81–88
2. Scharner J, Zammit PS (2011) The muscle satellite cell at 50: the formative years. Skelet Muscle 1:28
3. Mauro A (1961) Satellite cell of skeletal muscle fibers. J Biophys Biochem Cytol 9:493–495
4. Katz B (1961) The terminations of the afferent nerve fibre in the muscle spindle of the frog. Philos Trans R Soc Lond (Biol) 243:221–240
5. Schmalbruch H, Lewis DM (2000) Dynamics of nuclei of muscle fibers and connective tissue cells in normal and denervated rat muscles. Muscle Nerve 23:617–626
6. Schultz E, Gibson MC, Champion T (1978) Satellite cells are mitotically quiescent in mature mouse muscle: an EM and radioautographic study. J Exp Zool 206:451–456
7. Relaix F, Zammit PS (2012) Satellite cells are essential for skeletal muscle regeneration: the cell on the edge returns centre stage. Development 139:2845–2856
8. Negroni E, Vallese D, Vilquin JT, Butler-Browne G, Mouly V, Trollet C (2011) Current advances in cell therapy strategies for muscular dystrophies. Expert Opin Biol Ther 11:157–176
9. Seale P, Sabourin LA, Girgis-Gabardo A, Mansouri A, Gruss P, Rudnicki MA (2000) Pax7 is required for the specification of myogenic satellite cells. Cell 102:777–786
10. Gunther S, Kim J, Kostin S, Lepper C, Fan CM, Braun T (2013) Myf5-positive satellite cells contribute to Pax7-dependent long-term maintenance of adult muscle stem cells. Cell Stem Cell 13:590–601
11. Buckingham M, Relaix F (2015) PAX3 and PAX7 as upstream regulators of myogenesis. Semin Cell Dev Biol 44:115–25
12. Gnocchi VF, White RB, Ono Y, Ellis JA, Zammit PS (2009) Further characterisation of the molecular signature of quiescent and activated mouse muscle satellite cells. PLoS One 4:e5205
13. Halevy O, Piestun Y, Allouh MZ, Rosser BW, Rinkevich Y, Reshef R, Rozenboim I, Wleklinski-Lee M, Yablonka-Reuveni Z (2004) Pattern of Pax7 expression during myogenesis in the posthatch chicken establishes a model for satellite cell differentiation and renewal. Dev Dyn 231:489–502
14. Morrison JI, Loof S, He P, Simon A (2006) Salamander limb regeneration involves the activation of a multipotent skeletal muscle satellite cell population. J Cell Biol 172:433–440
15. Zammit PS, Partridge TA, Yablonka-Reuveni Z (2006) The skeletal muscle satellite cell: the stem cell that came in from the cold. J Histochem Cytochem 54:1177–1191
16. Zammit PS, Golding JP, Nagata Y, Hudon V, Partridge TA, Beauchamp JR (2004) Muscle satellite cells adopt divergent fates: a mechanism for self-renewal? J Cell Biol 166:347–357
17. Collins CA, Olsen I, Zammit PS, Heslop L, Petrie A, Partridge TA, Morgan JE (2005) Stem cell function, self-renewal, and behavioral heterogeneity of cells from the adult muscle satellite cell niche. Cell 122:289–301
18. Zammit PS, Heslop L, Hudon V, Rosenblatt JD, Tajbakhsh S, Buckingham ME, Beauchamp JR, Partridge TA (2002) Kinetics of myoblast proliferation show that resident satellite cells are competent to fully regenerate skeletal muscle fibers. Exp Cell Res 281:39–49
19. Rosenblatt JD, Parry DJ (1992) Gamma irradiation prevents compensatory hypertrophy of overloaded mouse extensor digitorum longus muscle. J Appl Physiol 73:2538–2543
20. van der Loos CM (2007) A focus on fixation. Biotech Histochem 82:141–154
21. Taylor CR, Shi SR, Chen C, Young L, Yang C, Cote RJ (1996) Comparative study of antigen retrieval heating methods: microwave, microwave and pressure cooker, autoclave, and steamer. Biotech Histochem 71:263–270
22. Shi SR, Key ME, Kalra KL (1991) Antigen retrieval in formalin-fixed, paraffin-embedded tissues: an enhancement method for immunohistochemical staining based on microwave oven heating of tissue sections. J Histochem Cytochem 39:741–748
23. van Essen HF, Verdaasdonk MA, Elshof SM, de Weger RA, van Diest PJ (2010) Alcohol based tissue fixation as an alternative for formaldehyde: influence on immunohistochemistry. J Clin Pathol 63:1090–1094
24. Collins CA, Zammit PS (2009) Isolation and grafting of single muscle fibres. Methods Mol Biol 482:319–330
25. Moyle LA, Zammit PS (2014) Isolation, culture and immunostaining of skeletal muscle fibres to study myogenic progression in satellite cells. Methods Mol Biol 1210:63–78
26. Fortier M, Figeac N, White RB, Knopp P, Zammit PS (2013) Sphingosine-1-phosphate receptor 3 influences cell cycle progression in muscle satellite cells. Dev Biol 382:504–516

Chapter 9

Isolation of Mouse Periocular Tissue for Histological and Immunostaining Analyses of the Extraocular Muscles and Their Satellite Cells

Pascal Stuelsatz and Zipora Yablonka-Reuveni

Abstract

The extraocular muscles (EOMs) comprise a group of highly specialized skeletal muscles controlling eye movements. Although a number of unique features of EOMs including their sparing in Duchenne muscular dystrophy have drawn a continuous interest, knowledge about these hard to reach muscles is still limited. The goal of this chapter is to provide detailed methods for the isolation and histological analysis of mouse EOMs. We first introduce in brief the basic anatomy and established nomenclature of the extraocular primary and accessory muscles. We then provide a detailed description with step-by-step images of our procedure for isolating (and subsequently cryosectioning) EOMs while preserving the integrity of their original structural organization. Next, we present several useful histological protocols frequently used by us, including: (1) a method for highlighting the general organization of periocular tissue, using the $MyoD^{Cre} \times R26^{mTmG}$ reporter mouse that elegantly distinguishes muscle ($MyoD^{Cre}$-driven GFP^{+}) from the non-myogenic constituents ($Tomato^{+}$); (2) analysis by H&E staining, allowing for example, detection of the pathological features of the dystrophin-null phenotype in affected limb and diaphragm muscles that are absent in EOMs; (3) detection of the myogenic progenitors (i.e., satellite cells) in their native position underneath the myofiber basal lamina using Pax7/laminin double immunostaining. The EOM tissue harvesting procedure described here can also be adapted for isolating and studying satellite cells and other cell types. Overall, the methods described in this chapter should provide investigators the necessary tools for entering the EOM research field and contribute to a better understanding of this highly specialized muscle group and its complex micro-anatomy.

Key words Ocular/periocular tissue, Extraocular muscles, Retractor bulbi muscle, $MyoD^{Cre} \times R26^{mTmG}$, mdx^{4cv}, Duchenne muscular dystrophy, Satellite cells, Pax7, Laminin, Antigen retrieval

1 Introduction

The extraocular muscles (EOMs) comprise a group of highly specialized skeletal muscles controlling eye movements [1]. The goal of this chapter is to provide detailed methods for the isolation and histological analysis of mouse EOMs. As described below, a number of features unique to EOMs, including their sparing in Duchenne muscular dystrophy, have drawn continuous interest, but

Michael Kyba (ed.), *Skeletal Muscle Regeneration in the Mouse: Methods and Protocols*, Methods in Molecular Biology, vol. 1460, DOI 10.1007/978-1-4939-3810-0_9, © Springer Science+Business Media New York 2016

knowledge about these hard to reach muscles is still limited. When we first began studying EOMs, we could find only very few resources to assist us in developing an approach for harvesting mouse periocular tissue for subsequent studies of isolated satellite cells [2, 3] and for histological analyses [3, 4]. We have been amazed by the degree of complexity of EOM micro-anatomy and we recognized the need for a comprehensive resource that includes relevant literature in the field and optimized methods for isolation and histological analysis of mouse EOMs. Here, we present our "EOM laboratory atlas" with the hope that it will provide useful tools for the study of EOMs and encourage researchers to enter the EOM research field.

EOMs represent a unique muscle phenotype based on a range of properties, including a diversity of expressed sarcomeric myosin isoforms, specialized patterns of innervation, and specific metabolic and contractile characteristics [5–8]. The developmental origin of EOMs adds another distinct facet to this muscle group. While body and limb muscles develop from the somites, EOMs are descended from prechordal and paraxial head mesoderm [9, 10]. The EOMs are also distinct from other skeletal muscles in their differential response to disease, being preferentially involved or spared in a variety of metabolic, mitochondrial and neuromuscular disorders [11–15]. Especially intriguing for muscular dystrophy research is the sparing of this muscle group in Duchenne muscular dystrophy, where EOMs remain anatomically and functionally spared even in late stages of the disease despite the severe pathology observed in other skeletal muscles [16, 17]. Likewise, EOMs are spared in animal models of muscular dystrophy resulting from the absence of dystrophin or other dystroglycan complex-related proteins [17–19]. While the sparing of EOMs in dystrophinopathy has made this muscle group a target for many investigations, the mechanism behind this EOM sparing remains unresolved [20–22]. Features that have been studied and proposed as potential protective mechanisms include the small myofiber caliber [23, 24], calcium homeostasis [17, 18, 25, 26], antioxidant capacity [27], and the compensatory role of utrophin (an autosomal dystrophin homologue) [28–30]. Specific properties of EOM myogenic progenitors have also been proposed as possible contributory factors of this EOM sparing [31–33]. Our recent study has indeed demonstrated that satellite cells isolated from EOMs harbor superior performance both in culture and in an in vivo engraftment model, compared to satellite cells from limb and diaphragm muscles [3]. We also showed that satellite cells isolated from dystrophin-null (mdx^{4cv}) mice retained their robust growth and renewal properties while satellite cells from muscles affected by dystrophin deficiency (i.e., limb and diaphragm muscles) performed poorly [3].

Preceding the protocol of EOM isolation detailed in the current chapter, we provide here a brief summary of the anatomical organization of EOMs as gathered from the literature, to assist with upcoming protocol details and nomenclature. For a comprehensive description of EOM anatomy readers can further refer to earlier reviews [1, 8] and to published detailed schematics [34] as well as additional online resources [35–37].

Each set of EOMs is composed of four recti (superior, inferior, medial, and lateral) and two obliques (superior and inferior), defined by their orientation relative to the eye and the range of motion for which they are responsible [8]. The four rectus muscles as well as the superior oblique muscle originate from the annulus of Zinn, a ring of fibrous tissue surrounding the optic nerve at its entrance at the apex of the orbit. The four rectus muscles then course anteriorly to insert directly into the sclera, anterior to the equator of the globe. The superior oblique muscle also courses forward passing through the trochlea (a rigid fibrocartillaginous ring attached to the orbit), before turning laterally to insert into the sclera on the superior part of the globe. The inferior oblique muscle originates from the maxillary bone in the medial wall of the orbit and inserts into the sclera on the lateral part of the globe, medial to the insertion of the lateral rectus. In amniotes, the EOMs are characterized by a distinctive compartmentalized organization where each EOM has an outer orbital layer and an inner global layer, facing toward the orbit and the eye, respectively [8, 10, 38]. As clearly demonstrated for the rectus muscles, while the global layer extends the full muscle length inserting into the sclera, the orbital layer terminates earlier, anchoring into the pulley, a specialized connective tissue band (containing smooth muscle) that encircles the globe near its equator [39–41]. The orbital and global layers which typically can be distinguished by their respective smaller and larger myofiber diameter [8, 42], also differ in their isoform expression profiles of a range of sarcomeric protein families including myosin, tropomyosin, troponin, and alpha-actinin [43–45].

In addition to the six principal EOMs, accessory extraocular muscles can be found, including: (1) the levator palpebrae superioris, which is involved in elevating of the upper eyelid. It originates from the orbital surface of the sphenoid bone and runs above the superior part of the globe to attach onto the upper eyelid and the superior tarsal plate [8]; (2) the retractor bulbi, which is involved in retracting the eyeball into the orbit, forcing the nictitating membrane (third eye lid) across the cornea (not present in humans and monkeys). The retractor bulbi is composed of several (1–4, depending on species) muscular slips, originating from the back of the orbit, positioned internally to the four rectus muscles, surrounding the optic nerve, and inserted onto the posterior hemisphere of the globe behind the attachments of the recti muscles in a nearly complete muscular cone [46–48]. Although, in common with the EOMs, these accessory muscles develop from the head mesoderm [4, 10], unlike the EOMs, they do exhibit dystrophin-null related pathology [3, 21, 29, 49], which further underscores the uniqueness of the EOMs even within the periocular tissue.

A summary of all methods for the isolation and histological analysis of the mouse EOMs described in this chapter is schematically depicted in Fig. 1. We first provide a detailed description and

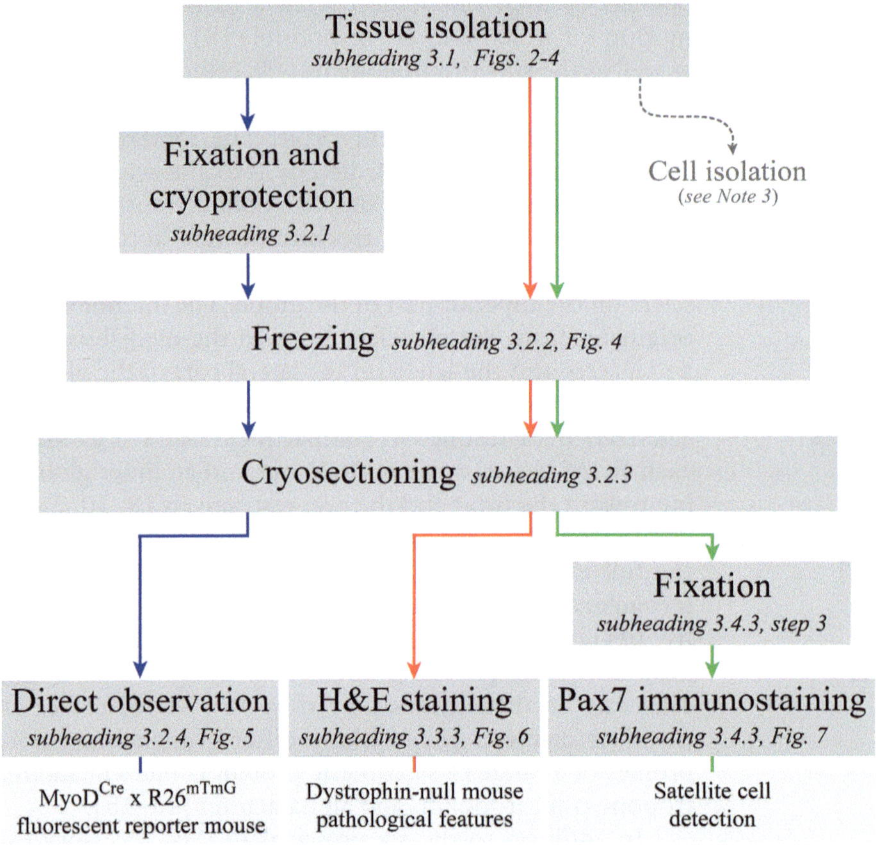

Fig. 1 A schematic summary of all the methods described in this chapter. After tissue isolation, three different tissue processing routes can be followed depending on the downstream application desired. For a direct study of a fluorescent reporter (MyoD^{Cre} × R26^{mTmG} reporter mouse, "blue route"), tissue is fixed and cryoprotected before freezing and is observed following cryosectioning. This route is also used by us when performing an X-Gal reaction on tissue isolated from LacZ-based reporter mice. If H&E staining is desired ("orange route"), tissue is freshly frozen after isolation, and H&E staining is performed after cryosectioning. If Pax7 immunostaining is desired ("green route"), tissue is also freshly frozen after isolation but a fixation is done after sectioning, before starting the immunostaining procedure. Notably, this latter route that has been optimized specifically for the detection of Pax7 includes an antigen retrieval step (Subheading 3.4.3, **steps 1**, **5–9**), which is not required for the detection of many other antigens. Furthermore, immunodetection of many antigens can be achieved on unfixed tissue or tissue that has been fixed prior to freezing ("orange route" and "blue route," respectively). The tissue harvesting procedure detailed here has been developed for the histological analyses of EOMs, but can also be adapted for isolating and studying satellite cells and other cell types as detailed in **Note 5**

figures with step-by-step images of our procedure for isolating EOMs while preserving as much as possible the integrity of their original structural organization within the orbit (Figs. 2 and 3, for a view of the final isolated ocular/periocular preparation *see* Fig. 4). We then detail our freezing and cryosectioning protocols, including fixation steps which are performed either prior to tissue freezing or after sectioning, according to the requirements of the specific downstream analysis. To highlight the general organization of the EOMs, we present a histological analysis of EOMs from the $MyoD^{Cre} \times R26^{mTmG}$ reporter mouse that provides an elegant tool for distinguishing between muscle ($MyoD^{Cre}$-driven GFP^{+}) and the non-myogenic tissue ($Tomato^{+}$) constituents (Fig. 6). We also describe our protocol for H&E staining, one of the principal stains used in histology, that allows us to detect, for example, pathological features in the dystrophin-null phenotype in affected limb and diaphragm muscles that are absent in EOMs (Fig. 6). Lastly, we provide our method for the detection of the myogenic progenitors (i.e., satellite cells) in their native position underneath the myofiber basal lamina using double immunostaining for Pax7, an established satellite cell marker [50, 51] and laminin, which highlights the myofiber basal lamina (Fig. 7). The EOM tissue harvesting procedure described here can also be used for isolating and studying satellite cells and other cell types as detailed in our recent publications [3, 4].

Overall, the methods described in this chapter should provide investigators with the necessary tools for entering the EOM research field and contribute to a better understanding of this highly specialized muscle group and its complex micro-anatomy.

2 Materials

2.1 General Comments

Throughout the tissue isolation and processing protocols described in this chapter, we use aseptic conditions as much as possible to minimize tissue degradation by potential bacterial contamination. For this:

1. All solutions are prepared using Milli-Q water and are typically sterilized by filtering through 0.22-μm filter. Glassware is sterilized by autoclaving, and the dissection tools are placed for few seconds into a hot glass bead dry sterilizer before use.
2. All the muscle isolation steps are carried out with the help of a stereo dissecting microscope that is placed in an isolation box/ clean area dedicated for this procedure in order to limit contamination.
3. All the slide-processing steps are done in the tissue culture hood to minimize potential contamination and permit long-term storage.

2.2 General Equipment

1. Stereo dissecting microscope equipped with a free-standing light source for adjustable illumination; to avoid excess heat during tissue isolation we recommend a fiber optic light source delivered through flexible gooseneck light pipes.
2. Isolation box/clean area dedicated for the stereo dissection microscope.
3. Tissue-culture hood.
4. Chemical fume hood.
5. Hotplate magnetic stirrer.
6. Cryostat.
7. –20 and –80 °C freezers.
8. Phase contrast/fluorescent microscope equipped with digitized photography system for analyzing tissue sections.
9. Bottle top 0.22-μm filters for sterile filtration of solutions.

2.3 Animals

Procedures described in this chapter have been performed by us with mice ranging in age from ~3 weeks to ~2 years old. All mouse strains listed below are on C57BL/6 background and from colonies maintained by us at the University of Washington. Animal care and experimental procedures were approved by the Institutional Animal Care and Use Committee at the University of Washington.

1. Wildtype mice.
2. $MyoD^{Cre/+} \times R26^{mTmG/+}$ double heterozygous mice [4] were obtained by crossing $MyoD^{Cre}$ knockin heterozygous males [$MyoD1^{tm2.1(icre)Glh}$ [52]] with knockin reporter females $R26^{mTmG}$ [Gt(ROSA) $26Sor^{tm4(ACTB\text{-}tdTomato,\text{-}EGFP)Luo}$/J [53]]. Both mouse strains are available at Jackson Lab: $MyoD^{Cre}$ (JAX stock 014140, on original FVB background), $R26^{mTmG}$ (JAX stock 007676).
3. Dystrophin-null mdx^{4cv} [3, 54, 55]. The mdx^{4cv} allele is one of several mdx "cv" strains that were generated by point mutations upon mutagen-treatment of male mice [54–56]. The mdx^{4cv} strain has been preferred by many laboratories due to the reduced occurrence of revertants (i.e., spontaneously appearing $dystrophin^{+}$ myofibers [57]) compared to the "standard" mdx mice (spontaneous mutation) [55]. This mouse strain is available at Jackson Lab (JAX stock 002378).

2.4 Dissecting Tools and Supplies

1. Dissecting scissors: 12.5 cm long, straight, stainless steel.
2. Two Iris forceps: 0.8 mm tips, 10 cm long, straight, serrated jaws, stainless steel.
3. Two Dumont #1 forceps: 0.20 × 0.12 mm tips, 12 cm long, straight, stainless steel.

4. Microscissors: Vannas type, 8.5 cm long, straight, 7 mm blades, 0.025×0.015 mm tips, stainless steel (for delicate cutting and fine incisions).
5. Two fine point forceps: Dumont tweezers #5, 11 cm length, straight, 0.05×0.01 mm tips, biology tips, stainless steel.
6. Two Dumont #5/45 forceps: 11 cm length, 45° angle, 0.05×0.01 mm tips, medical biology tips, stainless steel.
7. Long dressing forceps: 25.5 cm long, straight, serrated jaws, stainless steel.
8. Dissecting board (covered with Parafilm) with tissue pins.
9. Glass bead dry sterilizer (to sterilize dissection tools).
10. 70% ethanol in a plastic bottle spray (to disinfect the mouse head before dissection).

2.5 Supplies and Solutions for Processing the Isolated Tissue Through Cryosectioning

1. 24-well multiwell tissue culture dishes.
2. Dulbecco's phosphate-buffered saline (PBS).
3. 4% PFA fixative solution: 4% paraformaldehyde in sodium phosphate buffer with 0.03 M sucrose. All preparation steps must be done in a fume hood (*see* **Note 1** for specific handling). For preparing 100 ml, weigh 4 g of paraformaldehyde in a glass Erlenmeyer flask and place it on a hotplate (set to medium heat) to heat the powder until it starts to sweat. Cool for 10 min at room temperature, then add 1.026 g of sucrose, 10 ml of 1 M sodium phosphate pH 7.4 and 80 ml of Milli-Q water. Warm the solution on a hotplate magnetic stirrer to 60 °C with continuous stirring to dissolve the powder. Allow the solution to cool to room temperature. Add 1–4 drops of 1 *N* NaOH, until the opaque color of the solution clears. Adjust the pH to 7.2–7.4 using concentrated HCl while checking pH using color pH strips. Bring the volume to 100 ml and filter through a 0.22-μm disposable filter unit into a bottle. Store at 4 °C in an aluminum foil-wrapped bottle (for protecting from light exposure) for no more than 1 month. Pre-warm to room temperature before use; pre-warm only the volume that is required for immediate use to maintain quality and effectiveness of fixative.
4. Ascending series of sucrose solutions for tissue cryoprotection: 10%, 20%, and 30% sucrose in PBS.
5. Gyrating platform rotator (to agitate tissue during cryoprotection).
6. Cryomold: disposable vinyl specimen mold, $10 \times 10 \times 5$ mm.
7. Tissue-Tek OCT compound.
8. Stainless steel mortar (275 ml capacity) suspended above a stainless steel bowl (filled with liquid nitrogen) housed in an insulated plastic container (for more details about the specific apparatus we are using see manufacturer website [58]).

9. Liquid nitrogen.
10. Dewar flask for transporting and holding liquid nitrogen as required for tissue freezing.
11. Isopentane (2 methyl-butane).
12. Styrofoam box containing dry ice.
13. Aluminum foil.
14. Freezer bags.
15. Slides: Superfrost Plus microscope slides (Fisherbrand, or similar). These slides are positively charged to enhance adherence of tissue sections to the glass.
16. Slide boxes to organize and store slides in the −80 °C freezer.
17. Freezer box, cardboard to store frozen tissues in the −80 °C freezer.
18. Fine-tipped paintbrush to manipulate frozen sections just after sectioning, before being transferred onto the slides.

2.6 Supplies and Solutions for Direct Observation (DAPI-Staining and Mounting)

1. PBS as in **item 2** in Subheading 2.5.
2. DAPI solution (4′,6-diamidino-2-phenylindole dihydrochloride): stock concentration 1 mg/ml and a working concentration of 1 μg/ml diluted in PBS (*see* **Note 2**).
3. Rectangular glass staining dish (105 × 85 × 68 mm).
4. Mounting medium: Vectashield (Vector Laboratories) anti-fade mounting medium (hardening formulation), or alternative antifade mounting medium that inhibits photobleaching of fluorescent dyes and fluorescent proteins.
5. Cover glass, 22 × 50 mm, #1 thickness (0.13–0.17 mm), rectangular.
6. Nail polish to seal cover glasses after mounting.

2.7 Supplies and Solutions for H&E Staining

1. Slide plastic staining dishes with adapted slide holders.
2. Filter paper circles, Q5 quantitative grade.
3. Harris hematoxylin (non-acidified). Solution is filtered before the staining procedure through Q5 quantitative grade filter paper.
4. Scott's tap water substitute (pH = 8): dissolve 2 g of sodium bicarbonate and 20 g of $MgSO_4 \cdot 7H_2O$ in 1 L of tap water.
5. Eosin Y stock solution (1 %): dissolve 10 g of Eosin Y in 200 ml of distilled water and store at room temperature.
6. Eosin Y working solution (0.25 %): mix 250 ml of eosin Y stock solution with 750 ml of 80 % ethanol and 5 ml of concentrated glacial acetic acid, store at room temperature.
7. Ascending series of ethanol solutions for dehydration: 70 %, 95 %, and 100 % ethanol in Milli-Q water.

8. Shandon (or an alternative) xylene substitute.
9. Shandon (or an alternative) xylene substitute mounting medium.

2.8 Supplies and Solutions for Immunostaining

1. Small Pyrex loaf dish (86 × 286 × 152 mm).
2. Hydrophobic barrier pen.
3. Staining slide tray.
4. Antigen retrieval buffer: 10 mM Tris Base, 1 mM EDTA Solution, 0.05 % Tween 20, pH 9.0 (*see* **Note 3**).
5. TBS (Tris-buffered saline): 0.05 M Tris–HCl; 0.15 M NaCl; pH 7.4.
6. Blocking solution: TBS containing 2 % normal goat serum (TBS–2% NGS).
7. Detergent solution: TBS containing 0.05 % Tween 20 (TBS–0.05% Tween 20).
8. Flexible plastic coverslip (or a piece of Parafilm cut to the correct size) for the primary antibody incubation step.
9. Primary antibody solution consisting of the following primary antibodies diluted together in TBS–2% NGS blocking solution: (1) mouse anti-Pax7, IgG1, Developmental Studies Hybridoma Bank, bioreactor supernatant (our preferred product form due to reduced background when working with mouse tissue), stock at ~1.3 mg/ml, diluted 1:100; (2) rabbit anti-laminin, polyclonal, EMD Millipore, stock at 1 mg/ml, diluted 1:100. This solution is prepared just before using it (Subheading 3.4.3, **step 12**).
10. Secondary antibody solution consisting of the following secondary antibodies (typically purchased from Molecular Probes now part of Thermo Fisher Scientific) diluted together in TBS–2% NGS blocking solution: (1) Alexa Fluor 488 goat anti-mouse IgG1, stock at 2 mg/ml, diluted 1:1000; (2) Alexa Fluor 568 goat anti-rabbit IgG (H + L), stock at 2 mg/ml, diluted 1:1000. This solution is prepared just before using it (Subheading 3.4.3, **step 14**).
11. DAPI solution as in **item 2** in Subheading 2.6.
12. Mounting medium, cover glass, and nail polish as in **items 4–6** in Subheading 2.6.

3 Methods

A schematic summary of all procedures described in this chapter for studying EOMs is presented in Fig. 1. These histological and immunostaining protocols can be also used for studies of other mouse muscles (*see* Fig. 6 in current chapter and our publications [3, 4, 59, 60]) and adapted for studies of muscles from other species.

3.1 Isolation of Extraocular Muscles (Figs. 2–4)

The final isolated ocular/periocular tissue preparation consists of the EOMs still attached to the eyeball together with the accessory retractor bulbi muscle, the optic nerve and associated connective tissue (for a overall view of the ocular/periocular preparation after tissue isolation *see* Fig. 4). Isolating the EOMs together with their associated structures permits to preserve as much as possible the integrity of the original structural organization. Nevertheless, during the isolation, the superior oblique muscle is broken due to its association with the trochlea, hence only a part of this muscle is present within our final preparation. The tissue harvesting procedure described here for the histological analyze of EOMs can also be used for isolating and studying satellite cells and other cell types, for details refer to **Note 4**.

1. Euthanize mouse according to institute regulations (*see* **Note 5**).
2. Spray the head lightly with 70% ethanol.
3. Use dissecting scissors to decapitate the mouse.
4. Remove skin and eyelids from the head (Fig. 2a).
5. Use dissecting scissors to make a small incision at the base of the skull and cut along the midline of the skull through the sagittal suture in caudal to rostral fashion, then use two iris forceps to broke and remove the parietal bones exposing the brain (Fig. 2b).
6. Use two iris forceps to gently remove the brain paying attention not to severe the optic nerve attached to the eye (Fig. 2c).
7. Use two iris forceps to pull apart the two halves of the skull that should remain connected by the lower jaw and the bottom of the cranium (Fig. 2d) and secure the opened skull to the dissecting board with tissue pins (Fig. 2e).
8. Use two iris forceps to clean the nasal cavity from cartilage and soft tissues (Fig. 2f, black arrows point to the cartilage and soft tissues to be removed).
9. Use two Dumont #1 forceps to break and remove the bones of the orbit along their suture lines, exposing the eyes (Fig. 2g–l). This step typically results in the tearing of the superior oblique muscle, *see* **Note 6**.
10. Use the microscissors to cut the proximal tendons of the levator palpebrae superioris muscle (it has two tendons leading back to the sphenoid bone just above the origin of the rectus muscles) then grab it with the fine point forceps and cut its connection by the upper eyelid using the microscissors and remove it (Fig. 3a–c).
11. Use the fine point forceps to remove the Harderian gland that surrounds the EOMs and fills much of the orbital cavity (Fig. 3d–f). If EOMs are harvested for cell isolation, *see* **Note 4** regarding how to proceed from this point; for histological preparations, continue as described here.

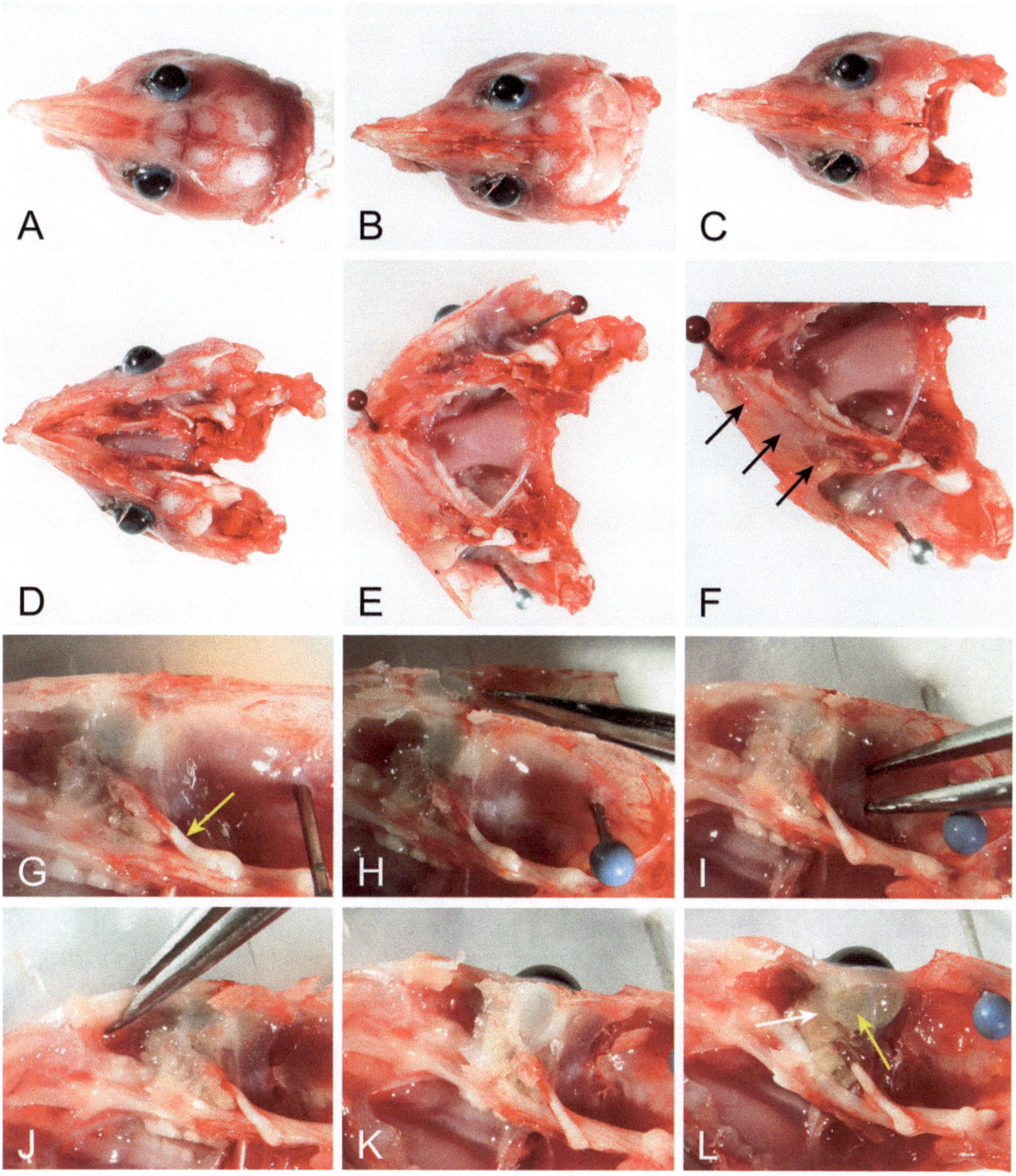

Fig. 2 Isolation of ocular/periocular tissue—First part: opening the skull and removing bony material to expose the ocular/periocular tissue. (**a**) Head with skin and eyelids removed. (**b**) Brain exposed by cutting the skull in two halves through the sagittal suture and by removing the parietal bones. (**c**) Brain removed. (**d**) Pulling apart the two halves of the skull and (**e**) securing the opened skull to the dissecting board with tissue pins. (**f**) Cleaning the nasal cavity from cartilage and soft tissues (indicated by *black arrows*). (**g**) The eye and associated tissues can be seen in transparency behind the orbit bones and the optic nerve is clearly visible exiting the orbital cavity through the optic foramen. (**h**–**k**) Removing of the orbital bones. (**l**) The intra-orbital tissues are now exposed and among them, the levator palpebrae superioris muscle (*white arrow*) and the Harderian gland (*yellow arrow*) can clearly by observed. Note that images in panels (**a**)–(**f**) were taken at a lower magnification and merged from a series of images to allow a full view of the mouse head or region of interest, before getting to finer details in panels (**g**)–(**l**), taken as single images

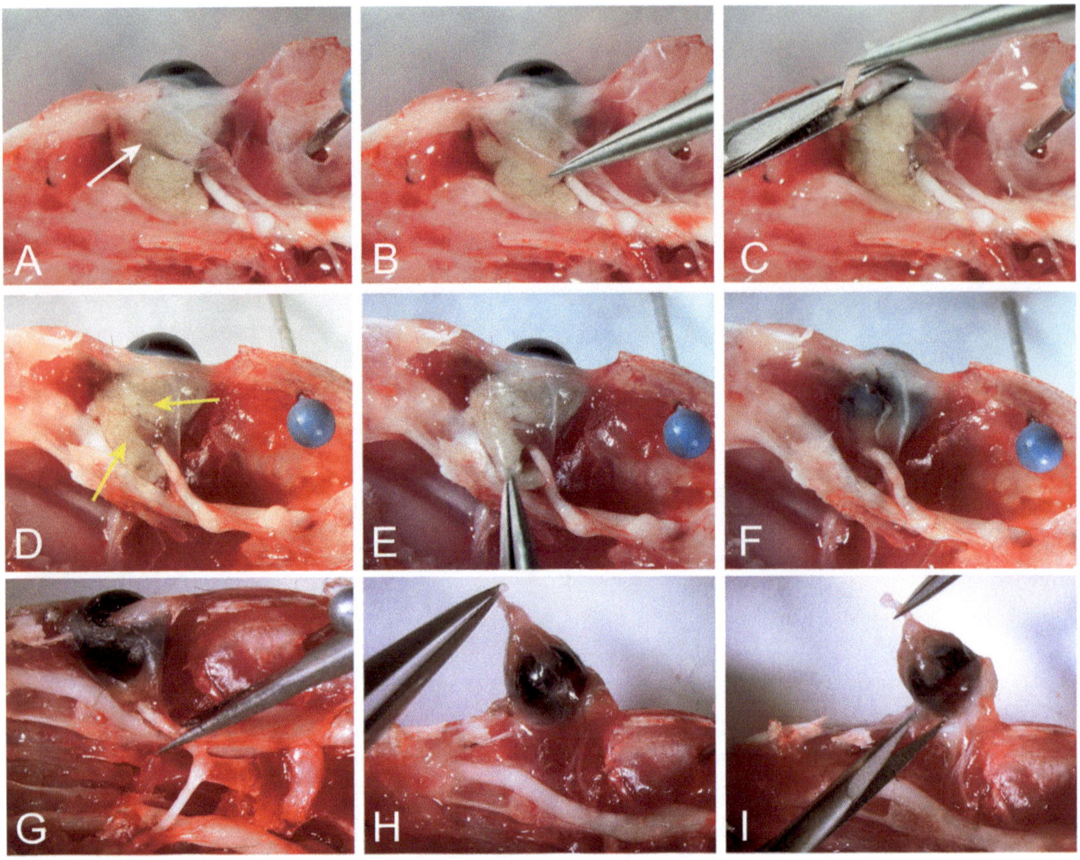

Fig. 3 Isolation of ocular/periocular tissue—Second part: exposing the extraocular muscles and final isolation of the ocular/periocular tissue. (**a**) Removal of the levator palpebrae superioris muscle (*white arrow*) by (**b**) cutting its proximal tendons and pulling it in order to (**c**) cut its insertion into the upper eyelid. (**d**) Removal of the Harderian gland (*yellow arrows*) by (**e**) grabbing it with a fine point forceps. (**f**) The extraocular muscles can now be seen. (**g**) Cutting the optic nerve between the annulus of Zinn and the optic chiasm. (**h**) Lifting the eye grabbing it by the optic nerve. (**i**) Cutting the remaining soft tissues and the proximal attachment of the inferior oblique

12. Use the microscissors to cut the optic nerve between the annulus of Zinn and the optic chiasm (Fig. 3g).
13. Lift the eye by grabbing the optic nerve with the fine point forceps (Fig. 3h) and release the eye from the orbit by cutting the remaining soft tissues and the proximal attachment of the inferior oblique (to the maxillary bone) with the microscissors (Fig. 3i).
14. Ocular/periocular preparations are placed in a 24-well multiwell tissue culture dish (two preparations per mouse per well, *see* Fig. 4) in PBS solution (typically 500 µl, at room temperature) and processed soon afterwards according to one of the three protocols described below.

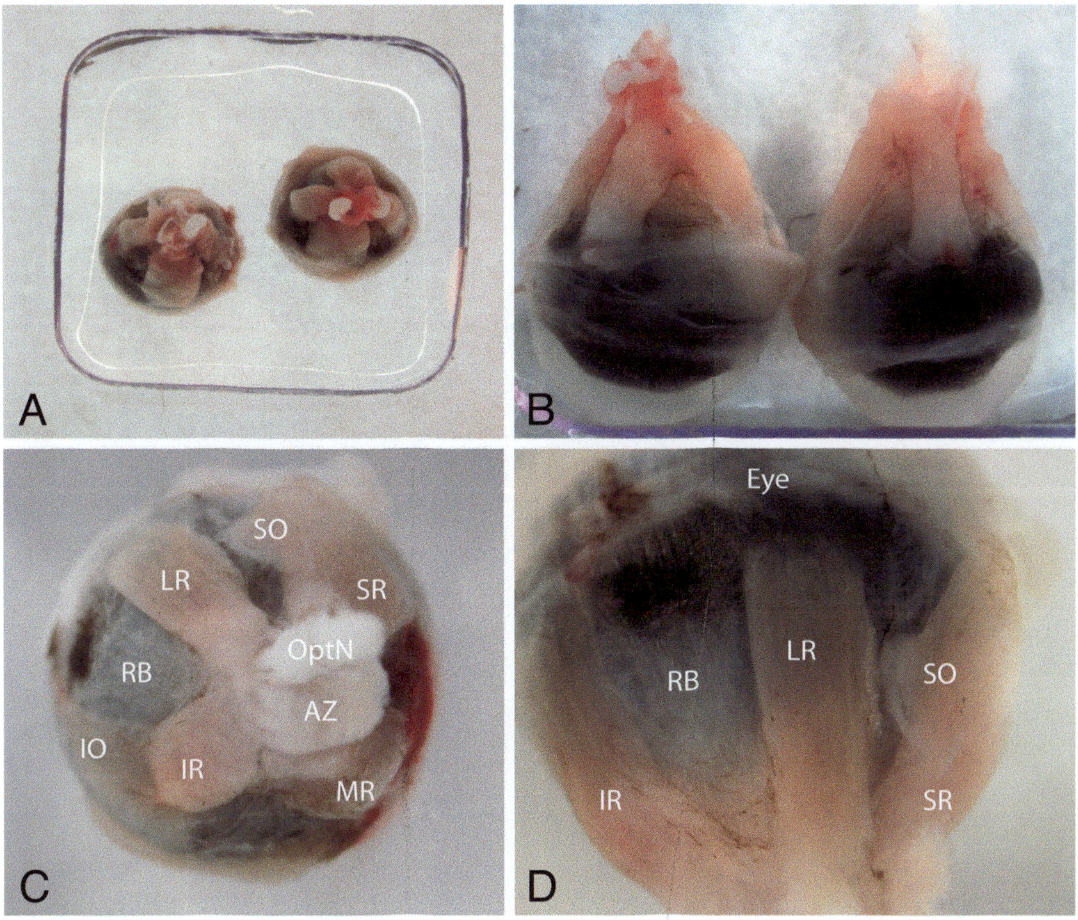

Fig. 4 Overall morphology of the ocular/periocular preparation at completion of the isolation procedure. The ocular/periocular specimen (a set of two per mouse) is already placed in a cryomold filled with OCT compound and is ready for freezing. (**a**, **b**) Lower magnification views from the top and the side of the preparation, respectively. (**c**, **d**) Higher magnification views from the top and the side, respectively, depicting finer details of the different muscles and the associated structures that can be observed. *SO* superior rectus, *IR* inferior rectus, *LR* lateral rectus, *MR* medial rectus, *SO* superior oblique, *IO* inferior oblique, *RB* retractor bulbi (darker appearance of this muscle is due to the dark pigmented eye visible through this thin muscle), *OptN* optic nerve, *AZ* Annulus of Zinn

3.2 Preparation of Extraocular Muscles Isolated from MyoD^{Cre} × R26^{mTmG} Reporter Mice for Direct Observation (Fig. 5)

The R26^{mTmG} reporter operates on a dual membrane-localized fluorescent system where all cells express Tomato until Cre-mediated excision of the Tomato coding sequence allows for GFP expression [53]. Consequently, when this R26^{mTmG} allele is combined with MyoD^{Cre} allele, all skeletal muscles and their resident satellite cells are GFP^{+} [4] due to ancestral MyoD expression in the myogenic lineage [52]. While we have been able to retain the membrane-bound GFP (and Tomato) signal from this R26$^{mT/mG}$ reporter even without fixation, in our hands, signal preservation with regard to

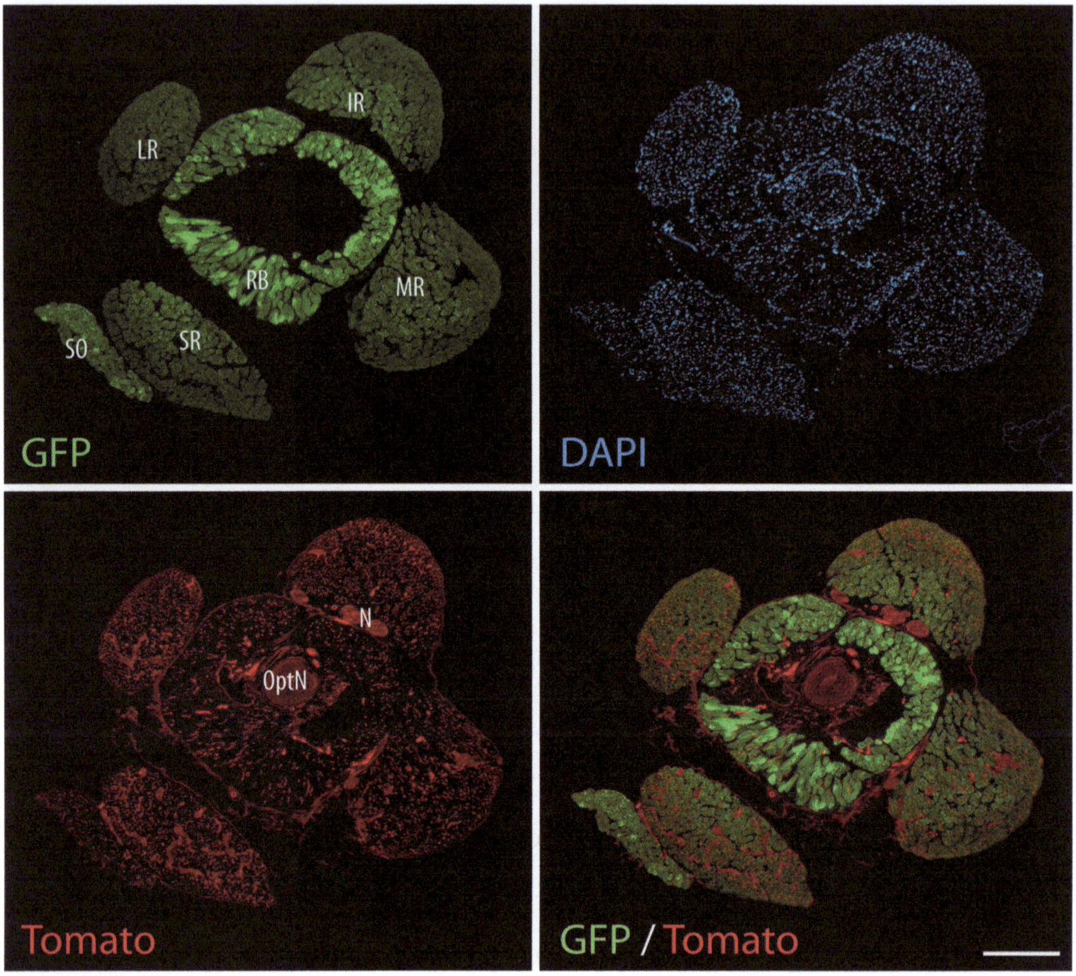

Fig. 5 Histological overview of the ocular/periocular tissue. Tissue processing followed the "blue route" as illustrated in Fig. 1. (**a–d**) Parallel fluorescent images (GFP, Tomato, DAPI, and merged GFP/Tomato) of a cross section from the MyoD^{Cre} × R26^{mTmG} reporter mouse (9 months old). The R26^{mTmG} reporter operates on a dual membrane-localized fluorescent system where all cells express Tomato until Cre-mediated excision of the Tomato coding sequence allows for GFP expression [53]. Consequently, when this R26^{mTmG} allele is combined with MyoD^{Cre} allele, all skeletal muscles and their resident satellite cells are GFP^{+} [4] due to ancestral MyoD expression in the myogenic lineage [52]. Images shown are from a cross section at the level of the optic nerve, immediately posterior to the eye (location is indicated by a *vertical dotted line* in the schematic shown in Fig. 6c). All muscle masses harbor strong GFP expression, while all connective tissues and nerve structures within and between muscles express the red fluorescent Tomato protein. The compartmentalization of the EOMs is clearly evident in the GFP-fluorescent image where an outer orbital layer and an inner global layer, harboring characteristic smaller and larger myofiber diameter, respectively, can be distinguished in the four rectus muscles. *SR* superior rectus muscle, *IR* inferior rectus muscle, *MR* medial rectus muscle, *LR* lateral rectus muscle, *SO* superior oblique muscle, *RB* retractor bulbi muscle, *OpN* optic nerve, *N* peripheral nerves. Scale bar, 500 μm

finer structural details is optimal if the tissue is fixed before freezing with 4 % PFA as described below (for fixative solution *see* **item 3** in Subheading 2.5). This tissue fixation step prior to freezing is however mandatory for retaining soluble forms of GFP, like that expressed for example in the frequently used by us Nestin-GFP and Sca1-GFP reporter mice [59, 61], where GFP would be lost if tissue was not fixed prior to freezing. Furthermore, tissue fixation prior to freezing is also our method of choice when performing X-Gal staining on tissue isolated from LacZ-based reporter mice [4]. Notably, as detailed below in Subheading 3.2.1, when fixing is performed prior to freezing, we also include a step of cryoprotection through an ascending series of sucrose solutions [62].

3.2.1 Fixation and Cryoprotection

1. Fix the isolated tissue (which is held in PBS following harvesting as described in Subheading 3.1, **step 14**) by adding 4% PFA fixative solution at 1:1 ratio to the PBS for 2 h at room temperature (for fixative solution *see* **item 3** in Subheading 2.5).
2. Remove fixative solution and rinse once with PBS.
3. Immerse the isolated tissue successively in 10% and 20% sucrose solutions, each for 30 min at room temperature (gently agitate the sample during this process using a gyrating platform rotator) followed by an overnight incubation in 30% sucrose solution at 4 °C. At the end of each sucrose step of this cryoprotection procedure, the tissue should sink at the bottom of the well when correctly infiltrated with the sucrose solution.

3.2.2 Freezing

1. Remove the isolated tissue from the 30% sucrose solution, grabbing it with Dumont #5/45 forceps by the optic nerve. Remove residual fluid by thoroughly blotting the isolated tissue onto a paper towel.
2. Fill the cryomold with OCT compound, being careful to avoid formation of bubbles. Remove any bubbles using a micropipette.
3. Using Dumont #5/45 forceps carefully position the eyes in the center of the OCT-filled cryomold, placing the preparation with the anterior part of the eye down as depicted in Fig. 4a and b. Carefully reposition the eye so the four rectus muscles are horizontally aligned when you look from the top (Fig. 4a and b), being careful to avoid the formation of bubbles. Remove any bubbles using a micropipette.
4. Fill the stainless steel bowl (housed in an insulated plastic container) approximately halfway with liquid nitrogen. Then, fill the stainless steel mortar with isopentane (~up to 1/3) and slowly place it over the stainless steel bowl lowering it into the liquid nitrogen. The liquid nitrogen level should be higher than the isopentane level (if needed regularly add more liquid nitrogen to maintain sufficient level).
5. Wait for the isopentane to reach the optimal freezing temperature (it usually takes less than 10 min), i.e., when the isopentane starts to thicken and small white crystals form at the base of the isopentane container (~−150 °C). If the isopentane entirely

freezes solid, thaw it and wait again until optimal freezing temperature is reached before subsequent use.

6. Hold the cryomold using a long dressing forceps and lower it into the isopentane until it is almost submerged. As the OCT freezing medium turns white and freezes, the cryomold is progressively further lowered into the isopentane until it is fully submerged. The whole process should take 20–30 s. From this point the tissue block should always be kept frozen as further detailed below (*see* **Note 7** for additional notes about specimen handling).
7. Immediately transfer the cryomold on top of dry ice placed into a Styrofoam box for short-time storage (less than 1 h) until all the isolated tissues have been frozen.
8. Rapidly wrap each cryomold in an aluminum foil and place it back immediately on top of the dry ice.
9. Place the cryomold into a freezer bag containing a paper-towel that has been wetted with deionized water (in order to avoid drying of the OCT compound and specimen) and keep the freezer bag on top of the dry ice, before transferring it soon afterwards into a freezer box placed into a –80 °C freezer for long-term storage.

3.2.3 Cryosectioning

1. Set the chamber temperature of the cryostat to –18 °C.
2. Using a Styrofoam box containing dry ice (*see* **Note 7** for additional details about specimen handling), transfer the frozen tissue block from the –80 °C freezer to the cryostat chamber. Wait about 15 min before starting to cut sections to allow the temperature of your sample(s) to equilibrate with the temperature of the cryostat chamber.
3. Mount the cryomold onto the specimen holder using OCT compound.
4. Place the specimen holder onto the orientable specimen head and remove the cryomold to expose the block. The eyeball part of the ocular/periocular preparation should now appear at the top of the block and will be cut first.
5. Start cutting 10 μm thick transverse sections and discard the initial sections until reaching the top of the frozen tissue (or the region of interest). While trimming the unwanted material, adjust the block orientation to ensure that the upper surface of the block is parallel to the cryostat blade (when the blade starts cutting through the block, it should hit all the block surface at once if perfectly parallel to the block).
6. Cut and collect 10 μm thick transverse sections onto cold slides one at a time (stored within the cryostat chamber). For more details about handling of the sections with the aid of a fine paint brush and collecting onto the slide *see* **Note 8**.

7. Examine the first few tissue sections under a microscope to evaluate (1) tissue orientation that can be readjusted if necessary, and (2) tissue quality (i.e., presence of tears, or freezing artifacts).
8. Collect an adequate amount of sections (number of sections depends on tissue size and goal of study) on each slide and let the slides air-dry at room temperature for at least 30 min (and up to 3 h).
9. If not processed immediately, store the slides in a slide box placed in an −80 °C freezer.

3.2.4 Mounting and Observation

1. If the slides were kept at −80 °C, warm up/air-dry the slides at room temperature for 10 min (*see* **Note 9** for details about slide handling in all subsequent steps).
2. Wash slides three times with PBS.
3. If visualization of nuclei is desired, incubate sections with DAPI (1 μg/ml) at room temperature for 5 min.
4. Briefly immerse slides into a rectangular glass staining dish filled with deionized water in order to get rid of the salts.
5. Allow slides to air-dry at room temperature for ~10 min.
6. Mount cover glass with mounting medium (*see* **Note 10**)
7. Seal the cover glass by applying nail polish around the edges of the cover glass and let the nail polish fully dry.
8. Observe the tissue sections using a fluorescent microscope and/or store at 4 °C protected from light for a later analysis; let the slide warm up few minutes at room temperature before later observation.

3.3 Preparation of Extraocular Muscles for H&E Staining (Fig. 6)

3.3.1 Freezing

1. Remove the isolated tissue from the PBS (following harvesting as described in Subheading 3.1) grabbing it with the Dumont #5/45 forceps by the optic nerve. Remove residual fluid by thoroughly blotting the isolated tissue onto paper towel.
2. Follow **steps 2–9** as described in Subheading 3.2.2.

3.3.2 Cryosectioning

1. *See* Subheading 3.2.3.

3.3.3 H&E Staining

Before starting the procedure, fill and arrange the necessary staining plastic dishes in the following order: tap water (x1); Harris hematoxylin (×1); warm (~37 °C) tap water (×3); Scott's tap water substitute (×1); tap water (×2); 70% ethanol (×1); Eosin Y (×1); 95% ethanol (×2); 100% ethanol (×3); xylene substitute (×3).

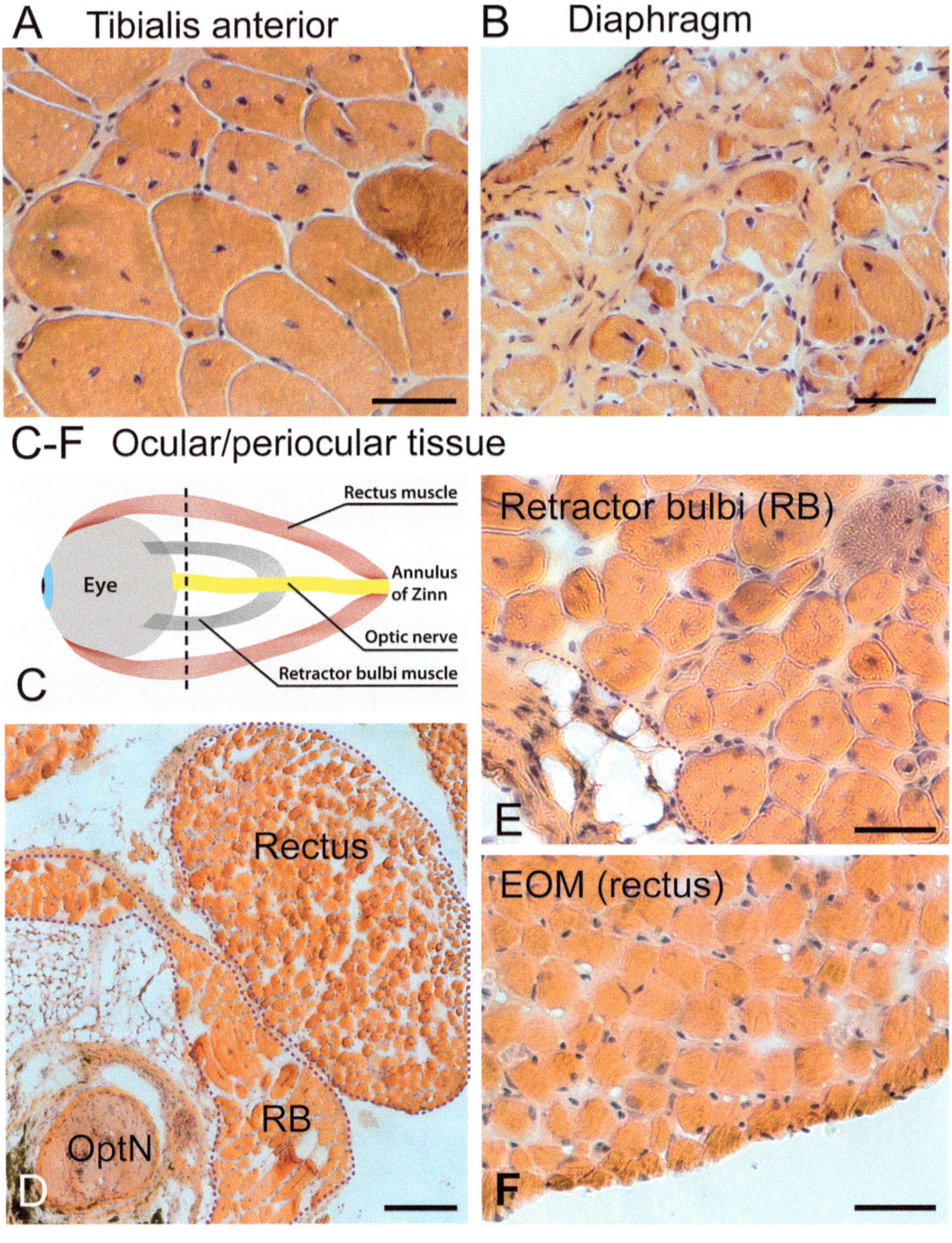

Fig. 6 H&E-stained cross sections of skeletal muscles from an mdx[4cv] mouse. The mdx[4cv] mouse is used as a animal model for Duchenne muscular dystrophy and is one of several mdx "cv" strains that were generated by point mutations in the dystrophin gene upon mutagen-treatment of male mice (*see* Subheading 2.3, **item 3** for more information on this dystrophin-null mouse strain). Tissues were isolated from a 1-year-old mouse. Tissue processing followed the "orange route" as illustrated in Fig. 1. (**a**) Limb muscle (tibialis anterior), (**b**) diaphragm muscle, and (**c**–**f**) ocular/periocular tissue. (**c**) A schematic of an ocular/periocular tissue preparation

Place slides into a slide holder and subsequently transfer/immerse the slide holder through the different staining plastic dishes as described below. Notably, starting with the first dehydration step, all steps should be performed within a chemical fume hood.

1. If the slides were kept at −80 °C, warm up/air-dry at room temperature for 10 min. Make sure sections are completely dry, *see* **Note 11**.
2. Hydrate in tap water for 5 min.
3. Physically remove most of the water from the slides by vigorous shaking the slide holder.
4. Stain in Harris hematoxylin for 5 min (progressive staining method, *see* **Note 12**).
5. Rinse in three changes of warm (~37 °C) tap water (15 dips each).
6. Bluing in Scott's tap water substitute (15 dips).
7. Rinse in two changes of tap water (15 dips each).
8. Dip in 70% ethanol for 2 min.
9. Dip in Eosin Y (0.25% working solution) for 2 min.
10. Dehydrate through an ascending series of ethanols. Dehydration steps: 95% ethanol (two times, 15 dips each) and 100% ethanol (three times, 15 dips each).
11. Clear in xylene substitute. Clearing steps: xylene substitute (three times, 15 dips each).
12. Mount cover glass with xylene substitute mounting medium (*see* **Note 13**).

3.4 Preparation of Extraocular Muscles for the Detection of Satellite Cells by Pax7 and Laminin Immunostaining (Fig. 7)

The methods described here have been developed for the optimal detection of satellite cells by Pax7 immunostaining but it can be adapted for the detection of other antigens. Notably, our protocol includes an antigen retrieval step (Subheading 3.4.3, **steps 1**, **5–9**), which is not required for the detection of many other antigens. Furthermore, immunodetection of many antigens can be achieved on unfixed tissue or tissue that has been fixed prior to freezing (Fig. 1, "orange route" and "blue route," respectively).

Fig. 6 (continued) depicting two rectus muscles, the retractor bulbi (RB) muscle and the optic nerve (OptN), harvested together with the eye. The *vertical dotted line* indicates the level of the sections shown in (**d**–**f**). This series of histological images illustrates that while limb (**a**) and diaphragm (**b**) muscles display typical signs of muscle pathology associated with the absence of dystrophin (central myonuclei, split myofibers and fibrosis, with the latter more pronounced in diaphragm), EOMs (**f**) show no such pathology. Noticeably, the retractor bulbi (**e**), although residing adjacent to the EOMs and of the same head mesenchyme embryonic origin [4], is affected by dystrophin absence as demonstrated by the extensive presence of central myonuclei similar to limb muscles. Scale bars, 50 μm (**a**, **b**) and (**e**, **f**), 200 μm (**d**). From ref. [3]

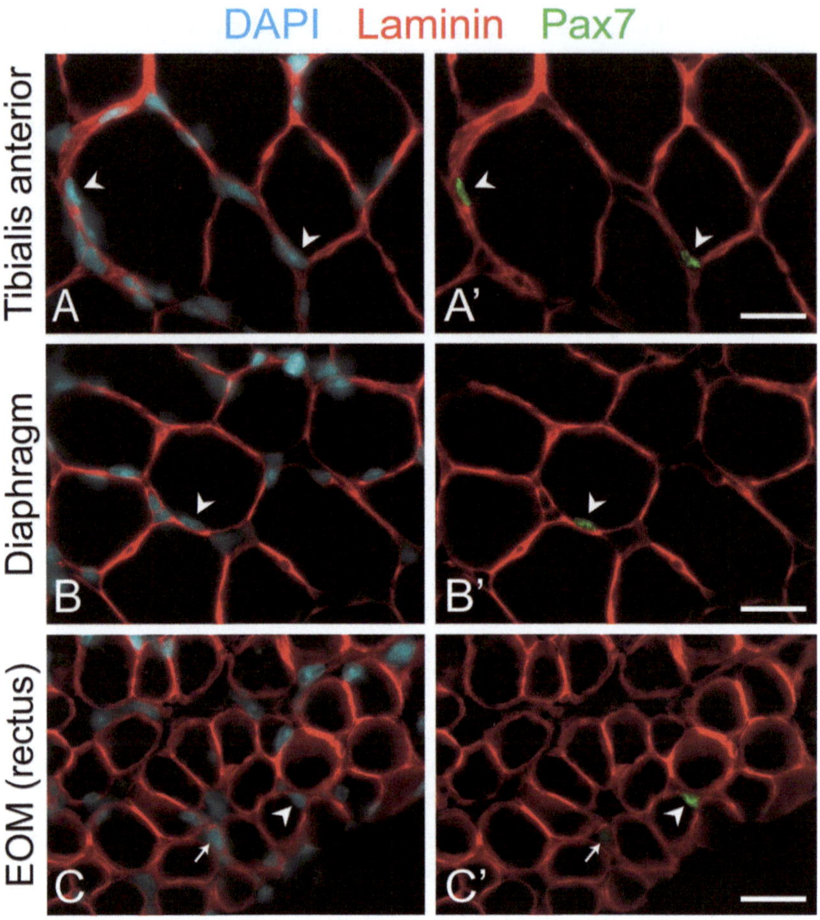

Fig. 7 Detection of satellite cells in skeletal muscles by Pax7 immunostaining. Tissues were isolated from a 4-month-old mouse. Tissue processing followed the "green route" as illustrated in Fig. 1. (**a**–**c**′) Immunostaining of cross sections from tibialis anterior, diaphragm and EOM (rectus muscle) identifies, in all three muscle groups, Pax7⁺ satellite cells (*arrowheads*) located beneath the basal lamina (delineated by laminin staining). The diameter of EOM myofibers is distinctively smaller than that of tibialis anterior and diaphragm myofibers, with some EOM myofibers being extremely narrow as highlighted by a Pax7⁺ nucleus occupying an entire cross-sectional myofiber width (*arrow*). Scale bars, 25 μm. Contributed by our former lab member Andrew Shearer, from ref. [3]

3.4.1 Freezing

1. Remove the isolated tissue from the PBS (following harvesting as described in Subheading 3.1) grabbing it with the Dumont #5/45 forceps by the optic nerve. Remove residual fluid by thoroughly blotting the isolated tissue onto paper towel.
2. Follow **steps 2–9** as described in Subheading 3.2.2.

3.4.2 Cryosectioning

1. *See* Subheading 3.2.3.

3.4.3 Immunostaining

1. Fill a slide staining dish with the retrieval buffer and place it into a small Pyrex loaf dish containing water heated at 95–100 °C. The water should by constantly stirred using a magnetic bar in order

to avoid bubbles formation underneath the staining jar and to ensure a homogenous water temperature.

2. If the slides were kept at −80 °C, warm up/air-dry at room temperature for 10 min.
3. Fix for 10 min at room temperature by gently adding on top of the slide enough volume of 4 % PFA fixative solution to cover all the sections (*see* **Note 9** for details about slide handling).
4. Wash three times by gently adding on top of the slide enough volume of TBS to cover all the sections. Then, place slides in slide holder and immerse in TBS for ~5 min.
5. Remove slides from TBS and immerse slides in heated retrieval buffer for 30 min (as described in **step 1**).
6. Remove the slide-staining dish from the boiling water (with the slide holder still in the retrieval buffer) and allow the solution to cool for 10 min.
7. Remove the slide holder from retrieval buffer and let the slides air-dry for ~5 min at room temperature.
8. Place the slide holder in TBS for ~5 min.
9. Remove slides from the slide holder and allow them to air-dry for ~10 min at room temperature, then use an hydrophobic barrier pen to draw on boundaries around sections (*see* **Note 9** for details about slide handling).
10. Gently add on top of the slides enough volume of TBS–2% NGS to cover all the sections and incubate overnight at 4 °C. To avoid drying of the tissue sections during this prolong incubation, slides are place in a covered and humidified staining slide tray.
11. Remove slides from 4 °C and wash three times by adding on top of the slides enough volume of TBS–0.05% Tween 20 to cover all the sections.
12. Incubate slides with the primary antibody solution (*see* Subheading 2.8, **item 9**) at 4 °C, overnight. Add 100 μl of primary antibody solution onto the slide. To ensure good spreading and avoid drying during the overnight incubation, gently cover with a flexible plastic coverslip (or a piece of parafilm cut to the correct size) making sure the solution sits evenly on the sections and avoiding bubbles. This approach has the advantage of minimizing the volume of primary antibody used, while still achieving optimal and reproducible results. Slides are place in a covered and humidified staining slide tray.
13. Remove slides from 4 °C and wash sections three times by gently adding on top of the slide enough volume of TBS–0.05% Tween 20 to cover all the sections.

14. Incubate slides with the secondary antibody solution (*see* Subheading 2.8, **item 10**) 2 h at room temperature (protected from light). Add enough secondary antibody solution to cover all the sections on the slide (i.e., ~250–500 μl) and place in a covered and humidified staining slide tray.
15. Wash slides three times by gently adding on top of the slide enough volume of TBS–0.05% Tween 20 to cover all the sections.
16. Wash slides one time by gently adding on top of the slide enough volume of TBS to cover all the sections and incubate for ~10 min.
17. Follow **steps 3–8** as described in Subheading 3.2.4.

4 Notes

1. Paraformaldehyde is a white powder with a formaldehyde-like odor. It is a rapid fixative and a potential carcinogen. When handling paraformaldehyde, wear gloves, mask, goggles and use a chemical fume hood. It is important to refer to the MSDS instructions and institutional regulations for further information regarding storage, handling and first-aid.
2. DAPI (4′,6-diamidino-2-phenylindole dihydrochloride) is potentially harmful. Avoid prolonged or repeated exposure. It is important to refer to the MSDS instructions and institutional regulations for further information regarding storage, handling and first aid. Using a chemical fume hood, we dissolve the entire powder stock, still in its original container, and generate a concentrated stock solution (1 mg/ml in Milli-Q water). This stock solution is then divided into 1 ml aliquots that are frozen at −20 °C for long-term (years) storage. Alternatively, to avoid preparation of DAPI solution and bypass the DAPI incubation step on slides (Subheading 3.2.4, **step 3**), a commercial anti-fade mounting medium that contains DAPI can be used, such as the Vectashield preparation available from Vector Laboratories.
3. While alternative retrieval buffers for Pax7 immunostaining have been described in the literature, we favor the one described here due to its simplicity and excellent outcome. For the full instructions of this antigen retrieval protocol refer to [63].
4. For cell isolation, each EOM is dissected from the ocular/periocular preparation by cutting the muscle at its proximal and distal attachments. The superior rectus as well as a fragment of the superior oblique (*see* **Note 6**) attached to the eye should be readily visible on top of the exposed eye (Fig. 3f). Start by grabbing the superior oblique by its free end using fine point

forceps and cut its connection to the eye using microscissors. Then, slide a fine point forceps underneath the superior rectus muscle and hold on, then using microscissors cut its connections to the annulus of Zinn and to the eye. Continue by repeating this procedure on the lateral and medial recti. Then cut the optic nerve between the annulus of Zinn and the chiasm and lift up the eye by the optic nerve using a fine point forceps. Dissect the inferior rectus repeating the procedure performed for the other rectus muscles. Finally, find the inferior oblique, grab it at mid belly using fine point forceps, and using microscissors cut its connections to the maxillary bone and to the eye. Place harvested tissues in a DMEM solution (at room temperature) for not more than 1 h (i.e., when processing multiple muscle groups or mice) for optimal tissue preservation, before being further processed to release single cell suspensions by enzymatic digestion and subsequent processing as detailed in our recent publications [3, 4].

5. When harvesting muscles either for histological preparation or for cell isolation, we prefer cervical dislocation for euthanizing mice as this method is more rapid and minimizes postmortem muscle stiffening.
6. The superior oblique muscle is broken into two parts during the isolation process due to its association with the trochlea. The trochlea is a cartilaginous ring attached to the frontal bone that is removed during the isolation procedure. Hence, only its portion attached to the eye is included in our final preparation.
7. It is important to never let the tissue block thaw once it has been frozen. Notably, precaution should be taken never to touch directly the frozen blocks with your fingers when handling them, and always place the frozen tissue on dry ice when transporting to and from −80 °C freezer storage.
8. Using a fine-tipped paint brush that has been precooled in the cryostat chamber, carefully position and flatten each freshly cut section so a slide can be placed just above it (positively charged side of slide facing down onto the section). Then, transfer the section to the slide by briefly pressing down the slide on the section with one of your finger. The number of sections per slide depends on tissue size and goals of the investigators.
9. All rinses and incubation steps described in Subheading 3.2.4 and in **steps 3, 4** and **10–16** of Subheading 3.4.3 are performed directly on the slide. This is achieved by just adding to the top of the slide enough solution to cover all sections, but not so much that it leaks off of the slide. The rinse/incubation solutions are removed by gentle aspiration. If slides are properly dried, the solution, gently added to the slide, should not spill over the slide borders. Alternatively, one can use a hydrophobic barrier pen to draw boundaries around tissue sections

to prevent liquid from running off the slides (as performed for our immunostaining procedure, Subheading 3.4.3, **step 9**).

10. Add a few drops of mounting medium on top of the slide, then place one edge of the cover glass at 45°, allowing the drops to spread along the edge of the cover glass. Then, gently lower the cover glass, allowing the mounting medium to spread slowly until it fully covers the surface of the cover glass, ensuring that no air bubbles are present under it.
11. To avoid shrinkage of the myofibers during H&E staining, it is essential to (1) dry the slides completely before the rehydration step and (2) never let the slides dry out in between the different steps once they have been rehydrated.
12. While the Harris hematoxylin stain is generally used in a "regressive" manner (i.e., the tissue is overstained and then destained until proper intensity is reached), it also can be used in a "progressive" manner (i.e., the tissue is left in the stain just long enough to reach the proper intensity). We have been using the latter approach which has yielded excellent staining results with 5–8 min staining time as depicted for example in Fig. 6.
13. After the last bath in the xylene substitute solution, the mounting medium should be applied quickly (i.e., within seconds) and amply (~1–2 ml) over the sections. Then, immediately place one edge of the cover glass at 45°, allowing the mounting medium to spread along the edge of the cover glass, and let go of the cover glass allowing the mounting medium to fully spread, ensuring no air bubbles are present under it. Drain the excess of mounting medium by blotting the slide on a paper towel. In case bubbles are nevertheless generated, the cover glass can be removed by placing the slide into the xylene substitute for few minutes followed by repeating the mounting process. After mounting, let the slide dry for about 1 h while kept under the chemical hood. The slide is then ready for viewing and can be stored long-term at room temperature.

Acknowledgements

We are grateful to Lindsey Muir for her helpful comments on the manuscript. We also express gratitude to our former lab members, Irina Kirillova and Andrew Shearer for their important contributions during the initial phase of our studies leading to the ocular/periocular isolation protocol described here. Our study of mouse extraocular muscles and their satellite cells is currently supported by grants to Z.Y.R. from the National Institute of health (AG035377 and NS088804) and by an AFM-telethon fellowship to P.S. (#18574). Z.Y.R. acknowledges additional support during the preparation of this chapter from the National Institutes of Health (NS090051).

References

1. Demer JL (2007) Mechanics of the orbita. Dev Ophthalmol 40:132–157
2. Kirillova I, Gussoni E, Goldhamer DJ, Yablonka-Reuveni Z (2007) Myogenic reprogramming of retina-derived cells following their spontaneous fusion with myotubes. Dev Biol 311:449–463
3. Stuelsatz P, Shearer A, Li Y, Muir LA, Ieronimakis N, Shen QW, Kirillova I, Yablonka-Reuveni Z (2015) Extraocular muscle satellite cells are high performance myo-engines retaining efficient regenerative capacity in dystrophin deficiency. Dev Biol 397:31–44
4. Stuelsatz P, Shearer A, Yablonka-Reuveni Z (2014) Ancestral Myf5 gene activity in periocular connective tissue identifies a subset of fibro/adipogenic progenitors but does not connote a myogenic origin. Dev Biol 385:366–379
5. Fischer MD, Gorospe JR, Felder E, Bogdanovich S, Pedrosa-Domellof F, Ahima RS, Rubinstein NA, Hoffman EP, Khurana TS (2002) Expression profiling reveals metabolic and structural components of extraocular muscles. Physiol Genomics 9:71–84
6. Lucas CA, Hoh JF (1997) Extraocular fast myosin heavy chain expression in the levator palpebrae and retractor bulbi muscles. Invest Ophthalmol Vis Sci 38:2817–2825
7. Porter JD, Khanna S, Kaminski HJ, Rao JS, Merriam AP, Richmonds CR, Leahy P, Li J, Andrade FH (2001) Extraocular muscle is defined by a fundamentally distinct gene expression profile. Proc Natl Acad Sci U S A 98:12062–12067
8. Spencer RF, Porter JD (2006) Biological organization of the extraocular muscles. Prog Brain Res 151:43–80
9. Couly GF, Coltey PM, Le Douarin NM (1992) The developmental fate of the cephalic mesoderm in quail-chick chimeras. Development 114:1–15
10. Noden DM, Francis-West P (2006) The differentiation and morphogenesis of craniofacial muscles. Dev Dyn 235:1194–1218
11. Valdez G, Tapia JC, Lichtman JW, Fox MA, Sanes JR (2012) Shared resistance to aging and ALS in neuromuscular junctions of specific muscles. PLoS One 7:e34640
12. Yu Wai Man CY, Chinnery PF, Griffiths PG (2005) Extraocular muscles have fundamentally distinct properties that make them selectively vulnerable to certain disorders. Neuromuscul Disord 15:17–23
13. Schoser BG, Pongratz D (2006) Extraocular mitochondrial myopathies and their differential diagnoses. Strabismus 14:107–113
14. Kaminski HJ, Richmonds CR, Kusner LL, Mitsumoto H (2002) Differential susceptibility of the ocular motor system to disease. Ann N Y Acad Sci 956:42–54
15. Ahmadi M, Liu JX, Brannstrom T, Andersen PM, Stal P, Pedrosa-Domellof F (2010) Human extraocular muscles in ALS. Invest Ophthalmol Vis Sci 51:3494–3501
16. Kaminski HJ, al-Hakim M, Leigh RJ, Katirji MB, Ruff RL (1992) Extraocular muscles are spared in advanced Duchenne dystrophy. Ann Neurol 32:586–588
17. Khurana TS, Prendergast RA, Alameddine HS, Tome FM, Fardeau M, Arahata K, Sugita H, Kunkel LM (1995) Absence of extraocular muscle pathology in Duchenne's muscular dystrophy: role for calcium homeostasis in extraocular muscle sparing. J Exp Med 182:467–475
18. Porter JD, Karathanasis P (1998) Extraocular muscle in merosin-deficient muscular dystrophy: cation homeostasis is maintained but is not mechanistic in muscle sparing. Cell Tissue Res 292:495–501
19. Porter JD, Merriam AP, Hack AA, Andrade FH, McNally EM (2001) Extraocular muscle is spared despite the absence of an intact sarcoglycan complex in gamma- or delta-sarcoglycan-deficient mice. Neuromuscul Disord 11:197–207
20. Porter JD (1998) Commentary: extraocular muscle sparing in muscular dystrophy: a critical evaluation of potential protective mechanisms. Neuromuscul Disord 8:198–203
21. Andrade FH, Porter JD, Kaminski HJ (2000) Eye muscle sparing by the muscular dystrophies: lessons to be learned? Microsc Res Tech 48:192–203
22. Porter JD, Merriam AP, Khanna S, Andrade FH, Richmonds CR, Leahy P, Cheng G, Karathanasis P, Zhou X, Kusner LL, Adams ME, Willem M, Mayer U, Kaminski HJ (2003) Constitutive properties, not molecular adaptations, mediate extraocular muscle sparing in dystrophic mdx mice. FASEB J 17:893–895
23. Karpati G, Carpenter S (1986) Small-caliber skeletal muscle fibers do not suffer deleterious consequences of dystrophic gene expression. Am J Med Genet 25:653–658
24. Karpati G, Carpenter S, Prescott S (1988) Small-caliber skeletal muscle fibers do not suffer necrosis in mdx mouse dystrophy. Muscle Nerve 11:795–803
25. Turner PR, Westwood T, Regen CM, Steinhardt RA (1988) Increased protein degradation results from elevated free calcium levels

found in muscle from mdx mice. Nature 335:735–738
26. Zeiger U, Mitchell CH, Khurana TS (2010) Superior calcium homeostasis of extraocular muscles. Exp Eye Res 91:613–622
27. Ragusa RJ, Chow CK, St Clair DK, Porter JD (1996) Extraocular, limb and diaphragm muscle group-specific antioxidant enzyme activity patterns in control and mdx mice. J Neurol Sci 139:180–186
28. Matsumura K, Ervasti JM, Ohlendieck K, Kahl SD, Campbell KP (1992) Association of dystrophin-related protein with dystrophin-associated proteins in mdx mouse muscle. Nature 360:588–591
29. Porter JD, Rafael JA, Ragusa RJ, Brueckner JK, Trickett JI, Davies KE (1998) The sparing of extraocular muscle in dystrophinopathy is lost in mice lacking utrophin and dystrophin. J Cell Sci 111(Pt 13):1801–1811
30. McDonald AA, Hebert SL, McLoon LK (2015) Sparing of the extraocular muscles in mdx mice with absent or reduced utrophin expression: a life span analysis. Neuromuscul Disord 25:873–887
31. Kallestad KM, Hebert SL, McDonald AA, Daniel ML, Cu SR, McLoon LK (2011) Sparing of extraocular muscle in aging and muscular dystrophies: a myogenic precursor cell hypothesis. Exp Cell Res 317:873–885
32. Porter JD, Israel S, Gong B, Merriam AP, Feuerman J, Khanna S, Kaminski HJ (2006) Distinctive morphological and gene/protein expression signatures during myogenesis in novel cell lines from extraocular and hindlimb muscle. Physiol Genomics 24:264–275
33. Pacheco-Pinedo EC, Budak MT, Zeiger U, Jorgensen LH, Bogdanovich S, Schroder HD, Rubinstein NA, Khurana TS (2009) Transcriptional and functional differences in stem cell populations isolated from extraocular and limb muscles. Physiol Genomics 37:35–42
34. Hughes MO (2007) A pictural anatomy of the human eye/anophthalmic socket: a review for ocularists. J Ophthal Prosthet 12:51–63
35. Lee SH, Wong M, Yap S (2015) Anatomy—muscles of the eye. http://rodsncones.blogspot.com/2014/05/anatomy-muscles-of-eye.html. Accessed 8 Dec 2015
36. LifeMap (2015) Extraocular muscles anatomy. http://discovery.lifemapsc.com/library/images/extraocular-skeletal-muscle-anatomy. Accessed 8 Dec 2015
37. Montgomery TM (2015) The extraocular muscles. http://www.tedmontgomery.com/the_eye/eom.html. Accessed 8 Dec 2015
38. Carry MR, O'Keefe K, Ringel SP (1982) Histochemistry of mouse extraocular muscle. Anat Embryol (Berl) 164:403–412
39. Demer JL, Oh SY, Poukens V (2000) Evidence for active control of rectus extraocular muscle pulleys. Invest Ophthalmol Vis Sci 41:1280–1290
40. Demer JL (2003) Evidence for a pulley of the inferior oblique muscle. Invest Ophthalmol Vis Sci 44:3856–3865
41. Ruskell GL, Kjellevold Haugen IB, Bruenech JR, van der Werf F (2005) Double insertions of extraocular rectus muscles in humans and the pulley theory. J Anat 206:295–306
42. Kono R, Poukens V, Demer JL (2005) Superior oblique muscle layers in monkeys and humans. Invest Ophthalmol Vis Sci 46:2790–2799
43. Zhou Y, Liu D, Kaminski HJ (2010) Myosin heavy chain expression in mouse extraocular muscle: more complex than expected. Invest Ophthalmol Vis Sci 51:6355–6363
44. Rossi AC, Mammucari C, Argentini C, Reggiani C, Schiaffino S (2010) Two novel/ancient myosins in mammalian skeletal muscles: MYH14/7b and MYH15 are expressed in extraocular muscles and muscle spindles. J Physiol 588:353–364
45. Rubinstein NA, Porter JD, Hoh JF (2004) The development of longitudinal variation of Myosin isoforms in the orbital fibers of extraocular muscles of rats. Invest Ophthalmol Vis Sci 45:3067–3072
46. Zhu Q, Hillmann DJ, Henk WG (2000) Observations on the muscles of the eye of the bowhead whale, Balaena mysticetus. Anat Rec 259:189–204
47. Zhou JB, Ge S, Gu P, Peng D, Chen GF, Pan MZ, Qu J (2011) Microdissection of guinea pig extraocular muscles. Exp Ther Med 2:1183–1185
48. Clarkson C, Brown A, Ekenstedt K, Fletcher TF (2015) Orbit, eyeball & related structures. http://vanat.cvm.umn.edu/carnLabs/Lab24/Lab24.html. Accessed 12 Nov 2015
49. Marques MJ, Pertille A, Carvalho CL, Santo Neto H (2007) Acetylcholine receptor organization at the dystrophic extraocular muscle neuromuscular junction. Anat Rec 290:846–854
50. Seale P, Sabourin LA, Girgis-Gabardo A, Mansouri A, Gruss P, Rudnicki MA (2000) Pax7 is required for the specification of myogenic satellite cells. Cell 102:777–786
51. Yablonka-Reuveni Z (2011) The skeletal muscle satellite cell: still young and fascinating at 50. J Histochem Cytochem 59:1041–1059

52. Kanisicak O, Mendez JJ, Yamamoto S, Yamamoto M, Goldhamer DJ (2009) Progenitors of skeletal muscle satellite cells express the muscle determination gene, MyoD. Dev Biol 332:131–141
53. Muzumdar MD, Tasic B, Miyamichi K, Li L, Luo L (2007) A global double-fluorescent Cre reporter mouse. Genesis 45:593–605
54. Chapman VM, Miller DR, Armstrong D, Caskey CT (1989) Recovery of induced mutations for X chromosome-linked muscular dystrophy in mice. Proc Natl Acad Sci U S A 86:1292–1296
55. Im WB, Phelps SF, Copen EH, Adams EG, Slightom JL, Chamberlain JS (1996) Differential expression of dystrophin isoforms in strains of mdx mice with different mutations. Hum Mol Genet 5:1149–1153
56. Banks GB, Combs AC, Chamberlain JS (2010) Sequencing protocols to genotype mdx, mdx(4cv), and mdx(5cv) mice. Muscle Nerve 42:268–270
57. Lu QL, Morris GE, Wilton SD, Ly T, Artem'yeva OV, Strong P, Partridge TA (2000) Massive idiosyncratic exon skipping corrects the nonsense mutation in dystrophic mouse muscle and produces functional revertant fibers by clonal expansion. J Cell Biol 148:985–996
58. Belart.com (2015) mortar w/bowl & housing, liquid nitrogen-cooled (Bel-Art# 372600000) https://www.belart.com/bel-art-h37260-0000-liquid-nitrogen-cooled-stainless-steel-ladle-and-reservoir-6-diameter-4-height.html. Accessed 7 Dec 2015
59. Stuelsatz P, Keire P, Almuly R, Yablonka-Reuveni Z (2012) A contemporary atlas of the mouse diaphragm: myogenicity, vascularity, and the Pax3 connection. J Histochem Cytochem 60:638–657
60. Yablonka-Reuveni Z, Danoviz ME, Phelps M, Stuelsatz P (2015) Myogenic-specific ablation of Fgfr1 impairs FGF2-mediated proliferation of satellite cells at the myofiber niche but does not abolish the capacity for muscle regeneration. Front Aging Neurosci 7:85
61. Day K, Shefer G, Richardson JB, Enikolopov G, Yablonka-Reuveni Z (2007) Nestin-GFP reporter expression defines the quiescent state of skeletal muscle satellite cells. Dev Biol 304:246–259
62. Yablonka-Reuveni Z, Christ B, Benson JM (1998) Transitions in cell organization and in expression of contractile and extracellular matrix proteins during development of chicken aortic smooth muscle: evidence for a complex spatial and temporal differentiation program. Anat Embryol (Berl) 197:421–437
63. IHC-World (2015) Tris-EDTA buffer antigen retrieval protocol. http://www.ihcworld.com/_protocols/epitope_retrieval/tris_edta.htm. Accessed 7 Dec 2015

Chapter 10

Skeletal Muscle Tissue Clearing for LacZ and Fluorescent Reporters, and Immunofluorescence Staining

Mayank Verma, Bhavani SR Murkonda, Yoko Asakura, and Atsushi Asakura

Abstract

Skeletal muscle is a highly ordered yet complex tissue containing several cell types that interact with each other in order to maintain structure and homeostasis. It is also a highly regenerative tissue that responds to damage in a highly intricate but stereotypic manner, with distinct spatial and temporal kinetics. Proper examination of this process requires one to look at the three-dimensional orientation of the cellular and subcellular components, which can be accomplished through tissue clearing. While there has been a recent surge of protocols to study biology in whole tissue, it has primarily focused on the nervous system. This chapter describes the workflow for whole mount analysis of murine skeletal muscle for LacZ reporters, fluorescent reporters and immunofluorescence staining. Using this technique, we are able to visualize LacZ reporters more effectively in deep tissue samples, and to perform fluorescent imaging with a depth greater than 1700 μm.

Key words Skeletal muscle, Tissue imaging, tdTomato, Green fluorescence protein (GFP), Satellite cell, Tissue clearing, lacZ reporter

1 Introduction

Skeletal muscle is the largest tissue in the body and possesses incredible regenerative capacity. It is composed of several different types of cells including muscle fibers, satellite cells, blood vessels, fibroblasts, and immune cells [1]. These diverse cell types have a complicated interplay and changes in anatomy from the basal state during injury and disease. Understanding the structure–function relationships of the cells within the tissue is important for modeling the dynamic changes during muscle development and regeneration. Imaging is one of the most powerful tools to study this interplay of different cell types. Typically, cross sections or longitudinal sections, obtained from frozen or paraffin embedded tissue blocks, are immunolabeled and a single slice is examined under the microscope. While this is informative in the lateral plane, the

Michael Kyba (ed.), *Skeletal Muscle Regeneration in the Mouse: Methods and Protocols*, Methods in Molecular Biology, vol. 1460, DOI 10.1007/978-1-4939-3810-0_10, © Springer Science+Business Media New York 2016

information in the axial direction is lost leading to underrepresentation or misrepresentation of the actual data. As such, it is important to obtain information from all three dimensions in the tissue to study the anatomical and cellular interactions in the tissue. An example of such interaction is of that between satellite cells and blood vessel endothelial cells. It has been implicated that satellite cells are located preferentially close to blood vessels [2]. However, this interaction may be underestimated in typical two-dimensional analysis (Fig. 1). This scenario highlights the importance of analyzing cellular interactions in muscle tissue in its native three-dimensional state. The techniques in this chapter would lead to whole-tissue quantification, at a single cell resolution, to allow for the understanding of spatial interactions between cell types in the muscle during homeostasis, injury and repair.

One approach to studying the spatial interaction of cells in their three dimensional (3D) anatomical state is by obtaining serial sections, imaging each section and reconstructing the 3D representation through manual or automated methods [3]. However, this process is technically challenging, error prone and labor intensive. Optical sectioning using software deconvolution and confocal microscopy also allows for imaging in 3D, but is severely limited in muscle tissue due to the presence of abundant light-scattering molecules such as lipids, extracellular matrix, and endogenously fluorescent molecules such as myoglobin and NADH. In addition, due to poor and unreliable labeling reagents for populations such as satellite cells, genetic reporter mouse lines encoding fluorescent molecules [4–7] or β-galactosidase (LacZ) [6, 8, 9] have been developed that allow for reliable and consistent assessment of cellular populations with high fidelity. LacZ reporters have been extensively utilized in studies of skeletal muscle regeneration and transplantations. X-gal staining for LacZ reporter has the benefit of maintaining stability over time and has high signal to noise ratio. One limitation of LacZ staining is the penetration depth of the staining and opacity of the muscle tissue hinders the ability to detect the signal far from the surface of the tissue.

An ideal optical tissue clearing protocol for muscle would reduce light-scattering via refractive index mismatching, signal from pigmented molecules, preserve fluorescence reporter molecule and be amenable to immunolabeling. In this chapter we adapt several tissue clearing and staining protocols to optimize them for the study of murine adult skeletal muscle [10–15]. In addition, we expand the utility of tissue clearing protocols by utilizing them for clearing LacZ stained samples. To the best of our knowledge, this is the first report to show the dramatic improvement in visualization of LacZ signal post-tissue clearing. Using the current protocol we can image completely through thinner muscles such as the cremaster and diaphragm muscles using a confocal microscope and through an extensor digitorum longus (EDL) or soleus muscle using a two-photon microscope (*see* **Note 2**).

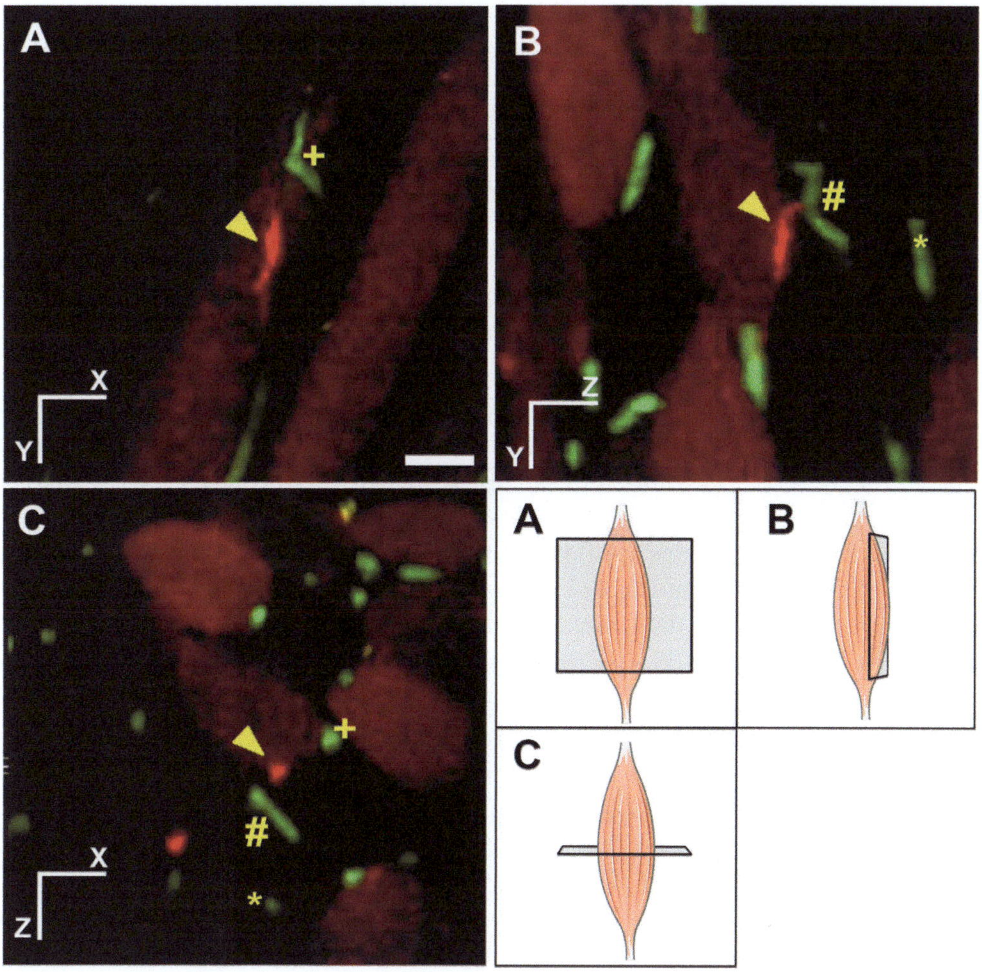

Fig. 1 Risk of under sampling when looking at cellular proximity in sections in muscle tissue. Spatial relationship is lost or misrepresented in thin section analysis. Orthogonal view of tdTomato labeled satellite cell (red) and green fluorescence protein (GFP) labeled blood vessels (green) from *Pax7CreERT2:R26R/tdTomato:Flk-1*$^{+/GFP}$ mice. Schematic representation of the orthogonal view is presented in the bottom right panel. Three different blood vessels are marked as (*plus*), (*hash*) and (*asterisk*). One satellite cell is marked as (*arrowhead*). Cross sectional analysis in panel (**A**) incorrectly shows the closest proximity between satellite cell (*arrowhead*) and blood vessels to be (*plus*). In addition, panel (**A**) and (**B**) greatly under-sample the amount of blood vessels in the proximity of the satellite cell, while panel (**C**) shows a better representation of the information. In traditional sectional analysis, only one section would be available for analysis thereby greatly underestimating the interactions between satellite cells and blood vessel. Tissue imaging sampling in three dimensions allows for more accurate quantification of cell proximity. Scale bar denotes 25 μm

2 Materials

1. Nutator mixer at 4 °C.
2. Bacterial shaker at 37 °C.
3. Rotating mixer at 37 °C.

4. 15 ml and 50 ml conical tubes.
5. Dissection tools.
6. Glass coverslips.
7. Silicone imaging spacers (Electron Microscopy Science).
8. 2 % paraformaldehyde (PFA) in phosphate buffered saline (PBS), pH 7.4.
9. PBST: phosphate buffered saline (PBS), pH 7.4, 0.01 % Triton.
10. X-gal wash buffer. 0.1 M sodium phosphate buffer, pH 7.2, 2 mM $MgCl_2$, 0.01 % sodium deoxycholate, and 0.02 % Nonidet-P40.
11. X-gal reaction buffer. 5 mM potassium ferrocyanide, 5 mM potassium ferricyanide, 10 mg/ml X-gal reagent (5-bromo-4-chloro-3-indolyl-β-D-galactoside) in X-gal wash buffer.
12. Polymerization solution. Dissolve 50 g of acrylamide and 1.25 g of VA-044 (Wako Pure Chemical) in 500 ml of PBS. Aliquot and freeze. Thaw solution overnight at 4 °C and degas the solution using nitrogen gas purge for 5 min on ice before use.
13. Clearing solution. 50 % *N*,*N*,*N*′,*N*′-Tetrakis(2-hydroxypropyl) ethylenediamine (87640, Sigma-Aldrich), 4 M urea in PBS.
14. Fluorescence wash buffer. 0.1 % Triton X-100, 0.1 % Tween 20, 0.1 % sodium deoxycholate, 10 % DMSO in PBS, pH 7.4.
15. Fluorescence staining buffer. 5 % donkey/goat/sheep serum, 0.1 % Triton X-100, 0.1 % Tween 20, 0.1 % sodium deoxycholate, 10 % DMSO in PBS, pH 7.4.
16. Refractory index matching solution (RIMS). 10 g of Histodenz (D2158 Sigma-Aldrich) in 7.5 ml of PBS, 0.1 % Tween 20, 0.1 % sodium azide (or alternative antibacterial agent).
17. Immunolabeling reagents (Table 1).
18. Image acquisition. Confocal or two-photon microscope with long working distance objective.
19. Deconvolution. Fiji (www.Fiji.sc) [16] (ImageJ) with DeconvolutionLab plugin [17]. Huygens Professional (Scientific Volume Imaging. www.svi.nl).
20. Image processing and visualization. Fiji (www.Fiji.sc) [16] with 3D viewer plugin [18], Icy (www.icy.bioimageanalysis.org) [19], Imaris (Bitplane www.bitplane.com), and Amira 3D (FEI www.fei.com).

3 Methods

3.1 Preparation of Tissue for Clearing

For proper quantification analysis, it is important to have homogenous signal throughout the sample. For this purpose, reporter lines are far superior to immunolabeling. For labeling satellite cells,

Table 1
Reagents for fluroscent labeling of skeletal muscle

Antigen	Company	Cat #	Concentration	Comment
CD31-Alexa 488	Biolegend	102514	5–10 μg for intravital	Marks blood vessels. Does not work on fixed tissue
DyLight 488 or 649 labeled tomato lectin	Vector staining	DL-1174, DL-1177	70 μg for intravital 1:50 for whole mount	Marks blood vessels. Fluorescently labeled fixable dextran or Isolectin B4 may also be used
Laminin	Sigma-Aldrich	L-9393, L-0663	1:100 for whole mount	Marks muscle membrane and blood vessels
Dystrophin	Abcam	Ab15277	1:100 for whole mount	Marks skeletal muscle membranes
Alexa Fluor® conjugated secondary antibodies	Thermo Fisher Scientific		1:200	Classical fluorescence dyes are not recommended
Donkey/goat/sheep serum	Sigma-Aldrich		5 %	Block based on the species the secondary antibody is from (*see* **Note 1**)

we utilized the *Pax7CreErT2* mice [20] crossed with (*loxP-stop-loxP-tdTomato*) reporter mice [21], which are currently the strongest conditional reporter mice available for satellite cells (*see* **Note 2**). For assessing LacZ signals, we utilized the *Myf5+/nLacZ* mice [22] for satellite cells and *Flt-1+/LacZ* [23] mice for blood vessels.

1. Sedate mice according to the institutionally approval protocol.
2. Administer fluorescent-labeled antibody or lectin through either tail vein or retro-orbital sinus for intravital staining of the vasculature or circulating cells. Allow the reagents to circulate and bind for 10 min.
3. Prepare cold 2 % PFA in PBS for perfusion fixation of the mouse (*see* **Notes 3** and **4**).
4. Perfuse the animal with 50 ml of fixative at a rate of 2.5 ml/min, as previously described [24] (*see* **Note 5**).
5. Dissect the tissue required for the study. For initial screening, we recommend the male cremaster muscle, which is thin, readily cleared and easily imaged. Steps for proper dissections are detailed in Fig. 2. Other suitable muscle includes the EDL and the soleus muscle. Their dissections have been previously described [25].

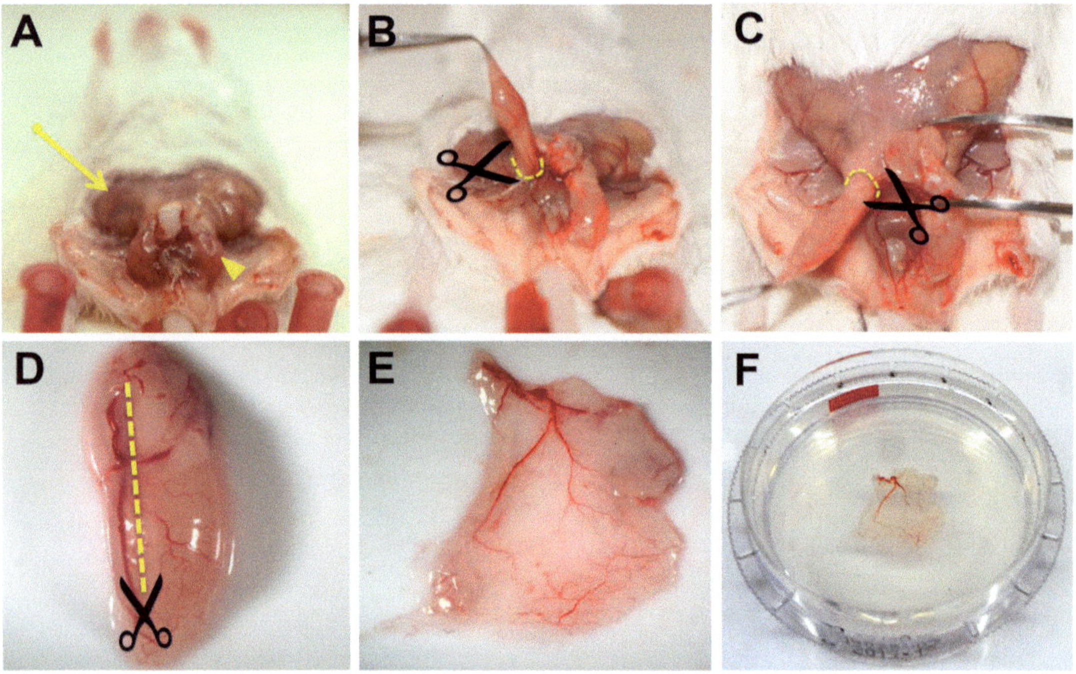

Fig. 2 Preparation of mouse cremaster muscle. (**A**) Remove the overlying skin (*arrow*) over the cremaster muscle (*arrowhead*). Gently apply pressure on the abdomen to allow the testicles to descend into the cremaster sac and grasp the muscle at the distal tip using blunt tweezers. (**B**, **C**) Retract the cremaster sac up and away from the body, gently removing the connective tissue as the muscle is freed. Make an incision around the cremaster sac (*broken lines*) and obtain the muscle along with the gonads. (**D**) Make an incision longitudinally along the muscle (*broken line*) taking care to avoid the primary artery feeding the muscle. (**E**) Flatten the muscle on a soft surface and remove the internal soft tissue. (**F**) Post-fix the muscle flat by submerging it underneath a coverslip to maintain the flat shape

3.2 LacZ Staining

Tissue clearing allows for greater transparency compared to traditional preparations following LacZ staining. Cleared tissue can be visualized all the way through to allow for proper assessment of signal deep within the tissue that would be traditionally missed (Fig. 3).

1. Wash the tissue samples using X-gal wash buffer three times at 37 °C for 2 h (*see* **Note 6**).
2. Light shield with aluminum foil and incubate the tissue in X-gal reaction buffer overnight at 37 °C in a bacterial shaker. Large pieces of tissue such as the adult gastrocnemius muscle should be cut into smaller pieces to allow to sufficient penetration of the substrate.
3. Wash the tissue three times in ice-cold PBST for 30 min each with rotation at 4 °C.
4. Incubate the tissue in polymerization solution for 6 h with rotation at 4 °C.

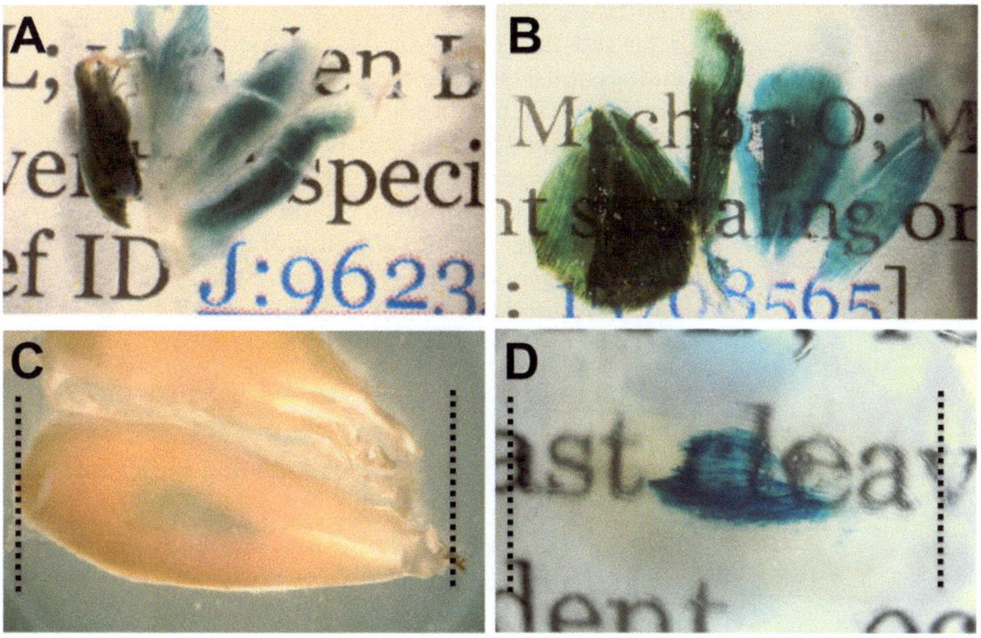

Fig. 3 Tissue clearing following LacZ staining allows for greater transparency in muscle allowing for visualization of deep tissue staining. Skeletal muscle from *Flt-1+/LacZ* mice marking blood vessels lacZ staining before (**A**) and after (**B**) tissue clearing. Tibialis anterior (TA) muscle before tissue clearing (**C**) shows moderate lacZ staining. TA muscle after tissue clearing (**D**) shows extensive signal in the middle of the tissue

5. Transfer the tissue into fresh nitrogen degassed polymerization solution. Overfill a 15 ml conical with the polymerization solution and place a plastic wrap over the tube before capping it to avoid introducing any air into the filled conical tube. Place the tube on a rotation mixer at 37 °C for 3 h. Discard the acrylamide waste appropriately.
6. Wash the tissue in clearing solution for 30 min and discard the residual acrylamide waste appropriately.
7. Incubate the tissue in clearing solution for 1–3 day at 37 °C in a bacterial shaker. Replace clearing solution each day.
8. Wash the tissue in PBST three times for 30 min.
9. Incubate the tissue in RIMS solution overnight.
10. Image the tissue using a stereoscope while it is immersed in RIMS. Flat muscle such as the cremaster muscle may be submerged in RIMS and flattened by a glass coverslip. Thicker muscle may be imaged by placing them into silicone isolator surrounded by coverslips (*see* **Note 7**).
11. The samples can remain in RIMS indefinitely and stored at room temperature.

3.3 Tissue Clearing for Fluorescent Reporter and Immunofluorescence

For deep tissue fluorescent imaging, it is recommended that the red or far-red fluorescent probes be used to avoid interference from endogenously fluorescent molecules. Skeletal muscle presents an additional challenge in selecting fluorophores due to the presence of myoglobin, NADH and FAD and non-homogenous concentration of these molecules. These molecules are especially problematic in the same spectral region as GFP [26]. As such, weak markers or myofiber markers should not be visualized in the green spectrum. GFP reporters may also be red-shifted using anti-GFP antibody. If multi-color-imaging spanning the spectra is desired, the green spectra should be reserved for non-myogenic markers.

1. Wash the tissue in fluorescence wash buffer three times for 60 min at 37 °C to permeabilize the tissue (*see* **Note 8**).
2. Block the tissue in fluorescence staining buffer for 4 h at 37 °C.
3. Incubate the tissue with primary antibody in fluorescence staining buffer for 24–48 h depending on the thickness of the tissue at 37 °C (*see* **Note 9**).
4. Transfer the tissue to 4 °C overnight to allow the antibody to bind.
5. Wash the tissue in fluorescence wash buffer three times at 37 °C in a bacterial shaker for 8 h total.
6. Incubate the tissue in 1:200 dilution of fluorescently conjugated secondary antibody (*see* **Note 10**) for 24–48 h at 37 °C.
7. Wash the tissue in fluorescence wash buffer three times at 37 °C in a bacterial shaker for 8 h total.
8. Incubate the tissue in RIMS overnight.
9. Mount the tissue in fresh RIMS between coverslips in silicone gaskets or 3D printed slide holders.
10. For long term storage, transfer the tissue into PBS and store at 4 °C for a few months. Samples should be incubated in RIMS again prior to imaging.

3.4 Image Acquisition

Image acquisition is highly dependent on the type of instrument used and the manufacturer or core-facility should be consulted for the best use case for a particular instrument. In this section, several tips will be discussed to allow deep tissue imaging. Typically, depth of 450 μm can be achieved using a confocal microscope system. Depth of more than 1700 μm can be obtained using a two-photon microscope system on cleared tissue. Two-photon excitation can be particularly challenging to predict and therefore proper planning is required. The spectra database hosted at the University of Arizona (http://www.spectra.arizona.edu) is a great resource for obtaining the proper excitation of various florescent dyes and molecules and the two-photon spectra is labeled as "-2P". The regular periodic organization of the sarcomeres also allows for label free second harmonic generation (SHG) detection of the myofibers [27] at an excitation wavelength of 870 nm. This can be particularly useful to delineate between intact and injured muscle fibers (Fig. 4). However, the emission will be in the same range as common DNA binding dyes such as DAPI or Hoechst 33342 and thus require separate acquisitions.

3.5 Image Processing

Some image processing is required to obtain good quality images suitable for analysis. While optical sectioning through confocal allows for excitation at a single point or layer, there is still some inherent scattering of light that causes noise and scattering of the signal, particularly in the axial direction. Software deconvolution can mathematically correct for this by redistributing the scattered light signal to its original state. A detailed discussion of deconvolution is out of the scope of this chapter. Some practical considerations have been addressed in previous reports [17, 28]. We utilize Huygens Professional, which has a user-friendly

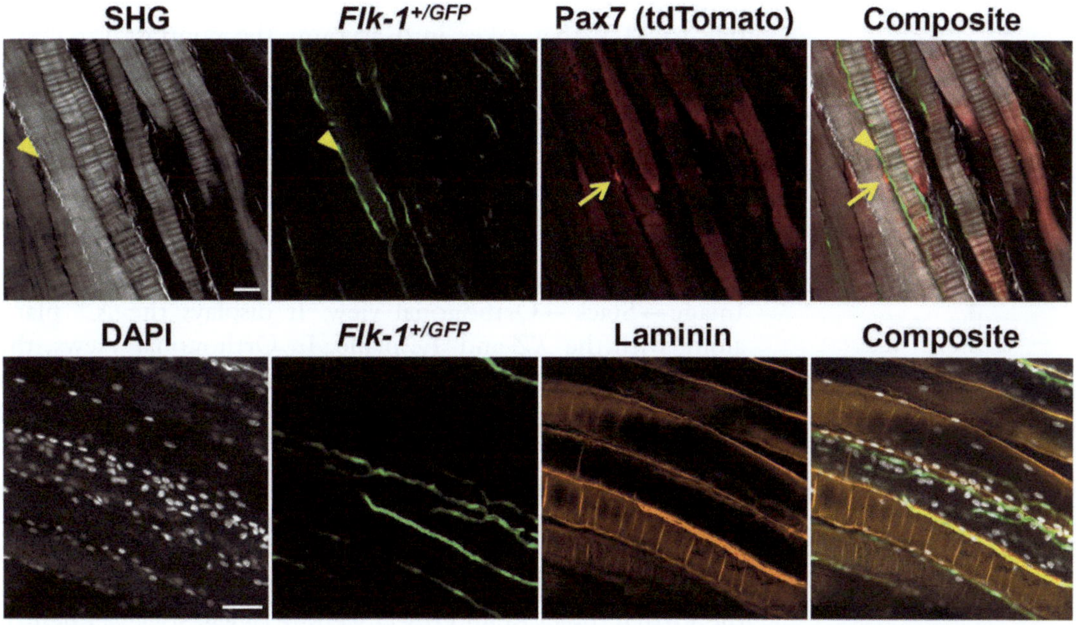

Fig. 4 Examples of cleared tissue with fluorescent reporter and immunostaining. (*Top*) Maximum intensity projection of 3×1 μm slices of cleared tissue with fluorescent reporters acquired using a two-photon microscope system. Second harmonic generation (SHG) for label free detection of collagen and muscle fiber sarcomere. tdTomato positive satellite cells (*arrows*) and Flk-1$^{+/GFP}$ positive blood vessels. Collagen fibers are also marked by SHG signal (*arrowheads*). (*Bottom*) Maximum intensity projection of 15×1 μm slices of cleared tissue with immunofluorescence and fluorescent reporter acquired using a confocal microscope system. DAPI labels all nuclei, Flk-1$^{+/GFP}$ marks the blood vessels and laminin immunofluorescence marks the myofiber and blood vessel membrane. Scale bar denotes 50 μm

interface and allows for theoretical calculation of a point spread function (PSF). Detailed tutorial on utilizing the software can be found on the manufacturers website and elsewhere [28].

3.6 Visualization of Data

Deconvoluted images can be visualized in several manner including intensity projection, orthogonal view, and 3D render. This can be accomplished in several program including Imaris, Amira 3D, Icy, and FIJI. FIJI [16] and Icy [19] are free open source software. For this chapter instructions for obtaining these views in FIJI are provided. Open the image file using the Bio-Formats Import plugin (File → Import → Bio Formats) as a Hyperstack.

1. Intensity projection is a convenient method to visualize data from many different focal planes in a 2D format. It is more suited for visualizing rare populations over multiple planes or complex vascular anatomy. To view an intensity projection, open the Z-Projection dialog from Image → Stacks → Z Project. Select the range for the projection and select the desired projection type. Maximum Intensity Projection (MIP) is the most popular projection algorithm for visualizing 3D data. However it is very susceptible to imaging artifacts and saturated hotspots. Standard Deviation Intensity Projection (SDIP) is an alternative to MIP that takes into account the contrast variability across the entire data set. SDIP is suited for samples with good signal to noise ratios (SNR) and membrane staining. Additional projections such as minimum intensity, average intensity, sum slices and median are less useful for use in projection of fluorescent samples.
2. Orthogonal view is a simple method to visualize 3D information. To view the hyperstack in orthogonal view go to Image → Stack → Orthogonal view. It displays the *XY* plane along with the *XZ* and *YZ* plane. In Orthogonal Views, the two alternative planes will dock along the respective dimension of the *XY* plane. Two intersecting yellow lines will live update in the alternative views automatically.
3. 3D Render. 3D rendering is possible but limited in FIJI. It requires the 3D viewer plugin [18]. Images can be 3D rendered by Plugin → 3D viewer. Detailed explanation of the 3D viewer is out of the scope of this chapter but can be found at the Virtual Insect Brain Project website (www.3dviewer.neurofly.de).

4 Notes

1. Animal serum may be substituted by 1 % bovine serum albumin (BSA). However, IgG free BSA should be used if anti-goat secondary antibodies are used due to similarities between bovine and goat IgG.

2. As with any Cre based system, it is important to validate the specificity of the Cre using a strong reporter system. For labeling satellite cells, we tested two popular tamoxifen inducible Cre mouse lines, *Pax7tm2.1(cre/ERT2)Fan/J(Fan)* and *Pax7tm1(cre/ERT2)Gaka(Gaka), available in Jackson Laboratory.*
3. Fresh PFA is important for imaging of fluorescently labeled tissue. Commercially available PFA solutions may contain methanol as preservative. Methanol can bleach fluorescent proteins and dampen β-gal activity.
4. If immunostaining is not required, 0.05 % or 1.0 % glutaraldehyde may be added the fixative solution for fluorescently labeled and β-gal labeled tissue, respectively.
5. Perfusion fixation followed by overnight immersion fixation in best for preserving fluorescent molecules. However, if immunostaining is susceptible to PFA fixation, tissue may be immersion fixed for only 30 min (with perfusion) or 4 h (without perfusion) 4 °C.
6. All washes are done in a bacterial shaker in 15 ml or 50 ml conical tubes filled 2/3 of the way and laying at a 45°. The vigorous agitation is required for proper penetration of the labeling regents. Prolong fixation time if the samples tend to break apart during the wash steps.
7. Silicone spacers may be washed, dried and reused. Use adhesive tape to remove dust particles prior to adhering the spacer onto a glass coverslip.
8. Light shield all samples using aluminum foil during the wash and incubation period to prevent loss of fluorescence.
9. Highly abundant epitope will require higher amounts of antibody as compared to immunostaining in cross sections. For example, anti-laminin staining requires tenfold higher antibody concentration as compared to immunostaining in cross sections.
10. It is important to use Alexa Fluor or DyLight-conjugated antibodies versus classical dyes such as FITC, rhodamine, Cy5, or other such dyes as they are much more susceptible to photobleaching and changes in pH.

References

1. Bentzinger CF, Wang YX, Dumont NA, Rudnicki MA (2013) Cellular dynamics in the muscle satellite cell niche. EMBO Rep 14:1062–1072. doi:10.1038/embor.2013.182
2. Christov C, Chretien F, Abou-Khalil R et al (2007) Muscle satellite cells and endothelial cells: close neighbors and privileged partners. Mol Biol Cell 18:1397–1409
3. Yoshida S, Sukeno M, Nabeshima Y-I (2007) A vasculature-associated niche for undifferentiated spermatogonia in the mouse testis. Science 317:1722–1726. doi:10.1126/science.1144885

4. Keefe AC, Lawson JA, Flygare SD et al (2015) Muscle stem cells contribute to myofibres in sedentary adult mice. Nat Commun 6:7087. doi:10.1038/ncomms8087
5. Bosnakovski D, Xu Z, Li W et al (2008) Prospective isolation of skeletal muscle stem cells with a Pax7 reporter. Stem Cells 26:3194–3204. doi:10.1634/stemcells.2007-1017
6. Sambasivan R, Gayraud-Morel B, Dumas G et al (2009) Distinct regulatory cascades govern extraocular and pharyngeal arch muscle progenitor cell fates. Dev Cell 16:810–821. doi:10.1016/j.devcel.2009.05.008
7. Relaix F, Rocancourt D, Mansouri A, Buckingham M (2005) A Pax3/Pax7-dependent population of skeletal muscle progenitor cells. Nature 435:948–953. doi:10.1038/nature03594
8. Asakura A, Seale P, Girgis-Gabardo A, Rudnicki MA (2002) Myogenic specification of side population cells in skeletal muscle. J Cell Biol 159:123–134. doi:10.1083/jcb.200202092
9. Tajbakhsh S, Bober E, Babinet C et al (1996) Gene targeting the myf-5 locus with nlacZ reveals expression of this myogenic factor in mature skeletal muscle fibres as well as early embryonic muscle. Dev Dyn 206:291–300. doi:10.1002/(SICI)1097-0177(199607)206:3<291::AID-AJA6>3.0.CO;2-D
10. Renier N, Wu Z, Simon D, Yang J et al (2014) iDISCO: a simple, rapid method to immunolabel large tissue samples for volume imaging. Cell 159(4):896–910. doi:10.1016/j.cell.2014.10.010
11. Tainaka K, Kubota SI, Suyama TQ et al (2014) Whole-body imaging with single-cell resolution by tissue decolorization. Cell 159:911–924. doi:10.1016/j.cell.2014.10.034
12. Susaki EA, Tainaka K, Perrin D et al (2014) Whole-brain imaging with single-cell resolution using chemical cocktails and computational analysis. Cell 157:726–739. doi:10.1016/j.cell.2014.03.042
13. Yang B, Treweek JB, Kulkarni RP et al (2014) Single-cell phenotyping within transparent intact tissue through whole-body clearing. Cell 158:945–958. doi:10.1016/j.cell.2014.07.017
14. Tomer R, Ye L, Hsueh B, Deisseroth K (2014) Advanced CLARITY for rapid and high-resolution imaging of intact tissues. Nat Protoc 9:1682–1697. doi:10.1038/nprot.2014.123
15. Hama H, Hioki H, Namiki K et al (2015) ScaleS: an optical clearing palette for biological imaging. Nat Neurosci 18:1518–1529. doi:10.1038/nn.4107
16. Schindelin J, Arganda-Carreras I, Frise E et al (2012) Fiji: an open-source platform for biological-image analysis. Nat Methods 9:676–682. doi:10.1038/nmeth.2019
17. Vonesch C, Unser M (2008) A fast thresholded Landweber algorithm for wavelet-regularized multidimensional deconvolution. IEEE Trans Image Process 17:539–549. doi:10.1109/TIP.2008.917103
18. Schmid B, Schindelin J, Cardona A et al (2010) A high-level 3D visualization API for Java and ImageJ. BMC Bioinformatics 11:274. doi:10.1186/1471-2105-11-274
19. de Chaumont F, Dallongeville S, Chenouard N et al (2012) Icy: an open bioimage informatics platform for extended reproducible research. Nat Methods 9:690–696. doi:10.1038/nmeth.2075
20. Murphy MM, Lawson JA, Mathew SJ et al (2011) Satellite cells, connective tissue fibroblasts and their interactions are crucial for muscle regeneration. Development 138:3625–3637. doi:10.1242/jcs098228
21. Madisen L, Zwingman TA, Sunkin SM et al (2010) A robust and high-throughput Cre reporting and characterization system for the whole mouse brain. Nat Neurosci 13:133–140. doi:10.1038/nn.2467
22. Tajbakhsh S, Rocancourt D, Cossu G, Buckingham M (1997) Redefining the genetic hierarchies controlling skeletal myogenesis: Pax-3 and Myf-5 act upstream of MyoD. Cell 89:127–138. doi:10.1016/S0092-8674(00)80189-0
23. Kearney JB, Ambler CA, Monaco KA et al (2002) Vascular endothelial growth factor receptor Flt-1 negatively regulates developmental blood vessel formation by modulating endothelial cell division. Blood 99:2397–2407
24. Brazelton TR, Blau HM (2005) Optimizing techniques for tracking transplanted stem cells in vivo. Stem Cells 23:1251–1265. doi:10.1634/stemcells.2005-0149
25. Collins CA, Zammit PS (2009) Isolation and grafting of single muscle fibres. Methods Mol Biol 482:319–330
26. Jackson KA, Snyder DS, Goodell MA (2004) Skeletal muscle fiber-specific green autofluorescence: potential for stem cell engraftment artifacts. Stem Cells 22:180–187. doi:10.1634/stemcells.22-2-180
27. Liu W, Raben N, Ralston E (2013) Quantitative evaluation of skeletal muscle defects in second harmonic generation images. J Biomed Opt 18:26005. doi:10.1117/1.JBO.18.2.026005
28. Wimmer VC, Möller A (2010) High-resolution confocal imaging in tissue. Methods Mol Biol 611:183–191. doi:10.1007/978-1-60327-345-9_15

Chapter 11

Isolation, Culture, Functional Assays, and Immunofluorescence of Myofiber-Associated Satellite Cells

Thomas O. Vogler, Katherine E. Gadek, Adam B. Cadwallader, Tiffany L. Elston, and Bradley B. Olwin

Abstract

Adult skeletal muscle stem cells, termed satellite cells, regenerate and repair the functional contractile cells in adult skeletal muscle called myofibers. Satellite cells reside in a niche between the basal lamina and sarcolemma of myofibers. Isolating single myofibers and their associated satellite cells provides a culture system that partially mimics the in vivo environment. We describe methods for isolating and culturing intact individual myofibers and their associated satellite cells from the mouse extensor digitorum longus muscle. Following dissection and isolation of individual myofibers we provide protocols for myofiber transplantation, satellite cell transfection, immune detection of satellite cell antigens, and assays to examine satellite cell self-renewal and proliferation.

Key words Skeletal muscle, Satellite cells, Muscle stem cells, Myofiber isolation, Extensor digitorum longus, Immunostaining, Proliferation, Self-renewal, Transfection, AraC, EdU, CFDA-SE, Transplant, Tibialis anterior, Mouse

1 Introduction

The cellular building blocks of skeletal muscle tissue are post-mitotic, multinucleated, syncytial cells called myofibers. The myonuclei in myofibers are incapable of repair and when damaged, myofibers rely on resident muscle stem cells, called satellite cells, for repair and regeneration [1]. In addition to rebuilding damaged myofibers, satellite cells continuously fuse and contribute myonuclei to myofibers in mature skeletal muscle [2, 3]. Satellite cells reside between the basal lamina and the sarcolemma of myofibers [4] and upon receiving signals to expand, quiescent satellite cells activate, induce myogenic transcription factors, enter into the cell cycle and proliferate, becoming myoblasts or muscle progenitor cells. Quiescent and activated satellite cells express the paired-box

* Authors Thomas O. Vogler and Katherine E.Gadek contributed equally.

Michael Kyba (ed.), *Skeletal Muscle Regeneration in the Mouse: Methods and Protocols*, Methods in Molecular Biology, vol. 1460, DOI 10.1007/978-1-4939-3810-0_11,

transcription factor Pax7 and upregulate the myogenic transcription factors MyoD and Myf5 prior to cell cycle entry and proliferation [1]. The majority of myoblasts will then terminally differentiate to fuse and repair muscle, while a subset self-renews to replenish the satellite cell pool for future rounds of muscle regeneration [1]. Satellite cell self-renewal is a quintessential characteristic of stem cells and requires both cell autonomous and cell nonautonomous signals [5, 6].

Elucidating the signals that regulate the finely tuned balance of satellite cell self-renewal and proliferation is critical for understanding muscle regeneration and homeostasis. In vivo assays aimed at understanding satellite cell behavior are cumbersome, time-consuming, and expensive. Isolating single, mononuclear satellite cells from whole muscle is one alternative approach. Post-isolation, satellite cells are cultured on tissue culture plastic or Matrigel; however, these conditions do not accurately recapitulate the niche in which satellite cells thrive. Another alternative involves isolating and culturing myofibers with their associated satellite cells, thereby maintaining the satellite cell niche. The myofiber niche provides a specific substrate stiffness that influences satellite cell behavior [7] as well as provides growth factors such as FGF [8]. Due to the satellite cell's intimate association with a myofiber, satellite cells remain attached to their corresponding myofibers during myofiber isolation. The extensor digitorum longus (EDL) muscle of *Mus musculus* provides hundreds of individual myofibers with associated satellite cells when isolated by enzymatic and mechanical dissociation.

Isolating and culturing isolating individual myofibers from EDL muscles, pioneered in the 1980s [9], has been used extensively by several groups [10–13]. Here we outline (1) how to isolate and culture myofiber-associated satellite cells from a mouse EDL, (2) assays used to study and manipulate myofiber-associated satellite cell behavior, and (3) how to fix and stain myofiber-associated satellite cells. Our approach can be modified to isolate, assay, manipulate, and identify epitopes by antibody staining on myofiber-associated satellite cells from different muscle groups including the gastrocnemius and soleus. We provide a comprehensive reagent list and detailed methods, optimized by over a decade of experience, to probe the behavior of myofiber-associated satellite cells.

2 Materials

Prepare all cell culture reagents in a laminar flow hood with sterile technique. Follow all waste disposal regulations when disposing of chemical and other waste materials.

2.1 Myofiber Growth Media

The components of myofiber growth media are F12-C media, horse serum, and antibiotics. Some of the following protocols require F12-C media alone, F12-C media with antibiotics or F12-C media with 15% horse serum and no antibiotics. Here we describe how to make myofiber growth media; however, this protocol can be modified to make F12-C media with antibiotics and no serum, to make F12-C media alone or to make F12-C media with 15% horse serum and no antibiotics.

1. For 1 L of F12-C media add 700 mL of tissue culture grade water to an Erlenmeyer flask and mix with stir bar.
2. Slowly add one packet of Ham's F-12 Powdered Nutrient Mix (Gibco/Thermo Fisher Scientific, Grand Island, NY, USA) and rinse the F-12 container with a small amount of water and add to mixture.
3. Stir until powder is dissolved.
4. Add 1.176 g $NaHCO_3$ and stir until dissolved. With 10 mL water rinse any residual powdered F-12 and $NaHCO_3$ from the sides and neck of the flask.
5. Adjust the pH to 7.15–7.20.
6. Add 10 mL 100× penicillin–streptomycin stock solution (*see* Subheading 2.1.1) and 4 mL 0.2 M $CaCl_2$ (*see* Subheading 2.1.2). Bring final volume to 1 L.
7. Sterile filter the solution through a 0.22 μm filter and aliquot 425 mL into tissue culture bottles. Store at 4 °C (*see* **Note 1**).
8. Add 75 mL of lot tested horse serum to 425 mL F12-C media to make myofiber growth media (*see* **Note 2**).

2.1.1 Penicillin–Streptomycin 100× Stock Solution

1. Dissolve 10 M units of tissue culture grade sodium salt penicillin G (Sigma-Aldrich, St. Louis, MO, USA) and 500 mg of tissue culture grade streptomycin sulfate (Sigma-Aldrich) in 900 mL tissue culture grade water.
2. Stir with stir bar until completely dissolved.
3. In tissue culture hood, filter the solution through a 0.22 μm bottle top filter, aliquot into 50 mL plastic conical vials and store at −20 °C until needed.

2.1.2 0.2 M $CaCl_2$ Stock Solution

1. Dissolve 11.1 g $CaCl_2$ in 450 mL tissue culture grade water and bring volume to 500 mL with tissue culture grade water.
2. In a tissue culture hood, filter the solution through a 0.22 μm bottle top filter into a sterile glass bottle. Store at room temperature.

2.2 Myofiber Isolation Components

1. F12-C media with antibiotics and no serum.
2. 2 mL microcentrifuge tubes.
3. 70% ethanol in a spray bottle.
4. Two pairs of Dumont #5 forceps (Fine Science Tools, Foster City, CA, USA).
5. Extra fine Bonn scissors with 13 mm cutting edge (Fine Science Tools).

6. ToughCut scissors with 22 mm cutting edge (Fine Science Tools).
7. 2 mL Pipet pump (Bel-Art Products, Wayne, NJ, USA).
8. Low oxygen humidified tissue culture incubator 37 °C, 5% CO_2, 6% N_2.
9. Six-well tissue culture plates.
10. Stereoscope or dissection microscope with light base.

2.2.1 Recombinant Human Basic Fibroblast Growth Factor (FGF2) Stock Solution

1. Resuspend FGF2 (Promega Corporation, Madison, WI, USA) to a concentration of 25 μg/mL (10,000×) in 1% BSA in PBS.
2. Sterile filter through a 0.22 μm filter. Aliquot FGF2 into 20 μL aliquots and store at −20 °C. Once thawed store at 4 °C (*see* **Note 3**).

2.2.2 Fire-Polished Pasteur Pipettes

Light a Bunsen burner and gently rotate the tips of glass Pasteur pipettes in the flame (*see* **Note 4**). Polish the end until it becomes smooth and rounded so as not to damage myofibers during transfer.

2.2.3 20× Saline G Solution

1. Mix 600 mL dH_2O and 160 g NaCl, 8.0 g KCl, and 3.0 g KH_2PO_4 in the order listed.
2. Add 3.0 g Na_2HPO_4, 22.0 g dextrose and 100 mg Phenol Red.
3. Stir until dissolved then add 400 mL dH_2O to bring final volume to 1 L.
4. Sterile filter the solution through 0.22 μm filter and store at 4 °C until needed.

2.2.4 Collagenase Type I 10× Stock Solution

1. Calculate the amount of collagenase type I (Worthington Biochemicals, Lakewood, NJ, USA) required based upon activity provided (*see* **Note 5**). The final 10× concentration should be 40,000 U/mL.
2. Freshly dilute 20× Saline G to 1× with sterile water and add collagenase.
3. Stir on ice until collagenase is dissolved.
4. Sterile filter the solution through a 0.44 μm filter then through a 0.22 μm filter (*see* **Note 6**).
5. Aliquot the collagenase into 100 μL aliquots and store at −20 °C. Thaw aliquots at room temperature when needed (*see* **Note 7**).

Fixing and Staining Components for Antigen Detection by Immunofluorescence

1. Phosphate buffered saline (PBS, 10× stock; dilute to 1× before using): 1.37 M NaCl, 27 mM KCl, 100 mM Na_2HPO_4, 18 mM KH_2PO_4.
2. PBS containing 0.5% Triton X-100.
3. Paraformaldehyde: 4% paraformaldehyde diluted in PBS from 10% EM grade paraformaldehyde stock solution (Electron

Microscopy Sciences, Hatfield, PA, USA). For every well of myofibers in a six well plate to be fixed, prepare 2.5 mL of 4 % paraformaldehyde. Mix and use immediately.

4. Hydrogen peroxide quench solution: 3 % H_2O_2 diluted with dH_2O from 30 % H_2O_2 stock solution (Fisher Scientific/ Thermo Fisher). Mix and use immediately.
5. Blocking solutions: 0.25 % Triton X-100, 3 % bovine serum albumin fraction V (BSA; Sigma-Aldrich) in PBS for initial permeabilization and blocking. 0.125 % Triton X-100, 3 % BSA in PBS for primary and secondary antibodies. For EdU Click-It chemistry 0.5 % Triton X-100, 3 % BSA in PBS (*see* **Note 8**).
6. Superfrost Plus Pre-cleaned Microscope Slides (Fisher Scientific/Thermo Fisher).

2.2.5 Myofiber Staining Baskets

1. Make basket top by carefully removing the lid and bottom half of a 1.5 mL tube making the edge as flat as possible (*see* **Note 9**).
2. Cut the Nylon Mesh (80 μm pore size, 46-μm thread diameter, Small Parts/Amazon.com Inc, Seattle, WA, USA) into a square slightly larger than the edges of the cut 1.5 mL tube (*see* **Note 10**).
3. Warm a hot plate loosely covered with aluminum foil to a medium heat setting (*see* **Note 11**).
4. Place the nylon mesh on the foil covered hot plate and place the basket top over the mesh. Press down gently and evenly until basket top melts into the nylon mesh.
5. Remove basket from hot plate and allow to cool.
6. Trim the edges of the extra nylon from the 1.5 mL tube using a razor blade or scalpel until the basket fits into a well of a 48 well plate easily (*see* **Note 12**).

2.2.6 Mowiol Mounting Medium

1. Place 6 g of glycerol into a 50 mL glass beaker with a stir bar.
2. While stirring, add 2.4 g Mowiol 4-88 (Calbiochem/EMD Millipore, Darmstadt, Germany).
3. Add 6 mL dH_2O and stir for 2 h at room temperature.
4. Add 12 mL 0.2 M Tris-HCl (pH 8.5) and incubate at 50–60 °C for 10 min using a bead or water bath followed by mixing on a stir plate for 10–15 min.
5. Repeat the heating and stirring for 1–2 h until all the Mowiol goes into solution.
6. Transfer the solution into 50 mL conical tubes and centrifuge at $5000 \times g$ for 15 min to remove any un-dissolved Mowiol from the solution.
7. Transfer the supernatant into a new 50 mL glass container with a stir bar and mix again. Add 0.569 g (2.5 % weight/volume

concentration) anti-fade agent 1,4-Diazabicyclo[2.2.2]octane (DACO) (Sigma-Aldrich).

8. Mix for 5–10 min until DACO is dissolved.
9. Store 1 mL aliquots at −20 °C. Warm aliquots to room temperature before use (*see* **Note 13**).

2.3 Myofiber Transplantation Components

1. 27 ½G 0.40×13 mm ½cc Tuberculin syringes (Fisher Scientific/Thermo Fisher Scientific).
2. Sterile filtered 0.9 % Saline Solution (*see* **Note 14**).
3. Sterile filtered 1.2 % Barium Chloride in 0.9 % Saline Solution (*see* **Note 15**).
4. Six-well tissue culture plates.
5. Myofiber growth media and 2.5 ng/mL FGF2.

2.4 Myofiber Associated Satellite Cell Transfection Components

1. OptiMEM Reduced Serum Media (Gibco/Thermo Fisher Scientific).
2. Lipofectamine 2000 (Invitrogen/Thermo Fisher Scientific).
3. Plasmid DNA to transfect.
4. 25 μg/mL FGF2 stock solution (*see* Subheading 2.1.1).
5. Myofiber growth media.
6. F12-C media with 15 % horse serum and no antibiotics.

2.5 CFDA-SE Self-Renewal Assay Components

1. Myofiber growth media.
2. 25 μg/mL FGF2 stock solution (*see* Subheading 2.1.1).
3. 10 mM CFDA-SE (Life Technologies/Thermo Fisher Scientific) stock solution in DMSO (*see* **Note 16**).
4. 10 μM CFDA-SE diluted from 10 mM stock in pre-warmed sterile PBS (*see* **Note 17**).

2.6 EdU Proliferation Assay Components

1. Click-iT® EdU Alexa Fluor® Imaging Kit (Invitrogen/Thermo Fisher Scientific).

2.6.1 10 mM 5-Ethynyl-2′-Deoxyuridine (EdU; Carbosynth, Berkshire, UK) In Vitro Labeling Stock Solution

1. Dissolve EdU in DMSO to 10 mM final concentration to generate a stock solution.
2. Aliquot in 50 μL volumes and store at −20 °C.

2.6.2 0.5 mg/mL EdU Water for In Vivo Labeling

1. Dissolve 0.5 mg/mL EdU in sterile filtered H_2O with 1 % dextrose (*see* **Note 18**).

2.7 AraC Assay Components

1. Myofiber growth media.

2.7.1 100 mM Cyotsine β-d-Arabinofuranoside (AraC; Sigma-Aldrich) Stock Solution

1. Generate a 100 mM AraC stock solution in dH_2O and store at −20 °C (*see* **Note 19**).
2. Dilute 100 mM stock solution to previously determined optimal working AraC concentration (*see* **Note 20**).

3 Methods

All procedures are carried out at room temperature unless otherwise specified. Follow all waste disposal regulations when disposing of all chemical and other waste materials.

3.1 Isolation of Single Myofibers from Mouse EDL Muscle

In mice, the EDL is located between the knee and ankle joints in the hindlimb muscle in a ventral-lateral orientation. The origin of the EDL is the lateral condyle of the tibia and the insertion is split between the four extensor tendons of the 2nd–5th digits. Anatomically, the belly of the EDL is located lateral and inferior to the tibialis anterior (TA) and superficial to the tibia [14]. The function of the EDL is to extend the 2nd–5th digits [15].

3.1.1 Initial Preparation Prior to EDL Harvest (per Mouse)

1. Add 900 μL F12-C media with antibiotics to a 2 mL microcentrifuge tube.
2. On a single six-well plate, add 4 mL of myofiber growth media and 2.5 ng/mL FGF2 to a single well and 3 mL to two additional wells and place into a 37 °C tissue culture incubator.
3. Thaw a single 100 μL frozen 10× collagenase aliquot (*see* Subheading 2.2.4).

3.1.2 Dissection and Harvest of EDL Muscle

The EDL should only be handled by grasping the tendon. If the EDL is grasped by the muscle belly, the integrity of the myofibers will be compromised and the yield of intact myofibers will be low.

1. Euthanize mouse according to IACUC and institutional protocols.
2. Apply 70% ethanol to mouse hindlimb so that the fur is saturated (Fig. 1a).
3. Place mouse on its back under dissection lamp.
4. Pinch the skin above the knee joint and use ToughCut scissors to cut a superficial notch in the skin approximately 0.75 cm in length that is parallel to the tibia (Fig. 1b). Be careful not to cut too deeply into the skin in order to avoid damaging the underlying musculature.
5. Pinch the skin with fingers both proximal and distal to the superficial cut and quickly pull distal portion towards the paw (Fig. 1c). This forceful action will quickly and cleanly remove the skin from the mid-foot to the knee allowing access to the

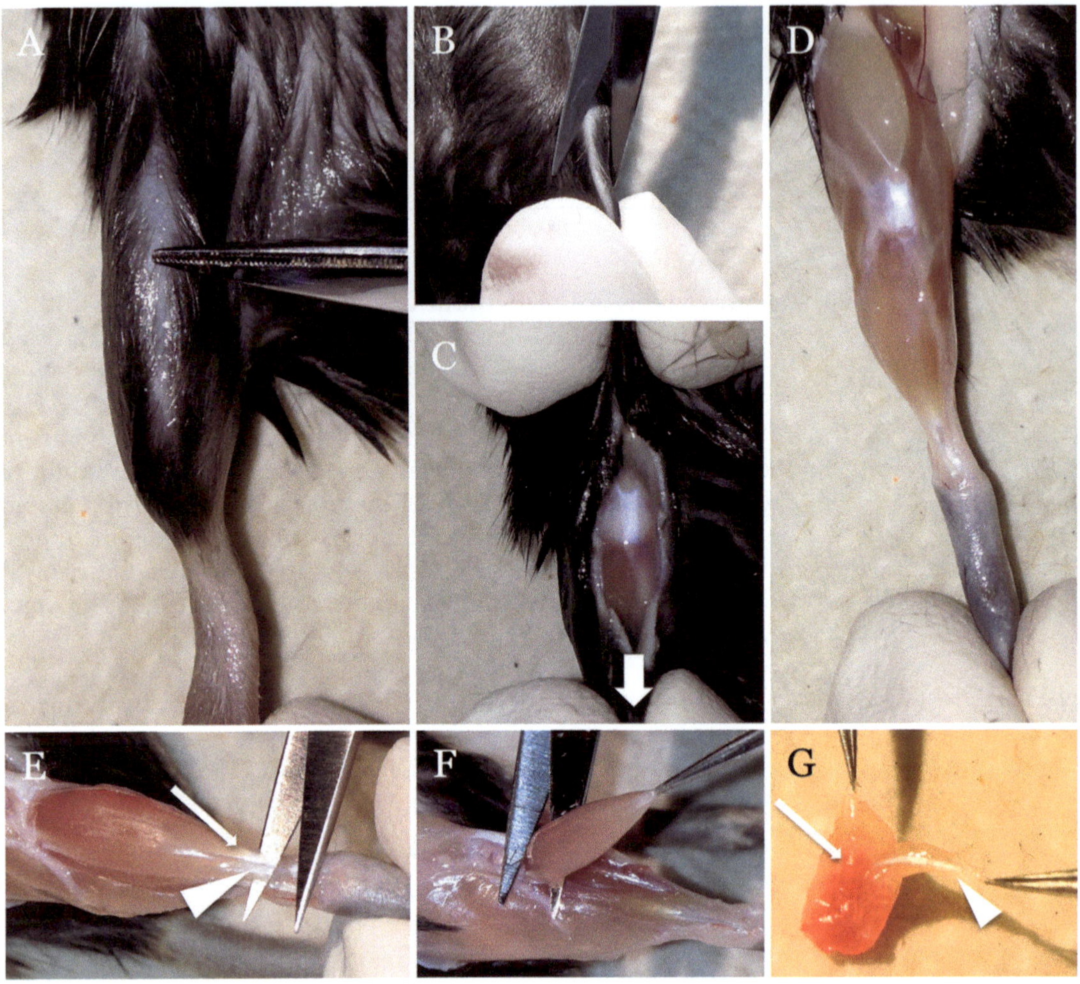

Fig. 1 Isolation of *Mus musculus* extensor digitorum longus (EDL) muscle. (**a**) The hind limb of *Mus musculus* is soaked with 70 % ethanol. The scissors are pointing to the knee joint. (**b**) The skin is pinched below knee joint and cut just above knee joint. (**c**) The distal portion of cut skin is pulled towards the paw (in the direction of the *white arrow*) to remove the skin. (**d**) The exposed musculature of the *Mus musculus* hindlimb after skin removal. (**e**) The tendons of the tibialis anterior (TA) (*white arrow*) and the EDL (*white arrowhead*) are cut at the ankle joint. (**f**) The TA and EDL are reflected and cut together at the origin point just below the knee joint (**g**) The EDL (*white arrowhead*) and TA (*white arrow*) are gently pulled apart with forceps to isolate the EDL. The isolated EDL is placed in collagenase solution for isolation of single myofibers and the associated satellite cells

EDL (Fig. 1d). Other methods to remove the skin require more tedious reflection of the skin using scalpel blades and forceps [15].

6. Using #5 forceps, gently remove the fascia that overlies the hind limb muscles (focusing on the fascia that is over the TA muscle). The fascia appears as a thin film surrounding the musculature. When the fascia is removed the identification and isolation of the EDL tendons is facilitated.

7. When the skin and fascia are removed there are two prominent tendons on the ventral side of the ankle joint. The medial tendon connects with the TA muscle and the lateral tendon is the confluence of the four insertion point tendons of the EDL. Using #5 forceps, gently isolate both tendons (Fig. 1e).
8. Gently isolate both the EDL and TA muscles by sliding #5 forceps underneath both isolated tendons. Gently slide the forceps up towards the EDL/TA origin. Be careful not to stretch the muscles.
9. Leaving the #5 forceps underneath the muscles, use the extra fine Bonn scissors to cut the two isolated tendons at the ankle joint (Fig. 1e). Be careful not to cut into any TA or EDL muscle.
10. With #5 forceps grasp the proximal portion of the cut tendons and reflect both the EDL and TA muscles proximally.
11. Carefully cut the reflected muscles at the insertion points on the tibia just below the knee joint (Fig. 1f). Be sure to not to stretch the muscle and cut as high as possible to avoid damaging or cutting the EDL musculature. Continue to grasp the distal part of the muscle by the tendons with #5 forceps. At this point the TA and EDL are separated from the mouse as a single unit.
12. Place the isolated muscles on the dissection pad and separate the two tendons on the distal part of the muscle.
13. Using two #5 forceps grab the TA and EDL tendons and gently separate the two by pulling apart. There should be very little resistance (Fig. 1g).

3.1.3 Collagenase Digestion of Isolated EDL Muscles

1. Immediately after isolation place both EDL muscles from a single mouse into prepared 2 mL microcentrifuge tube containing 900 μL of F12-C media with antibiotics. Be sure to only handle the EDL muscles by holding the tendons, even when placing into the media.
2. Add 100 μL of 10× stock collagenase (*see* Subheading 2.2.4) and incubate at 37 °C while inverting the tube ten times every 10 min for 70–90 min.
3. Retrieve the six well plate with pre-warmed myofiber growth media from the incubator. Stop the enzymatic digestion by decanting the entire contents of the 2 mL microcentrifuge tube (2 EDLs, F12-C media with antibiotics, collagenase) into a well with 4 mL of the pre-warmed myofiber growth media and 2.5 ng/mL FGF2. This well will be referred to as the "bulk well."

3.1.4 Isolation of Single Myofibers from Bulk EDL Muscle

Throughout the isolation process do not allow the bulk muscle tissue or single myofibers outside of an incubator or off a 37 °C warming plate for more than 15 min at a time. If performing isolations from two mice, work on one plate for 15 min and then switch to the other plate while the first re-warms in a 37 °C low O_2 incubator.

1. Under a stereo dissection scope, examine the collagenase digested EDL. The bulk muscle should resemble a loose bundle of translucent fishing line. If the myofibers appear white/opaque then they were likely damaged in the dissection and are not viable. Live myofibers are shiny and translucent under the light of the dissecting scope.
2. Transfer any completely detached individual myofibers into the well containing 3 mL of myofiber growth media and 2.5 ng/mL FGF2 (referred to as the "pick well") using a fire polished Pasteur pipet (*see* Subheading 2.2.2) and pipet pump. Minimize the amount of media, debris and dead myofibers moved to the pick well. This will minimize the amount of contaminating substances and non-myogenic cells as well as maximize myofiber survival (*see* **Note 21**).
3. Release any remaining, loosely attached myofibers in the bulk well by pipetting the whole muscle into a fire polished glass pipet and expelling it into the media with a pipet pump. This can be done a number of times with some force while still maintaining live myofibers. Be sure not to pinch or trap the bulk muscle of myofibers against the plastic of the six-well dish as this will kill or damage the myofibers.
4. After mechanical separation of the myofibers, transfer all detached individual myofibers to the pick well while minimizing debris transfer.
5. Transfer the individual myofibers from the first pick well to the second pick well. We have found two transfers are sufficient to isolate between 100–300 myofibers per EDL (*see* **Note 22**).

3.1.5 Culturing Isolated EDL Myofibers

Myofibers and their associated satellite cells can be cultured for extended periods of time (greater than 5 days). During this time a number of assays permit monitoring of myofiber-associated satellite cell behavior in culture; a number of these techniques will be described in the following sections (*see* Subheadings 3.2–3.6). Alternatively, myofibers and their associated satellite cells can be fixed immediately post-isolation to examine the in vivo activity of satellite cells and myofibers.

1. Myofibers are grown in myofiber growth media and 2.5 ng/mL FGF2 at 37 °C in a low (5 %) O_2 tissue culture incubator. Rock six-well plate back-and-forth gently every 24 h to redistribute recombinant FGF2. Add additional FGF2 stock solution (*see* Subheading 2.2.1) to maintain the 2.5 ng/mL concentration every 48 h.

2. Every 24 h examine the culture dish and remove hypercontracted/dead myofibers from the culture along with any remaining debris. If done carefully, this procedure will not considerably decrease the volume of myofiber growth media. If planning a short culture (0–96 h) the myofiber growth media will not need to be changed.
3. If planning a longer culture experiments (greater than 5 days) refer to other published protocols that describe methods to change media and maintain isolated myofibers [15]. For the following assays, longer myofiber cultures are unnecessary.

3.2 Transplantation of Myofiber-Associated Satellite Cells

Transplantation of myofiber-associated satellite cells into skeletal muscle results in lifelong hypertrophy, increased satellite cell numbers and resistance to age-related muscle degeneration [16]. The technique has been used to assess age-related muscle degeneration and the effects of the aged environment on young muscle stem cells [11, 16]. The procedure is technically challenging.

3.2.1 Initial Steps to Prepare for Transplantation

1. Pre-warm a six-well plate with 2.5 mL myofiber growth media and 2.5 ng/mL FGF2 in a 37 °C tissue culture incubator. One well of myofiber growth media and 2.5 ng/mL FGF2 will be needed per donor mouse.
2. Gather 27½ G tuberculin syringes (one for each injection) and recipient mice.
3. Add either sterile filtered 1.2 % $BaCl_2$ in 0.9 % saline or sterile filtered 0.9 % (*see* Subheading 2.4) to a 10 cm plate (8–12 mL) and let equilibrate to room temperature.

3.2.2 Transplantation into Recipient Mice

1. For myofiber-associated satellite cell transplantations, pick the cleanest, longest, straightest, translucent myofibers from a fresh (0 h) myofiber isolation (*see* Subheading 3.1) and transfer them into a clean six-well plate containing 2.5 mL of pre-warmed myofiber growth media and 2.5 ng/mL FGF2. Make sure there is no debris.
2. Let the myofibers culture in myofiber growth media and 2.5 ng/mL FGF2 for 3–5 h in a 37 °C, low O_2 tissue culture incubator.
3. Seal the FGF2 treated myofiber six-well plate with Parafilm before moving to the location of transplantation (mouse facility).
4. Anesthetize recipient mice according to IACUC and institutional guidelines.
5. Apply 70 % ethanol to the mouse hind limb and push aside hair so you can clearly see bare skin over the tibialis anterior (TA) muscle. The TA is the most superficial muscle in the ventral hind limb between the knee and ankle joint [14].

6. Once the mouse is anesthetized use a 27½ G tuberculin syringe to draw up ten myofibers while minimizing the media in the syringe if possible. Eject the myofibers into the 10 cm plate with either 0.9 % saline or 1.2 % $BaCl_2$ in 0.9 % saline. Quickly, void the syringe of any excess media and withdraw the myofibers that were just moved into the 10 cm plate. Make sure to count the number of myofibers you have withdrawn into the syringe. Avoid withdrawing excessive media/saline/$BaCl_2$ into the syringe (*see* **Note 23**).
7. Once withdrawn, let the myofibers settle to the bottom of the syringe and clear air bubbles to a final volume of 50 μL. Keep needle tip up until ready to inject mouse. When ready, move syringe so myofibers are floating in middle of solution closer to syringe needle tip. Place needle bevel point up into the belly of the TA muscle by going parallel to the tibia along the muscle while being careful to not go through the muscle (too deep) or just under the skin (too superficial). Advance the needle until there is minor resistance from the knee joint, at which point retreat the needle slightly (less than 0.5 cm).
8. Once the needle is placed correctly into the TA, quickly eject the solution containing the isolated myofiber-associated satellite cells. Be sure not to retreat the needle all the way out before all the contents are ejected.
9. After the injection into the mouse is complete, withdraw 100–200 μL 1.2 % $BaCl_2$ or 0.9 % saline into the syringe from the 10 cm plate used earlier then quickly eject back into a clean plate to see if myofibers are still in needle. At this point the myofibers will be hyper-contracted but will come out if you eject quickly/hard several times. Count the number of myofibers remaining following injection. This will tell you how many myofibers you transplanted into the mouse.
10. Put the recipient mouse back into its cage and note recovery. Mice will limp on their injured/transplanted hind limb for a few hours post injection.

3.3 Transfection of Myofiber-Associated Satellite Cells

Transfection of myofiber-associated satellite cells results in transient expression of target constructs. Care should be taken as transfection can result in expression by both the myofiber and the associated satellite cells. Careful experimental design is necessary to assess effects on both satellite cells and myofibers. It is important to note that antibiotic free media should be used in all steps with the exception to **step 8**.

1. Isolate myofibers from EDL muscle as described in Subheading 3.1. This protocol works directly following myofiber isolation (0 h myofibers).

2. Immediately following digestion isolate at least 50 myofibers (with minimal debris) per condition into wells of a 24 well dish. Include a transfection only control (*see* **Note 24**).
3. Rinse myofibers with F12-C + 15 % horse serum + 2.5 ng/mL FGF2. Place them in a final volume of 500 μL F12-C + 15 % horse serum + 2.5 ng/mL FGF2.
4. Mix 1 μg DNA and OptiMEM media to 62.5 μL final volume.
5. Separately mix 2.5 μL Lipofectamine 2000 with 62.5 μL OptiMEM media and let sit at room temperature for 5 min (no more than 25 min).
6. Combine OptiMEM/DNA mixture with OptiMEM/Lipofectamine 2000 mixture to obtain 125 μL final volume. Incubate mixture at room temperature for 25 min, no more than 6 h.
7. Add Lipofectamine 2000/DNA mixture to each well containing myofibers and 500 μL F12-C + 15 % horse serum. Incubate myofibers with Lipofectamine 2000/DNA mixture in a 37 °C in low oxygen incubator overnight (*see* **Note 25**).
8. Following transfection, rinse myofibers and change media to myofiber growth media + 2.5 ng/mL FGF2. Culture myofibers until needed, supplementing with 2.5 ng/mL FGF2 every 48 h.
9. Fix myofibers and store at 4 °C as described in Subheading 3.7.

3.4 CFDA-SE Self-Renewal Assay

CFDA-SE is a cell-permeant dye, which transforms within the cell, dramatically slowing its outflow rate. CFDA-SE is diluted upon cell division, and thus, can serve as a reliable marker for self-renewal or terminal differentiation [12]. CFDA-SE, which is fluorescein based, is available in several colors with distinct properties from Invitrogen/Thermo Fisher.

1. Isolate myofibers from EDL muscle as described in Subheading 3.1.
2. Culture myofibers with myofiber growth media and 2.5 ng/mL FGF2 for 48 h prior to addition of CFDA-SE.
3. Add previously determined working concentration (*see* Subheading 2.6 and **Note 17**) of pre-warmed CFDA-SE mixture to myofibers and pulse for 15 min at 37 °C.
4. Wash CFDA from myofibers with pre-warmed PBS and then again with pre-warmed myofiber growth media. Keep cells and fresh CFDA-SE away from light, as CFDA-SE is light sensitive (*see* **Note 26**).
5. Continue culturing myofibers in myofiber growth media and 2.5 ng/mL FGF2 until needed then fix and store at 4 °C as described in Subheading 3.7.

3.5 EdU Proliferation and Fusion Assay for EDL Myofibers

To assay for actively dividing cells, [^{3}H]Thymidine and BrdU have been historically used but these techniques have drawbacks including generation of radioactive waste and difficulties in anti-BrdU antibody accessibility to incorporated BrdU, respectively. EdU labeling and detection eliminates both these drawbacks and is much more reliable than BrdU staining. Both in vitro/ex vivo satellite cell proliferation percentages as well as in vivo satellite cell proliferation and myonuclear contribution can be determined with these techniques [3].

3.5.1 Ex Vivo S-Phase Labeling of Myofiber-Associated Satellite Cells

1. Isolate myofibers from EDL muscle as described in Subheading 3.1.
2. Culture myofibers in myofiber growth media and 2.5 ng/mL FGF2 until desired time point for collection.
3. Prior to collection, add 10 μM EdU from 10 mM EdU stock solution (*see* Subheading 2.7.1) to the cultured myofibers and incubate in a low O_2 tissue culture incubator at 37 °C. EdU incubation time should be empirically determined (*see* **Note 27**).
4. After completion of the EdU incubation, wash myofibers with pre-warmed PBS and fix the myofibers as described in Subheading 3.7.1.
5. Transfer the fixed myofibers to a myofiber staining basket (*see* Subheadings 2.3.1, 3.7.2 and Fig. 2).
6. Block and permeabilize the myofibers with 300 μL 3% BSA, 0.5% Triton X in PBS added to each well containing a myofiber staining basket for 30 min.

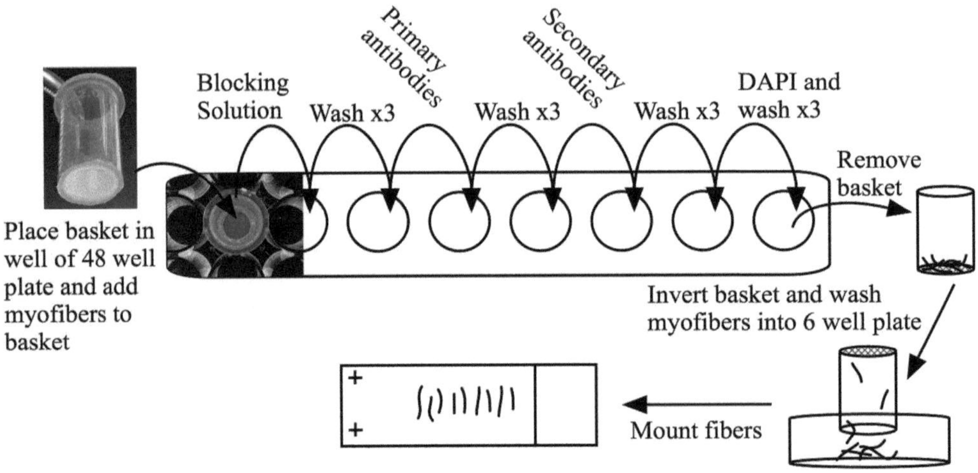

Fig. 2 Schematic of immunostaining procedure with myofiber staining baskets. Place a prepared myofiber staining basket (*see* Subheading 2.3.1) into a well of a 48 well plate and transfer myofibers into the basket. Move the basket into a new well and add blocking solution. Continue moving the basket into a new well for each step of the staining procedure. When complete, remove basket, invert basket and wash myofibers into six well dish with PBS then mount myofibers onto a charged slide (*see* Subheading 3.7.3)

7. Add 300 μL EdU Click-iT® cocktail (as described by the Click-iT® EdU Alexa Fluor® Imaging Kit listed in Subheading 3.7.1) to each well containing a myofiber staining basket and incubate for 30 min in the dark.
8. Continue immunolabeling for additional antigens (*see* Subheading 3.7.2) or mount myofibers (*see* Subheading 3.7.3).

3.5.2 In Vivo EdU Labeling of Satellite Cells and Myonuclei

1. Prepare appropriate volume of 0.5 mg/mL EdU drinking water as described in Subheading 2.7.2.
2. Place the sterile EdU drinking water in a mouse water bottle and keep the bottle full until the desired time point.
3. Sacrifice the mice and isolate myofibers as described in Subheading 3.1.
4. Fix the myofibers in 4% paraformaldehyde immediately post-isolation as described in Subheading 3.7.1.
5. Continue with the EdU Click-iT® chemistry as described in Subheading 3.5.1, **steps 5–8**.

3.6 AraC Treatment

AraC incorporates into DNA, inhibiting DNA replication and killing dividing cells. Myofibers, which are terminally differentiated are immune to AraC, but dividing satellite cells are affected [12]. While CFDA-SE (*see* Subheading 3.4) will label self-renewed cells brightest, AraC specifically selects for cells that have self-renewed or undergone terminal differentiation as neither of these cells are mitotically active. Immunostaining the satellite cells for Myogenin will distinguish between terminally differentiated cells and quiescent satellite cells.

1. Isolate myofibers from EDL muscle as described in Subheading 3.1.
2. Culture myofibers in myofiber growth media and 2.5 ng/mL FGF2 for 48–72 h.
3. After 48–72 h in culture, add previously determined concentration of 10 mM AraC stock solution (*see* Subheading 2.8.1 and **Note 20**) to media.
4. Change media on myofibers daily, gently washing with growth media to remove residual AraC before re-treating myofibers with AraC (*see* **Note 28**).
5. Culture myofibers in growth media containing AraC changing media and re-treating with AraC daily until fixation (*see* Subheading 3.7).

3.7 Fixation and Immunofluorescence of EDL Myofibers

EDL myofiber-associated satellite cells can be fixed and stained at different time points post-isolation depending on the experimental parameters to visualize proteins on or within both myofibers and satellite cells. Myofiber staining baskets, which we developed, provides simpler method for myofiber staining that minimizes myofiber loss and myofiber damage compared to currently used techniques.

3.7.1 Paraformaldehyde Fixation of Isolated EDL Myofibers

1. To fix isolated myofibers: make 2.5 mL of 4% paraformaldehyde in PBS for every well of myofibers to be fixed (*see* Subheading 2.3).
2. Transfer only single, straight and translucent myofibers to an empty well of a six well dish using a horse serum coated, fire polished, glass Pasteur pipet (*see* Subheading 2.3 and **Note 29**).
3. Remove as much growth media as possible (without losing myofibers) and add 2.5 mL of PBS to wash the myofibers.
4. Transfer the myofibers to an empty well of a six well dish using a horse serum coated, fire polished, glass pasture pipet.
5. Remove as much PBS as possible (without losing myofibers) and add 2.5 mL of 4% paraformaldehyde in a drop-wise manner to avoid myofiber clumping.
6. Incubate the myofibers in 4% paraformaldehyde for 10 min at room temperature. The myofibers should transform from shiny and translucent to white and opaque.
7. Transfer the fixed myofibers to an empty well of a six well dish and remove as much 4% paraformaldehyde as possible without losing myofibers. Then add 4.0 mL of PBS to the well to wash the myofibers.
8. Transfer the myofibers to an empty well and remove as much volume as possible without losing any myofibers and again add 4.0 mL of PBS. At this point the myofibers can be stored in PBS at 4 °C for up to 3 weeks or until you are ready to immunostain.

3.7.2 Immunostaining of Isolated EDL Myofibers

1. Transfer 4% paraformaldehyde fixed myofibers, using a horse serum coated, fire polished, glass pasture pipet, into a myofiber staining basket (*see* Subheading 2.3.1) sitting in a single well of 48 well dish (Fig. 2 and *see* **Note 30**).
2. Once 20–40 myofibers have been transferred, move the myofiber staining basket to an empty well and wash with 300 μL PBS as shown in Fig. 2 (*see* **Note 31**).
3. Add 300 μL 3% H_2O_2 (*see* Subheading 2.3) for 3–5 min and then move the myofiber staining basket to another well of the 48-well dish and rinse with PBS (*see* **Note 32**).
4. Move the myofiber staining basket to another well and add 300 μL 0.25% Triton X-100, 3% BSA in PBS and incubate for 60 min at room temperature.
5. Transfer the myofiber staining basket to an empty well and wash by adding 300 μL of PBS.

6. Transfer the myofiber staining basket to an empty well and add 300 μL primary antibody(s) diluted in 3% BSA; 0.125% Triton X in PBS. Incubate at 4 °C overnight or 1.5 h at room temperature.
7. After the primary antibody incubation, wash the myofibers three times in PBS by transferring the myofiber staining basket between wells (Fig. 2).
8. Incubate with the appropriate secondary antibodies diluted in 2% BSA; 0.125% Triton X in PBS for 1 h in the dark.
9. After the secondary antibody incubation, wash the myofibers three times in PBS by transferring the myofiber staining basket between wells.
10. Incubate the myofibers in 1 μg/μL DAPI for 10 min at room temperature.
11. Wash once with PBS by transferring the myofiber staining basket to an empty well.

3.7.3 Dry Mounting Myofiber-Associated Satellite Cells

1. To mount the myofibers, transfer the myofibers from the basket to a six-well dish or watch glass by turning the basket upside-down and then flushing the bottom of the basket with PBS to remove myofibers to the plate (*see* **Note 33**).
2. Under a dissection scope, transfer the myofibers individually to a charged glass slide by picking them up from one end with #5 forceps and placing them on the glass slide. Do not stretch the myofibers, as they will break. The charge from the slide will help the myofibers attach (Fig. 2).
3. Once all myofibers are transferred, mount with Mowiol (*see* Subheading 2.3.2 and **Note 34**), let stand at room temperature for at least 2–4 h to allow Mowiol to harden. Slides should be stored at room temperature in the dark until ready for imaging.

4 Notes

1. To make larger volumes of F12-C media adjust volumes of water, $NaHCO_3$, 100× penicillin–streptomycin, and 0.2 M $CaCl_2$ accordingly.
2. Multiple lots of horse serum should be tested to determine the lot that best supports myofiber and associated satellite cell growth. Typically, we test 8–15 lots of horse serum. Briefly, isolated satellite cells are cultured for 5 days then passaged and cultured for an additional 5 days. Each lot of horse serum is tested in quadruplicate 100 mm tissue culture plates. Cell numbers are counted each day to determine the doubling efficiency.

At the conclusion of the second round of 5 day culture, three plates from each lot are fixed and immunostained for skeletal muscle myosin heavy chain protein (MF20 antibody Hybridoma Studies Bank). For each lot, one plate is transferred to 1% serum supplemented with 1:100 ITS (Insulin, Transferrin, Selenium; Corning 25-800-CR) and cultured for 48 h to induce differentiation. The differentiated plates are stained for myosin heavy chain and the extent of differentiation determined for all lots. The lot that exhibits the best growth with the least differentiation in growth medium and the best differentiation when switched to low serum is purchased and stored at −80 °C. Prior to use each 0.5 L horse serum bottle is aliquoted into 75 mL volumes and stored at −20 °C. Aliquots are thawed at 37 °C when needed. Typically, one or two lots of horse serum of ten lots (10–20%) permit robust satellite cell growth and differentiation. Although laborious, this is a critical and key requirement for producing growth media.

3. FGF2 is very basic and will adhere to any glass surface. Always store FGF2 in plastic containers and use plastic pipette tips to add FGF2 to solutions and cell culture plates. If needed, aliquot media into a plastic conical tubes before adding FGF2.
4. Care should be taken to avoid overheating and sealing the pipette end. Once cool, rinse the glass-fired pipettes with media containing horse serum prior to their first use to prevent myofibers from sticking to the inside of the pipette during transfer.
5. Multiple lots of collagenase type I should be tested to determine the best lot for myofiber digestion and long term myofiber survival. We find there is significant lot to lot variability where some lots will dramatically reduce satellite cell viability.
6. Either syringe filters or bottle top filters will work depending upon volume filtered. Pre-filtering through the 0.44 μm filter allows lower applied pressure filtering through the 0.22 μm filter, especially if you are using syringe filters.
7. Do not freeze/thaw the collagenase 10× stock solution. Multiple freeze/thaws will decrease the enzyme efficiency.
8. PBS with Triton X-100 can be made in advance and stored at room temperature. Blocking solutions can be made in advance and frozen in 5–10 mL aliquots and stored at −20 °C. Thaw aliquots completely at room temperature prior to use.
9. Ideally, tubes are trimmed using a hand-held rotary tool with a cut-off wheel. Alternatively, a heated scalpel can be used to cut the tube. The ideal cut point on the tube is between the straight and tapered point of the tube.
10. Various mesh sizes were tested and 80 μm was found to be optimal. <80 μm mesh did not allow sufficient liquid flow-thru while >95 μm allowed myofibers to fall through.

11. The hot plate should be hot enough to melt basket tops but cool enough not to melt the nylon mesh. Foil covering serves two purposes—(1) allows easy cleanup of hot plate and (2) allows stuck tubes to be removed quickly without the risk of injury.
12. The tubes are examined for partial attachment or melted plastic within the mesh. Tubes with melted plastic in the mesh are discarded. The mesh, which is partially attached, can be remelted. Ideally tubes should sit slightly above the bottom of a 48 well plate to allow fluid flow.
13. The protocol can be scaled to make larger quantities. Aliquots can be freeze/thawed multiple times without consequence. Once added to slides, store in the dark to allow solution to harden. The solution will completely harden overnight and does not need any other sealing agents.
14. For 300 mL of 0.9 % saline solution, add 2.7 g NaCl to 250 mL dH_2O and stir until dissolved. Bring to final volume of 300 mL with dH_2O then sterile filter through 0.22 μm Steritop bottle filter into a sterile container and store at room temperature.
15. For 300 mL of 1.2 % $BaCl_2$ in 0.9 % saline solution, add 2.7 g NaCl and 3.6 g dehydrate $BaCl_2$ to 250 mL dH_2O, stir until dissolved. Bring to a final volume of 300 mL with dH_2O then sterile filter through 0.22 μm Steritop bottle filter into a sterile container and aliquot into 10 mL volumes for use with transplantation. Store at room temperature.
16. The CFDA-SE 10 mM stock solution should be made fresh for every experiment and stored in the dark, as CFDA-SE is light sensitive.
17. The optimal CFDA-SE concentration needs to be determined empirically. Satellite cells and myofibers, which autofluoresce green, uptake the dye at different rates. We recommend performing a titration curve of CFDA-SE concentrations ranging from 1 to 10 μM.
18. To dissolve EdU into sterile H_2O with 1 % dextrose place solution in 37 °C water bath for 1–2 h, occasionally mixing, until EdU goes into solution. Dextrose is added to the water to encourage the mice to drink the water.
19. Keep all AraC solutions away from light. A 100 mM stock solution is good for 1 year at −20 °C.
20. At high concentrations, AraC affects cellular processes other than DNA replication. An optimal working AraC concentration that kills as many cycling satellite cells as possible without killing the myofiber should be empirically determined. We recommend performing a titration curve of AraC concentrations ranging from 0.1 to 100 μM.

21. Other published protocols [15] suggest a number of rinses to remove excess collagenase from the bulk muscle. We have found that keeping the bulk muscle in diluted collagenase allows for a more complete digestion and less mechanical manipulation, ultimately leading to an increased single, live myofiber yield. Additionally, we have also found picking myofibers between multiple wells dilutes the collagenase sufficiently to avoid damage to the myofibers.
22. During the second transfer, be sure to only pick single, translucent, straight myofibers that have no associated debris or other live myofibers attached to them. Myofibers with bends, curls, kinks or hyper-contractions will often die within the first hours of culture.
23. To withdraw myofibers with minimal liquid, place the bevel of the syringe directly above a myofiber like a vacuum cleaner over a carpet. Withdraw the syringe minimally but swiftly.
24. Fifty myofibers is the minimum recommended for transfection. >100 myofibers should be done in a 12 well plate and amounts listed doubled. >200 myofibers should be done in a six well plate with all volumes quadrupled.
25. Myofibers and associated satellite cells should be monitored for death post transfection. Transfection time can be shortened to 4–6 h if necessary.
26. Wait at least 30 min before examining or fixing myofibers to determine if CFDA-SE pulse was effective.
27. The longer the myofibers spend cultured in the presence of EdU the higher the percentage of EdU labeled myofiber-associated satellite cells. The precise time-point used is under the discretion of the experimenter. When cultured in the presence of 2.5 ng/mL FGF2 for 48–72 h a 2-h EdU incubation will label approximately 70% of the myofiber-associated satellite cells.
28. AraC is only stable in culture for 24 h. For experiments lasting longer than 24 h where cycling cells need to be killed, myofibers will need to be washed with growth media and AraC replaced every 24 h.
29. It is critical to only transfer live myofibers during this step. Once myofibers are fixed it is difficult to distinguish between live and dead myofibers. This step is also a good opportunity to remove remaining debris or clumped myofibers.
30. If the volume inside the myofiber staining basket reaches capacity before all the myofibers are transferred move the myofiber staining basket to an empty well and finish transferring myofibers.
31. Baskets should be moved into empty wells and solution applied post move. Adding baskets to wells with solution present can

push myofibers to the top of the basket and can affect further staining steps. Depending upon basket length, volumes can be decreased if myofibers are covered with solution.

32. Do not leave quenching solution on myofibers longer than 3–5 min or it will hinder future staining. This step will reduce the autofluorescence of the myofibers during immunofluorescent microscopy.
33. Myofiber staining baskets can be reused for staining by autoclaving baskets in dH_2O and allowing them to air-dry. Alternatively, baskets can be rinsed thoroughly with dH_2O and allowed them to air-dry before their next use. However, be sure to check under a dissection microscope that no myofibers remain in the baskets before reuse.
34. For a 22×60 mm coverslip 40–60 μL of mowiol is sufficient. Do not add >100 μL mowiol as this will cause the coverslip to move and the myofibers will detach from the slide and become tangled. If too much mowiol is added, you can use a vacuum along the edges of the coverslip to remove the extra.

References

1. Yin H, Price F, Rudnicki MA (2013) Satellite cells and the muscle stem cell niche. Physiol Rev 93:23–67
2. Keefe AC, Lawson JA, Flygare SD, Fox ZD, Colasanto MP, Mathew SJ, Yandell M, Kardon G (2015) Muscle stem cells contribute to myofibres in sedentary adult mice. Nat Commun 6:7087. doi:10.1038/ncomms8087
3. Pawlikowski B, Pulliam C, Betta ND, Kardon G, Olwin BB (2015) Pervasive satellite cell contribution to uninjured adult muscle fibers. Skelet Muscle 5:1–13. doi:10.1186/s13395-015-0067-1
4. Mauro A (1961) Satellite cell of skeletal muscle fibers. J Biophys Biochem Cytol 9:493–495
5. Cheung TH, Rando TA (2013) Molecular regulation of stem cell quiescence. Nat Rev Mol Cell Biol 14:329–340. doi:10.1038/nrm3591
6. Doles JD, Olwin BB (2015) Muscle stem cells on the edge. Curr Opin Genet Dev 34:24–28. doi:10.1016/j.gde.2015.06.006
7. Gilbert PM, Havenstrite KL, Magnusson KE, Sacco A, Leonardi NA, Kraft P, Nguyen NK, Thrun S, Lutolf MP, Blau HM (2010) Substrate elasticity regulates skeletal muscle stem cell self-renewal in culture. Sci Signal 329:1078
8. Chakkalakal JV, Jones KM, Basson MA, Brack AS (2012) The aged niche disrupts muscle stem cell quiescence. Nature 490:355–360. doi:10.1038/nature11438
9. Bischoff R (1986) Proliferation of muscle satellite cells on intact myofibers in culture. Dev Biol 115:129–139
10. Rosenblatt JD, Lunt AI, Parry DJ, Partridge TA (1995) Culturing satellite cells from living single muscle fiber explants. In Vitro Cell Dev Biol Anim 31:773–779
11. Bernet JD, Doles JD, Hall JK, Kelly Tanaka K, Carter TA, Olwin BB (2014) p38 MAPK signaling underlies a cell-autonomous loss of stem cell self-renewal in skeletal muscle of aged mice. Nat Med 20:265–271. doi:10.1038/nm.3465
12. Troy A, Cadwallader AB, Fedorov Y, Tyner K, Tanaka KK, Olwin BB (2012) Coordination of satellite cell activation and self-renewal by Par-complex-dependent asymmetric activation of p38alpha/beta MAPK. Cell Stem Cell 11:541–553. doi:10.1016/j.stem.2012.05.025
13. Shefer G, Yablonka-Reuveni Z (2005) Isolation and culture of skeletal muscle myofibers as a means to analyze satellite cells. Methods Mol Biol 290:281–304
14. Delaurier A, Burton N, Bennett M, Baldock R, Davidson D, Mohun TJ, Logan MP (2008) The Mouse Limb Anatomy Atlas: an interactive 3D tool for studying embryonic limb patterning. BMC Dev Biol 8:83. doi:10.1186/1471-213X-8-83

15. Keire P, Shearer A, Shefer G, Yablonka-Reuveni Z (2013) Isolation and culture of skeletal muscle myofibers as a means to analyze satellite cells. Methods Mol Biol 946:431–468. doi:10.1007/978-1-62703-128-8_28

16. Hall JK, Banks GB, Chamberlain JS, Olwin BB (2010) Prevention of muscle aging by myofiber-associated satellite cell transplantation. Sci Transl Med 2:57ra83. doi:10.1126/scitranslmed.3001081

Chapter 12

Flow Cytometry and Transplantation-Based Quantitative Assays for Satellite Cell Self-Renewal and Differentiation

Robert W. Arpke and Michael Kyba

Abstract

In response to muscle damage, satellite cells proliferate and undertake both differentiation and self-renewal, generating new functional muscle tissue and repopulating this new muscle with stem cells for future injury responses. For many questions relating to the physiological regulation of satellite cells, quantitative readouts of self-renewal and differentiation can be very useful. There is a particular need for a quantitative assay for satellite cell self-renewal that does not rely solely upon sectioning, staining and counting cells in sections. In this chapter, we provide detailed methods for quantifying the self-renewal and differentiation potential of a given population of satellite cells using an assay involving transplantation into injured, regenerating muscle together with specific markers for donor cell identity and state of differentiation. In particular, using the Pax7-ZsGreen transgene as a marker of satellite cell state, self-renewal can be quantified by FACS on transplanted muscle to actually count the total number of resident satellite cells at time points following transplantation.

Key words Satellite cells, Pax7, Transplantation, Myogenesis

1 Introduction

The ability to quantify self-renewal and differentiation allows one to make statistical arguments about the effect of external or intrinsic variables on satellite cell physiology. Although satellite cells are often studied by extraction from muscle followed by ex vivo culture, when satellite cells are removed from muscle, they activate, irreversibly exit the stem cell state, and commit to myogenesis. Even a short period of ex vivo culture significantly ablates the ability of these cells to engraft the satellite cell compartment [1]. Therefore, such cultures should not be referred to as satellite cells and properties attributed to such cells should not be ascribed to satellite cells. Since true self-renewal only occurs in vivo, only an in vivo assay can quantify the native self-renewal potential of a given satellite cell population.

In order to track self-renewal, it is essential to have markers that allow the cell autonomous readout of differentiation state.

Michael Kyba (ed.), *Skeletal Muscle Regeneration in the Mouse: Methods and Protocols*, Methods in Molecular Biology, vol. 1460, DOI 10.1007/978-1-4939-3810-0_12, © Springer Science+Business Media New York 2016

The assay that we describe makes use of two such markers, one for the satellite cell and the other for the differentiated myofiber. Conveniently, within skeletal muscle, satellite cells uniquely express the transcription factor Pax7 [2] and this marker is often used to quantify satellite cells in sections (*see* Quiroga and Zammit, Chapter 8). We have exploited this unique expression pattern to develop a transgenic mouse in which quiescent satellite cells are labeled with green fluorescence. A BAC containing the *Pax7* locus was modified to replace the first coding exon of *Pax7* with *ZsGreen*, and introduced into the mouse genome by pronuclear injection. The resulting Pax7-ZsGreen mouse can be used for flow cytometric purification or flow cytometric quantification of satellite cells from any muscle [3]. A key feature of ZsGreen is that it serves as both a donor marker (provided the recipient does not carry the reporter) as well as a marker of satellite cell state—the green fluorescence is rapidly lost when cells satellite cells differentiate [3].

The second marker is the protein dystrophin. Dystrophin is strongly expressed in differentiated myofibers, detected just beneath the sarcolemma [4, 5] where it ensures integrity of the cell membrane during cycles of contraction and release [6], by linking the contractile apparatus to the extra cellular matrix [7]. The mdx mouse lacks dystrophin [8], thus when wild-type (WT) satellite cells are transplanted into the muscle of an mdx host, their contribution to skeletal muscle regeneration can be observed by immunostaining of sections with antibodies to dystrophin. Thus, as with Pax7 above, dystrophin serves as both a host marker and a marker for differentiated state.

Because experiments evaluating the function of specific genes in satellite cell physiology often require many genetic components, the mice whose satellite cell self-renewal and differentiation potential need to be quantified are typically outbred. Tissue from outbred mice transplanted into mdx mice, which are conventionally on a C56BL/10 background, would in most cases provoke a strong immune response and be rejected. To eliminate any immune system interference with the output of this assay, we have developed an immunodeficient version of the mdx mouse, *NSG-mdx4Cv*, which serves as an excellent and highly tolerant recipient of satellite cell grafts [9]. This strain carries the SCID mutation, which leads to a complete absence of mature B and T cells [10], together with the *IL2Rg* mutation, which ablates NK cells [11], important mediators of nonself MHC recognition. In addition, it utilizes an allele of *Dmd* referred to as *mdx4Cv* [12] which has a very low rate of spontaneous reversion [13]. Because the *dystrophin* gene has so many exons, there are many opportunities for alterations in splicing to produce a shortened protein with functional N- and C-termini but bypassing the mutation; these relatively infrequent reversion events also occur in Duchenne patients [14]. Spontaneous reversion leads to dystrophin positivity in revertant fibers, therefore its presence sets a level of background noise. With the 4Cv mutation, this background is close to zero [13].

In overview, the experiment involves purifying Pax7-ZsGreen-labeled cells from one or more donor mice, transplanting these cells into both TA muscles of an irradiated, pre-injured NSG-mdx^{4Cv} mouse, harvesting both TA muscles 1 month later, and processing one for FACS and the other for sectioning and dystrophin/laminin staining. The rate of self-renewal can be quantified as the number of ZsGreen cells in the muscle 1 month post-transplant per input ZsGreen cell. The rate of differentiation can be quantified as the number of dystrophin+ fibers one month post-transplant, per input ZsGreen cell. We typically find that an *n* of 6 is a good starting point for reliable output data in both arms.

2 Materials

2.1 Irradiation

1. X-ray irradiator.
2. NSG-mdx4Cv mice [9].
3. Ketamine [45 mg/mL]–xylazine [5 mg/mL].
4. Lead shields.

2.2 Cardiotoxin Injury

1. Ketamine [45 mg/mL]–xylazine [5 mg/mL].
2. 10 μM cardiotoxin from *Naja mossambica mossambica* solution: 5 mg lyophilized cardiotoxin (Sigma—C9759) dissolved in 73.25 mL PBS, aliquoted and frozen.
3. Hamilton syringe with 26-G needle.
4. 6-0 nylon suture.

2.3 Harvest of Satellite Cells for Transplant

1. Mice carrying the Pax7-ZsGreen reporter gene [3], *see* **Note 1**.
2. Digestion solution 1: 500 mL DMEM high glucose plus 4500 mg/L glucose without L-glutamine and sodium pyruvate supplemented with 1 % penicillin–streptomycin and 1 g collagenase type II.
3. Rinsing solution: Ham's/F-10 medium plus 1.00 mM L-glutamine supplemented with 10 % horse serum, 1 % 1 M HEPES buffer solution, and 1 % penicillin–streptomycin.
4. Digestion solution 2: 7 mL rinsing solution supplemented with 500 μL of digestion solution 1 and 1.25 mL of 0.4 % dispase in rinsing solution per sample.
5. Sorvall LEGEND RT centrifuge (Thermo Electron Corporation).
6. 100 × 15 mm petri dishes.
7. Razor blades.
8. Serological pipettes.

9. 50 mL conical tubes (sterile).
10. Pasteur pipettes.
11. Small pipette bulbs.
12. 10 mL syringes.
13. 18-G needles.
14. 16-G needles.
15. 40 μm cell strainers.
16. Collection medium: Ham's/F-10 medium plus 1.00 mM L-glutamine supplemented with 20% fetal bovine serum, 1% penicillin–streptomycin, 1% GlutaMAX, 10 ng/mL human basic fibroblast growth factor, and 114.4 μM 2-mercaptoethanol (2 μL in 250 mL).

2.4 Transplantation

1. Ketamine–xylazine mixture: ketamine [45 mg/mL]–xylazine [5 mg/mL].
2. Hamilton syringe with 26-G needle.
3. Phosphate-buffered saline without calcium and magnesium, 1×.
4. 6-0 nylon suture.

2.5 TA Harvest for FACS Analysis

1. Digestion solution 1: (same as Subheading 2.3 above).
2. Rinsing solution: (same as Subheading 2.3 above).
3. Digestion solution 2: 3.5 mL rinsing solution supplemented with 250 μL of digestion solution 1 and 625 μL of 0.4% dispase in rinsing solution per sample.
4. 60 × 15 mm petri dishes.
5. 15 mL conical tubes (sterile).
6. 3 mL syringes.

2.6 FACS Staining

1. FACS staining medium: phosphate-buffered saline without calcium and magnesium supplemented with 2% fetal bovine serum.
2. FACS staining medium with propidium iodide: as above, supplemented with 1 μg/mL propidium iodide.
3. Antibody mixture: (1 μL CD31-PE-Cy7, 1 μL CD45-PE-Cy7, 1 μL VCAM-Biotin, 1 μL Streptavidin-PE, 2 μL α7-integrin-647).

Specific antibodies for FACS Staining:

4. CD31-PECy7 Clone 390 (BD Biosciences—561410).
5. CD45-PE-Cy7 Clone 30-F11 (BD Biosciences—552848).
6. VCAM-BiotinClone429(MVCAM.A)(BDBiosciences—553331).
7. Streptavidin-PE (BD Biosciences—554061).

8. α7-Integrin Clone R2F2 (AbLab—67-0010-05).

2.7 Sectioning and Immunohistochemistry

1. O.C.T. compound.
2. Peel-A-Way embedding molds.
3. 2-methylbutane.
4. Liquid nitrogen.
5. Cryostat.
6. Microscope slides.
7. Acetone.
8. Liquid-repelling PAP Pen.
9. Bovine serum albumin.
10. Phosphate-buffered saline, 1×.
11. Rabbit polyclonal antibody to dystrophin (AbCam—ab15277).
12. Mouse monoclonal antibody to laminin Clone LAM-89 (Sigma).
13. Alexa Fluor 555 goat anti-rabbit IgG antibody (Invitrogen—A21429).
14. Alexa Fluor 488 goat anti-mouse IgG antibody (Invitrogen—A11029).
15. DAPI.
16. IMMU-MOUNT.
17. Glass coverslips.
18. Upright fluorescence microscope.

3 Methods

3.1 Irradiation Procedure

In order to limit the contribution of the endogenous satellite cells to muscle repair, 48 h prior to transplantation, both hind limbs of the *NSG-mdx4Cv* recipient mice are irradiated.

1. Sedate mice with ketamine–xylazine mixture.
2. Placed mice in a supine position on irradiator platform.
3. Protect the body of each mouse with lead shielding, leaving only the hind limbs exposed (Fig. 1).
4. Apply tape to immobilize the feet of each mouse. This maintains optimum exposure of the tibialis anterior (TA) during irradiation, as turning can result in shielding of the TA by the bone.
5. Tape is also used to prevent the exposure of the tails to the irradiation, by holding the tail in place under the lead shield (Fig. 1).
6. Expose mice to 1200 cGy irradiation (*see* **Notes 2** and **3**).

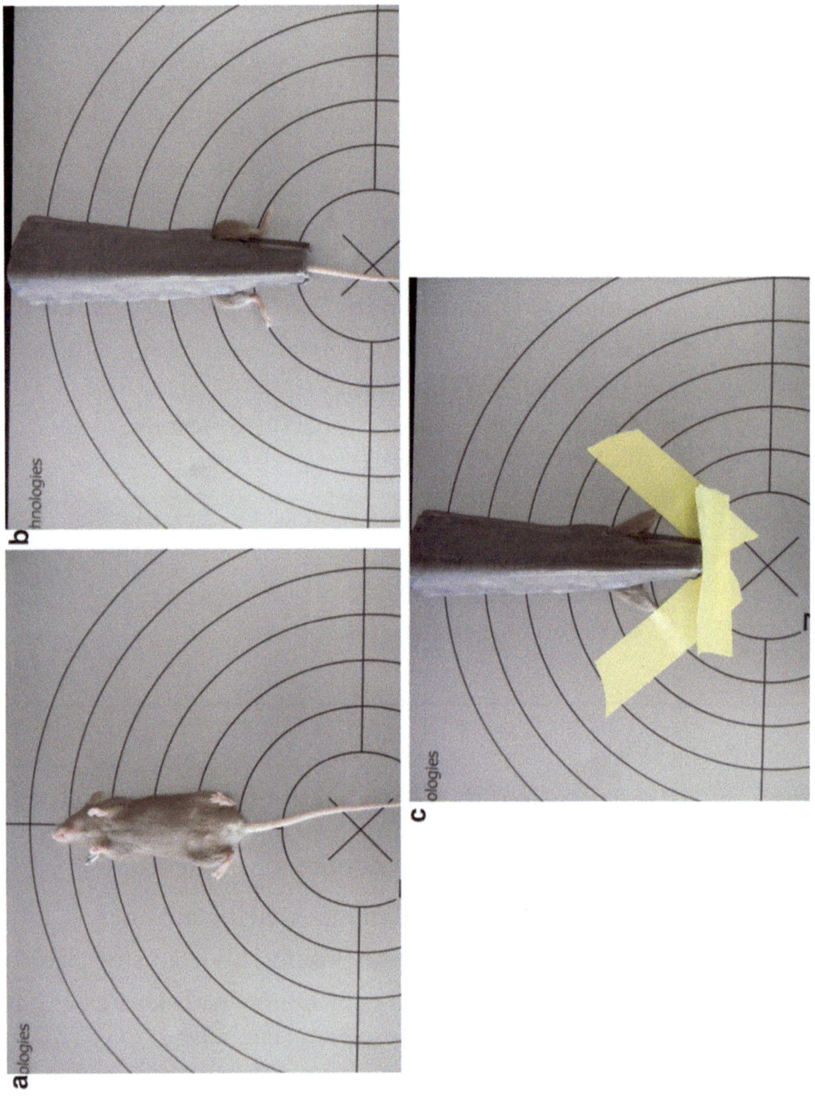

Fig. 1 Positioning of mice for irradiation. (**a**) Mice are placed on the irradiation platform in a supine position. (**b**) Except for their limbs, the mice are protected from the irradiation with lead shields. (**c**) Tape is applied to hold the hind limbs in place and to secure the tail under the lead shield

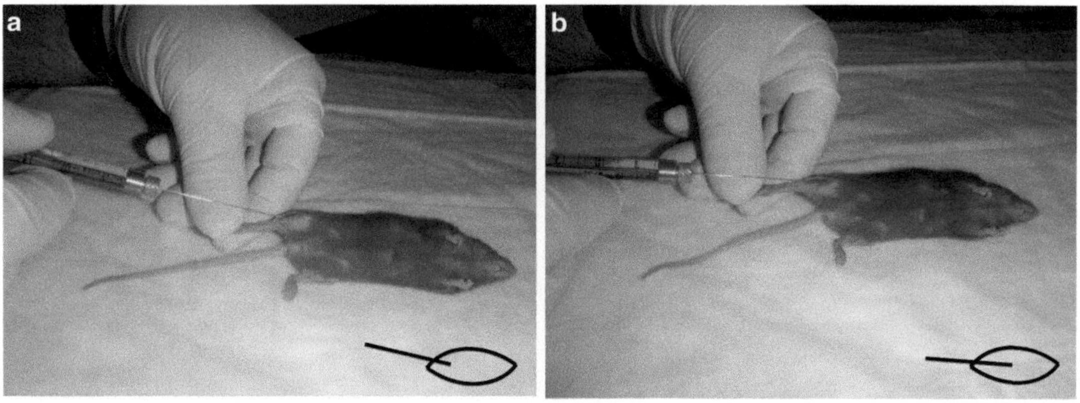

Fig. 2 Positioning of Hamilton syringe for the injection of cardiotoxin or cells. (**a**) To first pierce the muscle, the needle is brought in at an angle. Diagram at lower right indicates orientation of needle and muscle. (**b**) Prior to transplanting cells, the needle is oriented with the axis of the muscle, and inserted. Diagram at lower right indicates the desired orientation

3.2 Cardiotoxin-Injury Procedure

While the *NSG-mdx4Cv* mice lack full length dystrophin protein, they have a relatively mild muscle phenotype compared to humans with a similar mutation in the dystrophin gene. To further stimulate muscle regeneration, both TAs are injected with cardiotoxin. This induces hypercontraction, ultimately destroying muscle fibers throughout a large part of the TA muscle.

1. Sedate mice with ketamine–xylazine mixture.
2. Shave skin and swab with 70% ethanol and Betadine solution.
3. A 7–10 mm skin incision is made to expose the TA.
4. The TA is injected intramuscularly with a 26-G needle on a Hamilton syringe containing 15 μL of 10 μM cardiotoxin solution.
   (a) The bevel of the needle should be up when entering the TA, and the needle should be at an angle of approximately 30° relative to the axis of the muscle (Fig. 2a).
   (b) After piercing the muscle, change the angle of the needle to align with the axis, so the needle can be inserted through the core of the muscle. Rotate the needle slightly as it is pushed further into the TA prior to injecting the volume (Fig. 2b).
   (c) After volume is injected, hold the needle in place for approximately 30 s prior to withdrawing from the muscle. If the needle has become attached to the muscle, a gentle twist of the needle may be needed to withdraw the needle.
5. Close the skin incision with a 6-0 nylon suture.

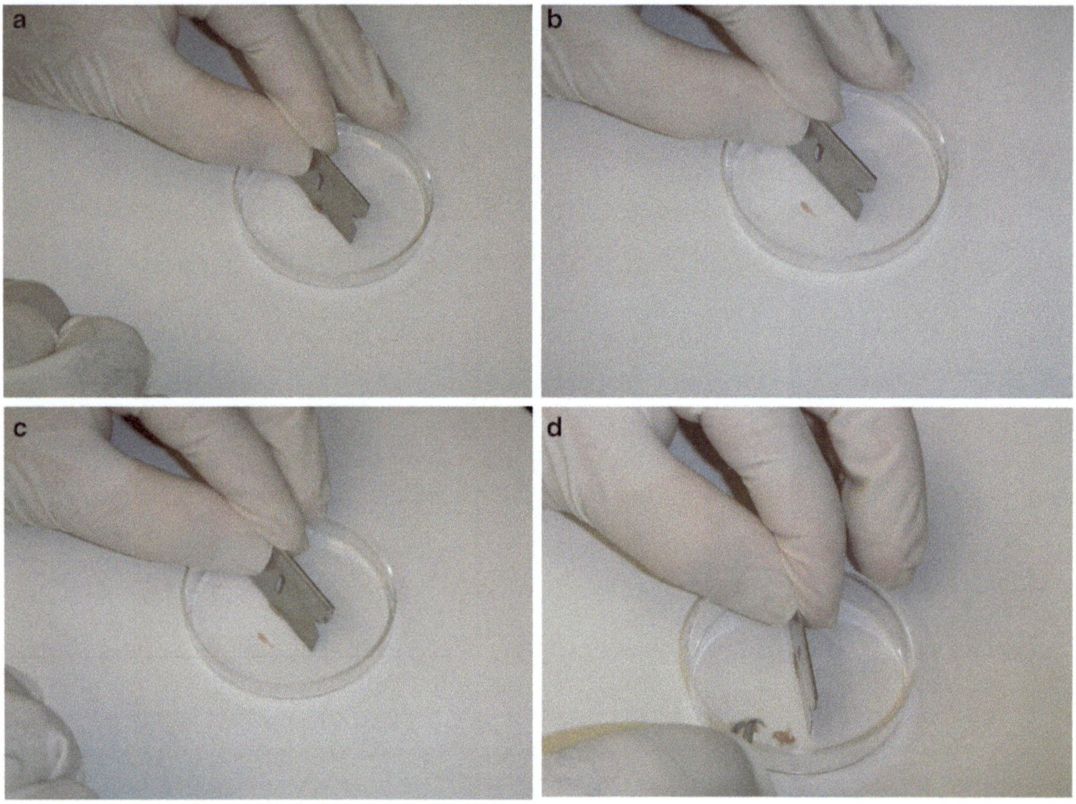

Fig. 3 Preparing muscle for digestion. (**a**) The razor blade is aligned parallel to the long axis of the muscle fibers. (**b**) Forceps are sliced along the razor blade to separate the muscle into approximately 2 mm wide pieces. (**c**) Steps **a** and **b** are repeated. (**d**) This process is continued for the entire muscle

3.3 Harvest of Satellite Cells for Transplant

1. Carefully dissect both hind limbs, triceps, and psoas muscles.
2. Chop the muscle parallel to the muscle fiber direction into approximately 2 mm thick pieces (use the razor blade as a straight edge and pull along the straight edge with a curved forceps) (Fig. 3).
3. Place the sample into a 50 mL tube containing 15 mL of digestion solution 1.
4. Keep the samples on ice until all samples have been harvested.
5. Incubate shaking for 75 min at 37 °C.
6. Invert the tubes five times, then let the sample gravity settle for approximately 5 min.
7. Aspirate the supernatant to the 10 mL mark of the 50 mL conical tube. Add 6 mL rinsing solution, invert the tubes five times; then let the samples settle for 5 min.
8. Repeat **step** 7.
9. Aspirate the supernatant, again leaving behind 10 mL.

10. Add 2 mL rinsing solution, and pour muscle solution into an inclined 100×15 cm petri dish.
11. Scrape the muscle solution against the bottom of the inclined petri dish and squeeze into and out of a sheared Pasteur pipette (with small bulb). This will mechanically disrupt the muscle fibers.
12. Add 2 mL rinsing solution to wash the petri dish, transferring the solution back to the 50 mL conical tube. Repeat this rinse to collect as much residue as possible.
13. Centrifuge the samples at 300×*g* for 5 min on a benchtop centrifuge.
14. Aspirate the supernatant, leaving behind 5 mL.
15. Add 10 mL rinsing solution, and resuspend the pellet.
16. Centrifuge the samples at 300×*g* for 5 min.
17. Aspirate the supernatant to the 5 mL mark of the tube.
18. Add 8.5 mL digestion solution 2, and resuspend the pellet.
19. Vortex for 30 s.
20. Incubate shaking for 30 min at 37 °C.
21. Vortex for 30 s.
22. Place a 40 μm cell strainer on a new 50 mL conical tube and wet with 2 mL rinsing solution.
23. Draw the sample into and out of a 10 mL syringe with a 16-G needle four times.
24. Draw into and out of a 10 mL syringe with an 18-G needle four times.
25. Apply the sample through the 40 μm cell strainer, collecting into the new 50 mL conical tube.
26. Collect any residue in the original 50 mL tube by adding 10 mL rinsing solution and pouring through the cell strainer into the new 50 mL tube.
27. Centrifuge the samples at 300×*g* for 5 min.
28. Place a 40 μm cell strainer onto a new 50 mL conical tube and wet with 2 mL rinsing solution.
29. Again, draw the sample into and out of a 10 mL syringe with an 18-G needle four times.
30. Apply the sample through the 40 μm cell strainer, collecting in the new 50 mL conical tube.
31. Collect any residue by adding 10 mL rinsing solution and adding to the the new 50 mL tube, through the strainer.
32. Centrifuge the samples at 300×*g* for 5 min.
33. Carefully aspirate until there is only approximately 100 μL volume left above the cell pellet.

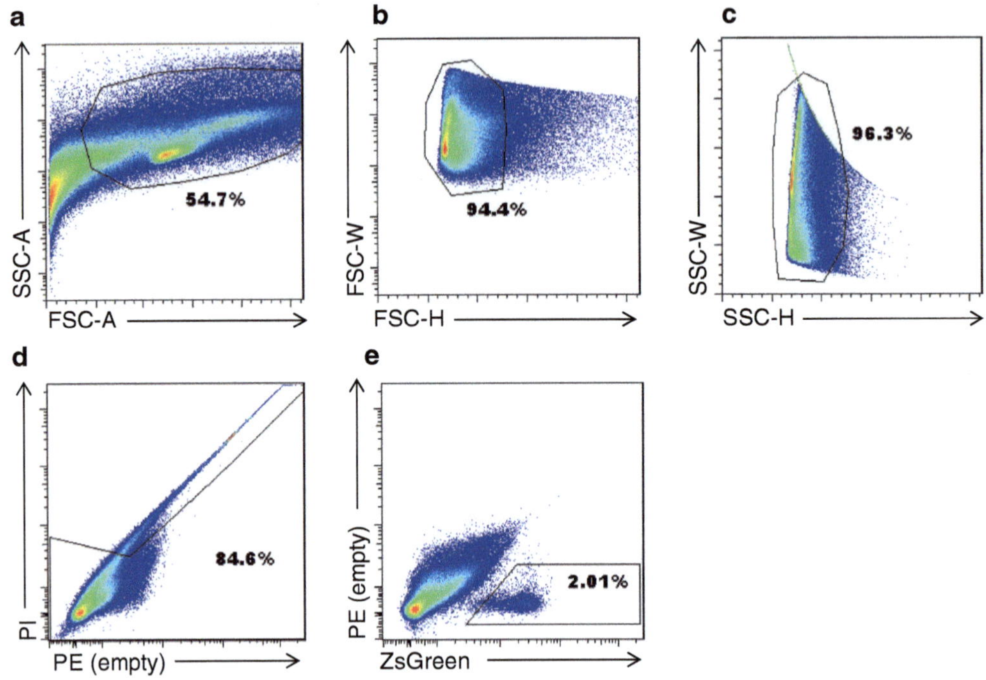

Fig. 4 Gating for FACS sorting of satellite cells bulk hind limb. (**a**) FSC-A × SSC-A. (**b**) FSC-H × FSC-W. (**c**) SSC-H × SSC-W. (**d**) PE × PI. (**e**) ZsGreen × PE

34. Add 2 mL of staining medium to the sample, resuspend the cells, and transfer to a new 15 mL conical tube for FACS sorting.
35. Sort the ZsGreen+ cells (Fig. 4), and collect into collection medium. Make sure to keep a count of cells as they are sorted.

3.4 Transplantation Procedure

Resuspend donor Pax7-ZsGreen+ cells from Pax7-ZsGreen mice in sterile saline at the desired cell number, typically a density allowing injections of 10 μL volume to contain 300 Pax7-ZsGreen cells (*see* **Note 4**). Cells are transplanted into both the right and left TAs of at least six *NSG-mdx4Cv* mice per condition.

1. Sedate mice with ketamine–xylazine mixture.
2. Reopen the incision made the previous day by cutting the nylon suture to expose the TA.
3. Inject 10 μL of the appropriate number of cells suspended in sterile PBS intramuscularly into each TA using a 26-G needle on a Hamilton syringe.
   (a) The bevel of the needle should be up when entering the TA, and the needle should be at an angle of approximately 30° relative to the axis of the muscle (Fig. 2a).
   (b) After piercing the muscle, change the angle of the needle to align with the axis, so the needle can be inserted through the core of the muscle. Rotate the needle slightly as it is

pushed further into the TA prior to injecting the volume (Fig. 2b).

(c) After volume is injected, hold the needle in place for approximately 30 s prior to withdrawing from the muscle. If the needle has become attached to the muscle, a gentle twist of the needle may be needed to withdraw the needle.

4. Close the skin incision with a 6-0 nylon suture.

3.5 Harvesting Mononuclear Cells from Individual TA Muscles

Transplanted TA muscles are harvested 4 weeks post-transplant. Utilizing the Pax7-ZsGreen mouse as the donor for the transplanted satellite cells allows for quantification of contribution to the satellite cell pool. Undifferentiated donor-derived satellite cells are identifiable by their green fluorescence. The procedure for harvesting cells from the TA muscle is similar to that for harvesting donor cells from total hind limb, but because the amount of tissue is so small, the volumes at each step are reduced, disruption through syringes is done only once and the number of wash steps is reduced in order to limit cell loss. In addition, it is important not to lose any satellite cells that might be released during the first digestion (which releases the fibers). Therefore, rather than letting fibers gravity settle, in the individual TA prep, all cells are centrifuged into the pellet after the first digestion.

1. Carefully dissect that TA muscles to be used for FACS analysis, saving the opposite TA muscles for histology (Subheading 3.7, below).
2. Chop the TA muscle parallel to the muscle fibers into approximately 2 mm thick pieces (use the razor blade as a straight edge and pull along the straight edge with a curved forceps).
3. Place the sample into a 15 mL tube containing 3 mL of digestion solution 1.
4. Place the samples on ice until all samples have been harvested.
5. Incubate shaking for 75 min at 37 °C.
6. Invert the tubes five or six times, then centrifuge for 5 min at 300 × *g*. Centrifugation at this step is different from the bulk prep, above.
7. Aspirate the supernatant to the 1 mL mark of the 15 mL conical tube. Add 3 mL rinsing solution, pipette to resuspend pellet with a P1000 pipette; then centrifuge for 5 min at 300 × *g*. Centrifugation at this step is different from the bulk prep, above.
8. Repeat **step** 7.
9. Aspirate the supernatant, again leaving behind 1 mL.
10. Add 3 mL rinsing solution, pull the pellet up and down eight times through a sheared Pasteur pipette (approximately 1.5 cm from the narrow end of the pipette).
11. Centrifuge the samples at 300 × *g* for 5 min.

12. Aspirate the supernatant, leaving behind 1 mL.
13. Add 3 mL rinsing solution, resuspend using a 1 mL pipette.
14. Centrifuge the samples at 300 × *g* for 5 min.
15. Aspirate the supernatant to the 0.5 mL mark of the tube.
16. Add 3 mL digestion solution 2, resuspend using a 1 mL pipette.
17. Vortex for 30 s.
18. Incubate shaking for 30 min at 37 °C.
19. Vortex for 30 s.
20. Draw into and out of a 3 mL syringe with a 16-G needle four times.
21. Draw into and out of a 3 mL syringe with an 18-G needle four times.
22. Using a 1 mL pipette, apply the sample through a 40 μm cell strainer, collecting in a new 15 mL conical tube.
23. Discard the cell strainer, then add 3 mL rinsing solution to each sample.
24. Centrifuge the samples at 300 × *g* for 5 min.
25. Aspirate until only approximately 50 μL remains in the conical tube.
26. Tap the tubes to resuspend the cells.
27. Add 250 μL of staining medium to each sample.

3.6 FACS Staining

In addition to determining the number of Pax7-ZsGreen+ satellite cells that have self-renewed based upon their green fluorescence, staining the cell samples for surface markers prior to analysis on the FACS allows for the enumeration of the total number of satellite cells in the TA. The cell samples are stained for CD31 and CD45 (Lineage-negative markers) as well as the satellite cell markers VCAM1 and α7-integrin [15–19].

1. Prepare the antibody mixture. Subheading 2.6, **item 3** provides the recipe per TA. Multiply these volumes by the number of TAs to be analyzed.
2. 6 μL of the antibody mixture is added to each TA sample (this is enough to stain approximately two million cells).
3. Incubate on ice for at least 20 min (*see* **Note 5**).
4. Add 5 mL staining medium to each sample.
5. Centrifuge at 300 × *g* for 5 min.
6. Aspirate the staining medium until only approximately 50 μL remains in the conical tube.
7. Tap the tubes to resuspend the cells.
8. Add 250 μL of staining medium with propidium iodide to resuspend the cells for FACS analysis.

3.7 FACS Gating Strategy

Cells can be analyzed/sorted on a variety of instruments, using the same strategies. The examples provided below pertain to our instrument of choice, a BD FACS Aria II, equipped with red (641 nm), blue (488 nm), and yellow-green (561 nm) lasers. The cells are analyzed sequentially through a series of gates. Below are the gates utilized for the analysis of satellite cell self-renewal (Fig. 5).

1. The Forward Scatter Threshold eliminates the signal from debris and setting this threshold appropriately is important to ensure speed of sort and sort efficiency (*see* **Note 6**).
2. Side Scatter Area × Forward Scatter Area is a plot of event complexity versus size of events. This plot allows for exclusion of debris and inclusion of mononuclear cells (Fig. 5a).
3. Side Scatter Height × Side Scatter Width plots exclude doublets (Fig. 5b).
4. Forward Scatter Height × Forward Scatter Width plots exclude doublets (Fig. 5c).
5. Live cells are determined by the ability of the cells to exclude propidium iodide (Fig. 5d). We plot PI against PE because of the strong spectral overlap of these two channels, using a triangular gate to include all PI− cells, including those with a strong PE signal that bleeds into the PI channel, and to exclude true PI+ cells.
6. Satellite cells do not express the lineage surface markers CD31 and CD45 (PE-Cy7 or lineage negative cells), therefore the negative fraction is gated (Fig. 5e).
7. Lineage negative cells are examined for the expression of VCAM and α7-integrin and a gate is placed on the well separated double-positive population (Fig. 5f).
8. Finally, the Pax7-ZsGreen status of the VCAM/α7-integrin double positive cells is determined using the green channel vs. PE-Cy7 (Fig. 5g and h).

3.8 Cryosectioning and Dystrophin/Laminin Staining

Quantification of the ability of the transplanted cells to differentiate and contribute to myofibers can be determined by staining TA cross-sections with antibodies to dystrophin and laminin (Fig. 6).

1. Place each TA into a cryoblock containing O.C.T. compound.
2. Cool 2-methylbutane with liquid nitrogen.
3. After the 2-methylbutane has become solid, place the cryoblocks on the surface of liquid nitrogen-cooled 2-methylbutane for approximately 2 min.
4. Store cryoblocks containing samples in −80 freezer until ready to section.

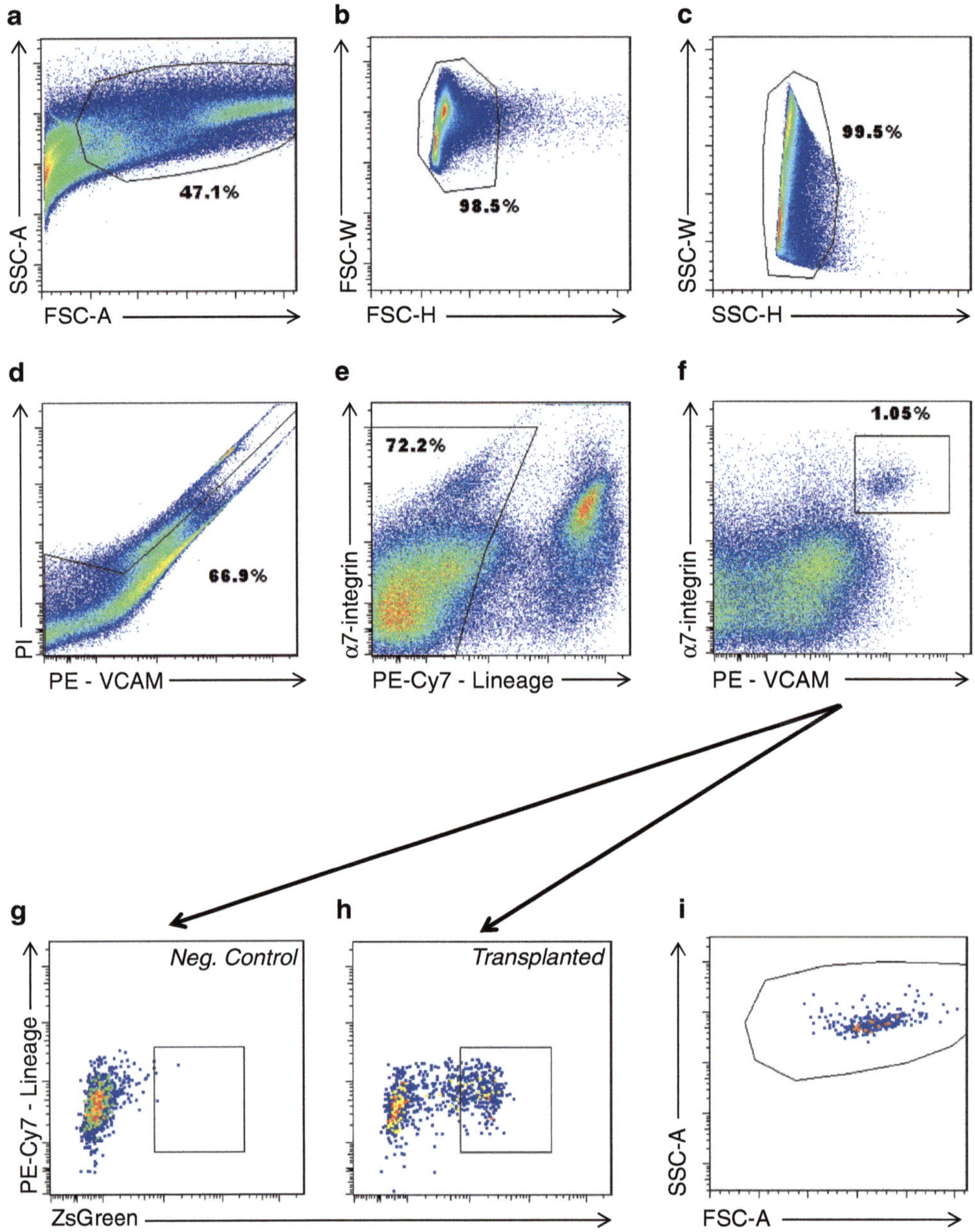

Fig. 5 Gating for FACS analysis of transplanted TAs. (**a**) FSC-A × SSC-A. (**b**) FSC-H × FSC-W. (**c**) SSC-H × SSC-W. (**d**) PE (VCAM) × PI. (**e**) PE-Cy7 (Lineage) × α7-integrin. (**f**) PE (VCAM) × α7-integrin. (**g**) ZsGreen × PE-Cy7 (Lineage) for a control sample that was not transplanted with any cells. (**h**) ZsGreen × PE-Cy7 (Lineage) for a sample transplanted with 300 transplanted cells. (**i**) Back gating of ZsGreen+ cells in H back to the forward/side scatter (FSC-A × SSC-A) plot. This should be done to verify that the FSC × SSC gate is collecting all of the ZsGreen+ cells

5. 10 μm sections are collected onto microscope slides; approximately 150 μm between sections, ten sections per slide. Each slide therefore includes ten sections at different places through a single TA muscle.

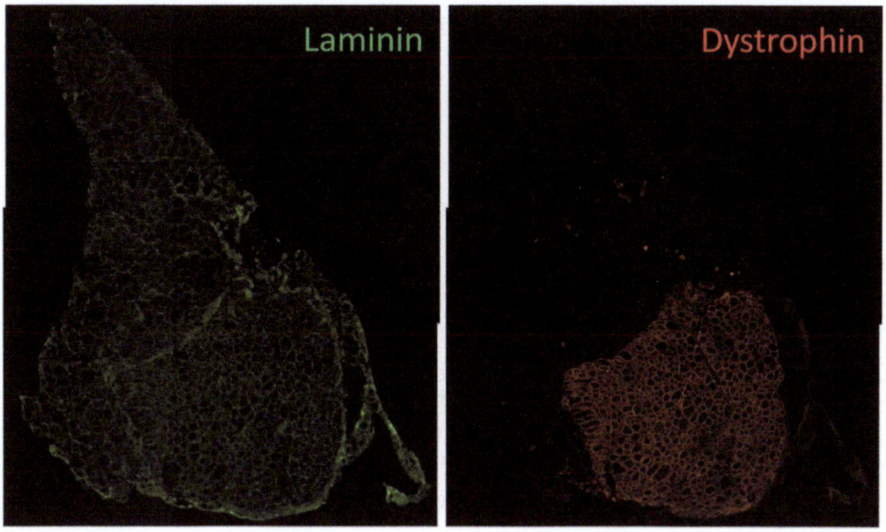

Fig. 6 Merged image for dystrophin and laminin staining of transplanted TA. *Left*: laminin staining; *right*: dystrophin staining

6. Slides are fixed with cold acetone for 5 min.
7. Allow the slides to air dry for 10 min and circle the samples with PAP Pen.
8. Rehydrate the samples with 1× PBS for 10 min.
9. Block with 3% BSA in 1× PBS for 1 h at room temperature.
10. Incubate with a 1:250 dilution of rabbit polyclonal antibody to dystrophin and 1:500 dilution of mouse monoclonal antibody to laminin in 3% BSA in 1× PBS for 1 h at room temperature.
11. Wash slides 3× for 5 min with 1× PBS.
12. Incubate with a 1:750 dilution of Alexa Fluor 555 goat anti-rabbit IgG and 1:750 dilution of Alexa Fluor 488 goat anti-mouse IgG in 3% BSA in 1× PBS for 45 min at room temperature.
13. Wash slides 3× for 5 min with 1× PBS.
14. Incubate with a 1:1000 dilution of DAPI in 1× PBS for 10 min at room temperature.
15. 24×60-1 glass coverslips are mounted using Immu-Mount.
16. Slides are photographed at 5× magnification with a Zeiss Axio Imager M1 Upright microscope with an AxioCam HRc camera.
17. Multiple images are needed to visualize the entire TA. Images can be merged using Adobe Photoshop.
18. Under File > Automate > Photomerge > Select Interactive Layout, browse to select the images to be merged (*see* **Note 7**), then save your merged file image.

19. Counting dystrophin + (donor-derived) and laminin + (total) fibers can be done using Adobe Photoshop: Select Analysis > Count Tool.
20. As you click on each fiber, Photoshop marks the site with a number (*see* **Note 8**).

4 Notes

1. The Pax7-ZsGreen transgene should never be allowed to become homozygous. We maintain this stock by back crossing to C57BL/6 and PCR genotyping. Carrier mice should never be bred with each other, as we have found that in the homozygous state, the transgene can undergo generational silencing. This does not occur if the transgene is kept heterozygous.
2. We routinely perform irradiation on a RS 2000 Biological Research Irradiator (Rad Source Technologies, Inc.) for 6 min 45 s with the shelf at level 3.
3. NSG-mdx^{4Cv} mice carry the SCID mutation, which makes them about 3× more sensitive to irradiation than WT mice. If performing transplantations into SCID+ mice, the dose would need to be increased accordingly.
4. When using the Pax7-ZsGreen mouse on a clean C57BL/6 background, 300 cells should result in approximately 300 donor fibers and 300 detectable donor-derived satellite cells after 1 month. Beyond about 500 cells, the correlation between cells in to cells/fibers out deviates from linearity. However, if the Pax7-ZsGreen cells are on a mixed background, the linear range will need to be determined empirically.
5. Twenty minutes is a minimum time. Incubation with antibodies on ice may be extended to 40 min.
6. On the BD FACS Aria II, we set the Forward Scatter Threshold to 5000.
7. Depending on memory and image size, Photoshop may only let you merge two images at a time. If this is the case, it will be necessary to perform multiple iterative merges.
8. Make sure to record the number of counted fibers at the end of each image analysis. The program does not save the count.

Acknowledgments

This work was supported by grants from the NIH (R01 AR055685) and the Muscular Dystrophy Association (MDA351022).

References

1. Montarras D et al (2005) Direct isolation of satellite cells for skeletal muscle regeneration. Science 309(5743):2064–2067
2. Seale P et al (2000) Pax7 is required for the specification of myogenic satellite cells. Cell 102(6):777–786
3. Bosnakovski D et al (2008) Prospective isolation of skeletal muscle stem cells with a Pax7 reporter. Stem Cells 26(12):3194–3204
4. Hoffman EP, Brown RH Jr, Kunkel LM (1987) Dystrophin: the protein product of the Duchenne muscular dystrophy locus. Cell 51(6):919–928
5. Watkins SC, Hoffman EP, Slayter HS, Kunkel LM (1988) Immunoelectron microscopic localization of dystrophin in myofibres. Nature 333(6176):863–866
6. Petrof BJ, Shrager JB, Stedman HH, Kelly AM, Sweeney HL (1993) Dystrophin protects the sarcolemma from stresses developed during muscle contraction. Proc Natl Acad Sci U S A 90(8):3710–3714
7. Ervasti JM, Campbell KP (1991) Membrane organization of the dystrophin-glycoprotein complex. Cell 66(6):1121–1131
8. Sicinski P et al (1989) The molecular basis of muscular dystrophy in the mdx mouse: a point mutation. Science 244(4912):1578–1580
9. Arpke RW et al (2013) A new immuno-dystrophin-deficient model, the NSG-Mdx mouse, provides evidence for functional improvement following allogeneic satellite cell transplantation. Stem Cells 31:1611–1620
10. Bosma GC, Custer RP, Bosma MJ (1983) A severe combined immunodeficiency mutation in the mouse. Nature 301(5900): 527–530
11. Cao X et al (1995) Defective lymphoid development in mice lacking expression of the common cytokine receptor gamma chain. Immunity 2(3):223–238
12. Chapman VM, Miller DR, Armstrong D, Caskey CT (1989) Recovery of induced mutations for X chromosome-linked muscular dystrophy in mice. Proc Natl Acad Sci U S A 86(4):1292–1296
13. Danko I, Chapman V, Wolff JA (1992) The frequency of revertants in mdx mouse genetic models for Duchenne muscular dystrophy. Pediatr Res 32(1):128–131
14. Hoffman EP, Morgan JE, Watkins SC, Partridge TA (1990) Somatic reversion/suppression of the mouse mdx phenotype in vivo. J Neurol Sci 99(1):9–25
15. Chan SS et al (2013) Mesp1 patterns mesoderm into cardiac, hematopoietic, or skeletal myogenic progenitors in a context-dependent manner. Cell Stem Cell 12(5):587–601
16. Blanco-Bose WE, Yao CC, Kramer RH, Blau HM (2001) Purification of mouse primary myoblasts based on alpha 7 integrin expression. Exp Cell Res 265(2):212–220
17. Fukada S et al (2007) Molecular signature of quiescent satellite cells in adult skeletal muscle. Stem Cells 25(10):2448–2459
18. Jesse TL, LaChance R, Iademarco MF, Dean DC (1998) Interferon regulatory factor-2 is a transcriptional activator in muscle where It regulates expression of vascular cell adhesion molecule-1. J Cell Biol 140(5):1265–1276
19. Seale P, Ishibashi J, Holterman C, Rudnicki MA (2004) Muscle satellite cell-specific genes identified by genetic profiling of MyoD-deficient myogenic cell. Dev Biol 275(2):287–300

Chapter 13

Noninvasive Tracking of Quiescent and Activated Muscle Stem Cell (MuSC) Engraftment Dynamics In Vivo

Andrew T.V. Ho and Helen M. Blau

Abstract

Muscle stem cells play a central role in muscle regeneration. Most studies in the field of muscle regeneration focus on the unraveling of muscle stem cell biology to devise strategies for treating failing muscles as seen in aging and muscle-related diseases. However, the common method used in assessing stem cell function in vivo is laborious, as it involves time-consuming immunohistological analyses by microscopy on serial cryo-sections of the muscle post stem cell transplantation. Here we describe an alternative method, which adapts the bioluminescence imaging (BLI) technique to allow noninvasive tracking of engrafted stem-cell function in vivo in real-time. This assay system enables longitudinal studies in the same mice over time and reveals parameters, not feasible by traditional analysis, such as the magnitude and dynamics of engrafted muscle stem cell expansion in vivo in response to a particular drug treatment or muscle injury.

Key words Muscle stem cells, Muscle regeneration, In vivo tracking, Bioluminescence imaging, Engraftment dynamics

1 Introduction

Adult skeletal muscle has remarkable regenerative capacity due to the presence of muscle stem cells (MuSCs), also referred to as satellite cells [1], which are identified by their location in the muscle niche between the sarcolemma and basal lamina of the myofibers and by their expression of intracellular and surface markers, Pax7, c-Met, syndecan, calcitonin receptor, CD34, and integrin $\alpha7$ [2–4]. In healthy adult muscles, MuSCs reside as mitotically quiescent population that persists throughout life, but are rapidly activated in response to muscle injury to give rise to progenitors that differentiate and fuse to repair damaged myofibers [5, 6]. However, the function of MuSCs declines with age or stress due to changes both in intrinsic and extrinsic cellular factors [7–11], leading to progressive muscle weakness. Transplantation therapy with myoblasts (more committed muscle progenitors) from young mice has shown some beneficial effects in muscle regeneration, but

Michael Kyba (ed.), *Skeletal Muscle Regeneration in the Mouse: Methods and Protocols*, Methods in Molecular Biology, vol. 1460, DOI 10.1007/978-1-4939-3810-0_13, © Springer Science+Business Media New York 2016

~90% of the injected myoblasts die and these progenitors lack long-term regenerative capacity as they do not repopulate the niche [2, 12]. By contrast, MuSCs are capable of self-renewal and extensive repair of entire muscle if freshly isolated from murine tissue and transplanted into damaged tissues [2]. Thus, MuSCs comprise an essential source for muscle repair and cell-based therapy. The identification of factors and conditions that enhance or restore stem cell function in vivo have been shown to be efficacious in promoting muscle functions and reversing muscle weakness [7].

Here we describe a protocol, combining the MuSC isolation, transplantation and BLI imaging techniques, to assess for the MuSC regeneration potential non invasively in vivo without the need for serial sectioning and immunofluorescence analysis of the engrafted muscles. An additional advantage of this method is that it allows longitudinal studies of MuSC functions in vivo to be conducted in the same mouse over time.

The MuSC isolation protocol is based on a modification of the published myofiber isolation protocol [13] and a flow cytometry protocol [4, 14, 15]. Briefly, a gentle collagenase digestion and mincing by the MACs Dissociator enables numerous single fibers to be dissociated, followed by dispase digestion to release mononucleated cells from their niche. Subsequently, the cells with hematopoietic lineage markers and any non-muscle cells ($CD45^+$/$CD11b^+$/$CD31^+$) are depleted using a magnetic bead column selection approach. The remaining cell mixture is then subjected to FACS analysis to sort out MuSCs co-expressing CD34 and α7-integrin markers [2]. We have previously reported that the prospectively isolated population by FACS based on CD34 and α7-integrin identify bona fide stem cells and shown that they are capable of self-renewal and differentiation by single MuSC transplantation [2].

To enable in vivo tracing of engrafted cells, MuSCs are prospectively isolated from transgenic GFP/luciferase donor mice [2] and transplanted by intramuscular injection into the tibialis anterior (TA) muscle of irradiated immunocompromised NOD/SCID recipients [2]. Alternatively, MuSCs from other donors without the GFP/luciferase transgene can be labeled with GFP/luciferase by infecting the sorted MuSCs with a lentiviral vector encoding these two reporter genes (e.g., *luc*-IRES-*GFP* lentivirus) in vitro overnight in culture [16]. Regeneration is assayed as positive engraftment evaluated using noninvasive bioluminescence imaging (BLI). BLI is a dynamic readout conducted in live mice and therefore allows repetitive measurements of engraftment dynamics noninvasively over a period post-transplantation. BLI provides high sensitivity due to its high signal-to-noise ratio, as excitation light used in fluorescence imaging (which generates background noise) is not required [17, 18]. The BLI signal emitted from the luciferin catalytic reaction by the luciferase in the engrafted cells is directly recorded by the cooled charge-coupled device (CCD) camera. To standardize the signal measurements across different samples and experiments, a same size region of interest (ROI) encompassing the areas with BLI signals are placed over each hindlimb (Subheadings 3.3 and 3.4) to define the local background (noise) and therefore enable the calculation of the radiance (signal) as a unit of measurement in photons/s/cm^2/sr.

Typically, transplantation of 10–100 MuSCs in an 18-Gy irradiated tibialis anterior muscle generates a steady increase in BLI signal output by 2 log-fold that plateaus by week 4, indicative of stable engraftment [2, 16, 19]. The long-term regeneration potential can be demonstrated by analyzing the dynamics of BLI signals over a specific duration (typically 1–3 months). Additionally, snake venom toxin such as Cardiotoxin or Notexin can be injected into the engrafted muscle to induce a second injury. This subsequent injury challenges the engrafted stem cells that have reconstituted the stem cell reservoir to produce a second wave of MuSC activation and expansion, which can be detected by a further increase in BLI signal [16]. Engraftment, differentiation and homing of MuSCs to muscle fibers to the satellite cell niche underneath the basal lamina of myofibers can be readily monitored by GFP immunohistochemistry in fixed tissue sections.

Coupling MuSC isolation and transplantation to luciferase bioluminescence imaging system enables noninvasive repetitive measurements of the engrafted muscles and reveals stem cell functions in vivo, which cannot be achieved by the endpoint serial section and immunofluorescence analysis method. Moreover, the relationship of the BLI magnitude and dynamics of engrafted muscle cell expansion in vivo over time can be further inferred to assign in vivo cell fates and functions elicited by a particular drug treatment or muscle injury.

2 Materials

2.1 Muscle Stem Cell Preparation

1. Digestion buffer

   2% collagenase, diluted in low glucose DMEM (*see* **Note 1**).
2. Lineage depletion cocktail (*see* **Note 2**)

   CD45-biotin 2 μl (BD Pharmingen).

   CD31-biotin 5 μl (eBioscience).

   CD11b-biotin 5 μl (BD Biosciences).

   Sca1-biotin 5 μl (BD Pharmingen).
3. Muscle stem cell isolation cocktail

   Streptavidin Microbeads 100 μl (Miltenyi Biotec).

   Streptavidin-APC-Cy7 5 μl (Invitrogen).

Anti-ITGA7 (α7-integrin) antibody conjugated to PE 5 μl (Miltenyi Biotec).

Anti-CD34-eFluor 660 15 μl (eBioscience).

Pacific Blue—live/death discrimination dye 2 μl (Invitrogen).

4. Myoblast media

   40 % F10 media.

   44 % DMEM low glucose.

   15 % fetal bovine serum.

   1 % penicillin–streptomycin solution.

5. FACS buffer

   1× PBS.

   2.5 % goat serum or 1 % BSA.

   2 mM EDTA.

   Others

6. GentleMACS Dissociator (Miltenyl Biotec).
7. GentleMACS C-tube (Miltenyl Biotec).
8. 1× RBC lysis buffer (eBioscience).
9. 40 μm filters.
10. MACS Magnetic Columns (Miltenyi Biotec).

2.2 Transplantation

1. 26-G needle and 50 μl Hamilton syringe.
2. Isoflurane.
3. Anesthesia machine equipped with a vaporizer set with 1–4 % 1 L/O2 (VetEquip).

2.3 Bioluminescence Imaging

1. Ketamine 80–100 mg/kg + xylazine 5–10 mg/kg IP.

3 Methods

3.1 Isolation of Muscle Stem Cells

3.1.1 Muscle Isolation and Digestion

1. In a sterile hood, prepare 10 ml of 0.2 % collagenase (diluted in DMEM) into a purple capped gentleMACS C-tube.
2. Dissect and place all of the hindlimb muscles from a single mouse into the C-tube.
3. Tighten the cap, place the C-tube (lid face down) on the gentleMACS Octo Dissociator and start the "m_muscle_01" program preinstalled by the manufacturer (*see* **Note 3**).
4. Leave the cap loosened and incubate tubes at 37 °C in humidified incubator for 60 min.
5. Tighten the cap and perform another round of MACS homogenization with "m_muscle_01".

6. Add 100 μl of 2 % collagenase and 50 μl of dispase—pipette up and down to disperse and incubate with cap loosened at 37 °C for an additional 30 min.
7. Draw the cell mixture through a 21-G needle syringe ten-times to dissociate residual undigested fibers.
8. Pass the sample through a 40 μm filter (pre-wet with myoblast media) into a 50 ml conical tube.
9. Wash the filter with 10 ml Myoblast media to dislodge cells that remain attached to the filter.
10. Pellet cells with 350×*g* centrifugation and lyse the red blood cells by resuspending the cell pellet in 1 ml of 1× RBC lysis buffer.
11. After 5 min of incubation at room temperature, add 9 ml of FACS buffer and centrifuge tube at 350×*g* for 10 min to collect cell pellet.
12. Resuspend pellet in 1 ml FASC buffer.

3.1.2 Immunofluorescence Staining

1. Add Biotin lineage depletion cocktail and incubate with slow agitation at 4 °C for 20 min.
2. Wash with 10 ml FACS buffer and pellet cells with 350×*g* centrifugation for 10 min.
3. Resuspend with 1150 μl FACS Buffer.
4. Save five aliquots of 50 μl each to five new FACS tubes and set aside for FACS compensation (unstained control, Pacific blue-live/death discrimination dye, CD34- eFluor 660, α7-integrin-PE and Streptavidin-APC-Cy7).
5. Stain the 900 μl remaining cell mixture with muscle stem cell isolation cocktail and incubate on an end-to-end rotator in cold room for 30 min.
6. Wash with 10 ml FACS buffer and pellet cells with 350×*g* centrifugation at 4 °C for 10 min.
7. Resuspend with 1 ml FACS Buffer.

3.1.3 Cell Depletion and Sorting

1. Equilibrate Mac column placed on the magnet stand with 2 ml FACS buffer.
2. Run the 1 ml of stained cells (Subheading 3.1.2, **step** 7) through the column to deplete the lineage-positive cells and collect flow-through into a 15 ml conical tube.
3. Wash column with additional 9 ml FACS buffer.
4. Pellet the collected flow-through with 350×*g* centrifugation at 4 °C for 10 min.
5. Resuspend cell pellet with 3 ml FACS buffer.
6. Sort for stem cells via FACS. After setting up FACS compensation with the control tubes, set up the sorter with gates to

select for live single cell population (Pacific blue-negative), lineage negative (Streptavidin-APC-Cy7 negative) and collect muscle stem cell population in the double CD34-Alexa Fluor 660 and α-7 PE population.

3.2 Muscle Stem Cell Engraftment

3.2.1 18 Gy Irradiation of NOD/SCID

1. Recipient mice are anesthetized by intraperitoneal injection of Ketamine 80–100 mg/kg + xylazine 5–10 mg/kg IP.
2. Placed in a lead-jig (to protect the rest of the body from irradiation), and only hindlimbs remain exposed to γ-irradiation outside the jig (Fig. 1a).
3. 18 Gy of total γ-irradiation is given in one dose (*see* **Note 4**).

3.2.2 Transplantation

1. Anesthetize animals 24–48 h post irradiation by isoflurane inhalation via precision vaporizer 1–4% 1 L/O2 min. Use the toe or tail pinch to ensure the animal is completely anesthetized.
2. Inject 50–500 prospectively isolated muscle stem cells (freshly isolated or after culture) intra-muscularly into the irradiated tibialis anterior muscles of NOD/SCID recipient mice through a Hamilton syringe with 26-G needle and with consistent needle depth (~4mm) and angle (~50 degrees), as illustrated in Fig. 1b.

3.3 Bioluminescence Imaging

1. Anesthetize by isoflurane inhalation 1–4% 1 L/O2 until the mouse is unresponsive to palm pinching test.
2. Inject intraperitoneally with 100 μl of d-luciferin solution. This dose causes no known undesirable side effects or toxicity.

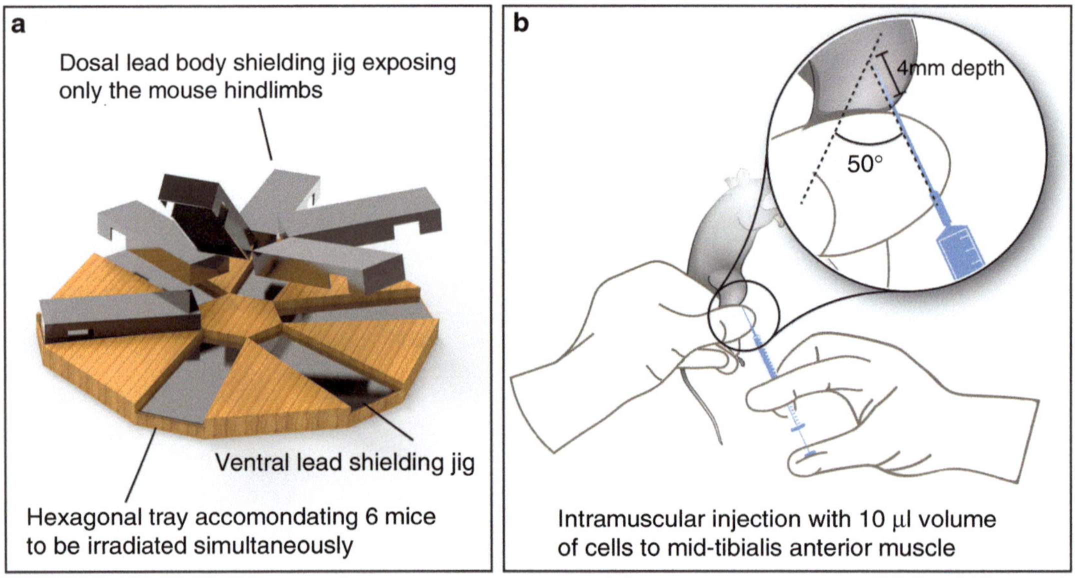

Fig. 1 Illustrations depict (**a**) the lead jig configuration to accommodate six mice for irradiating the lower limbs simultaneously while protecting the rest of the body from irradiation; and (**b**) the intramuscular injection performed by a single person with one hand stabilizing the mouse lower limb while the other hand delivers the muscle stem cells from a syringe into the tibialis anterior muscle

3. Place the mice in a light-tight box under a light-sensitive camera with the hindlimbs extended and keep palms down with tape.
4. The signal is captured via the living image software connected to the IVIS imaging system (Perkin Elmer) (*see* **Note 5**). Use the imaging wizard to guide you through a series of steps, prompting the user to enter the required information for setting up the acquisition in the control panel panel; generally, the setting of 60-sec exposure time and F=1 camera aperture is used (*see* **Note 6**).
5. Set up the acquisition to capture the bioluminescence from the animal over a 10 min period with a sequence of 10 images capturing at 1-min interval. (Fig. 2).

3.4 Analysis

1. BLI values in Radiance for each ROI is exported to either to excel or Prism application for analysis.
2. Using a graph-plotting program such as Prism, the BLI curve is generated using the XY scatter plot of radiance ($p/s/cm^2/sr$) versus time (weeks). A threshold of radiance = 1×10^4 $p/s/cm^2/sr$ (approximately 1×10^5 p/s in total flux) indicates above background signal and can be deemed positive for engraftment.

4 Notes

1. Prepare fresh and keep on ice.
2. Mouse hematopoietic lineage panel was reported to identify, enrich, and/or deplete cells committed to the T, B, NK, myeloid and erythroid lineages.
3. Alternatively, **steps 2–4** can be replaced by manual mincing with a sterile razor in the culture hood until the samples comprise fragments that can be easily passed though a 10 ml pipette tip.
4. The actual radiation dose set point varies based on the calculation that takes the correction factor specific to each radiator and jig design into consideration.
5. IVIS acquisition control panel is only available on a PC workstation that connects to the IVIS imaging system. The imaging wizard will determine items available in the control panel depending on the selected imaging mode (luminescent or fluorescent) and acquisition mode (image Setup or Sequence Setup).
6. The auto exposure setting allows the system to acquire an image at maximum sensitivity. This is useful in a scenario where signal strength is unknown. The software will adjust the parameters and retake another image.

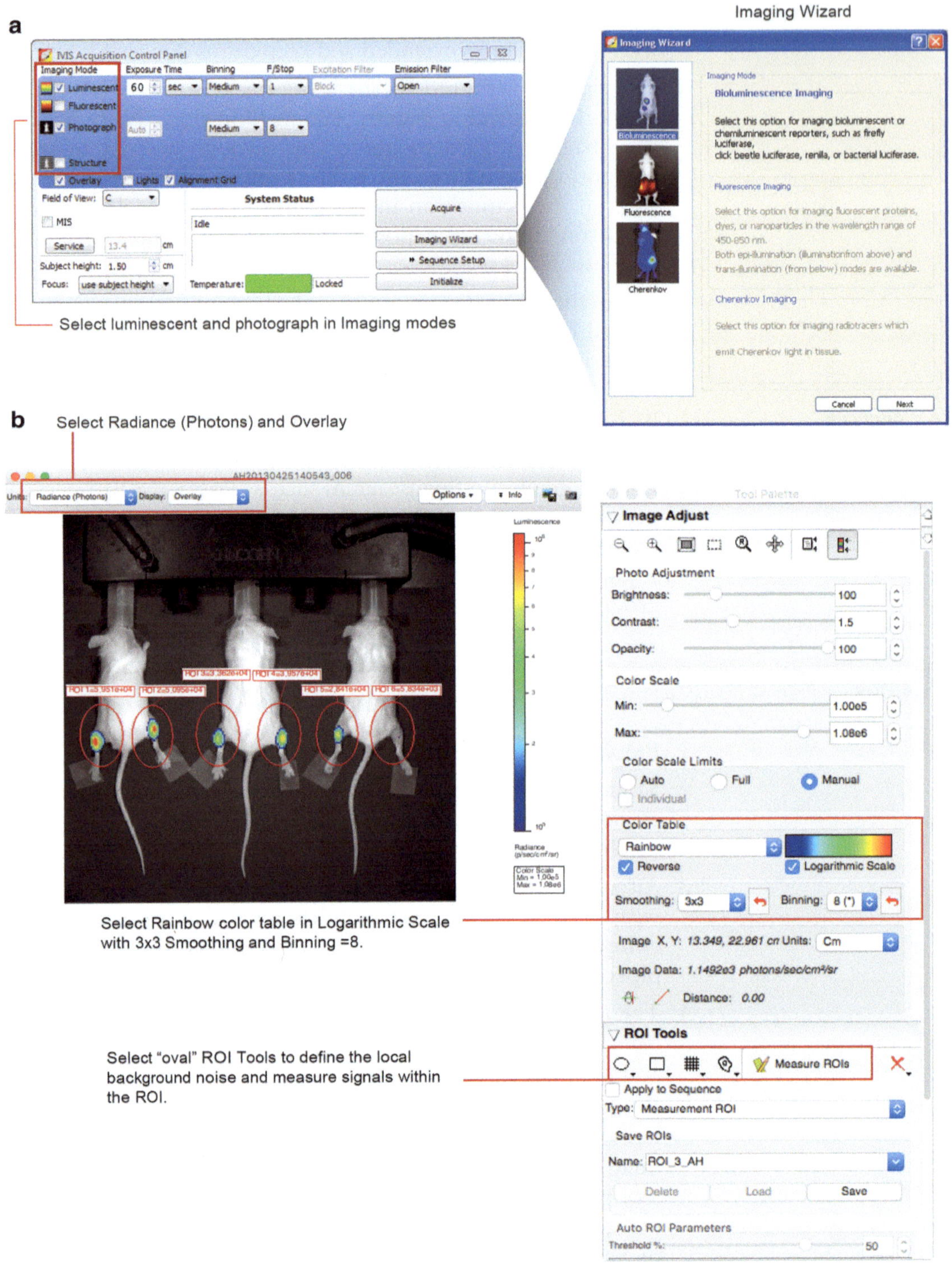

Fig. 2 Software interface from the Living Image application controlling the acquisition processes in IVIS imaging system (PerkinElmer). (**a**) The acquisition control panel highlighting the selection key to bioluminescent imaging. (**b**) Standardized settings in the tool palette panel, optimized for acquiring images and bioluminescent signals of the transplanted mouse limbs

References

1. Mauro A (1961) Satellite cell of skeletal muscle fibers. J Biophys Biochem Cytol 9: 493–495
2. Sacco A, Doyonnas R, Kraft P, Vitorovic S, Blau HM (2008) Self-renewal and expansion of single transplanted muscle stem cells. Nature 456(7221):502–506. doi:10.1038/nature07384, nature07384 [pii]
3. Fukada S, Uezumi A, Ikemoto M, Masuda S, Segawa M, Tanimura N, Yamamoto H, Miyagoe-Suzuki Y, Takeda S (2007) Molecular signature of quiescent satellite cells in adult skeletal muscle. Stem Cells 25(10):2448–2459. doi:10.1634/stemcells.2007-0019
4. Montarras D, Morgan J, Collins C, Relaix F, Zaffran S, Cumano A, Partridge T, Buckingham M (2005) Direct isolation of satellite cells for skeletal muscle regeneration. Science 309(5743):2064–2067
5. Dhawan J, Rando TA (2005) Stem cells in postnatal myogenesis: molecular mechanisms of satellite cell quiescence, activation and replenishment. Trends Cell Biol 15(12): 666–673
6. Partridge TA (2002) Cells that participate in regeneration of skeletal muscle. Gene Ther 9(11):752–753. doi:10.1038/sj.gt.3301764
7. Blau HM, Cosgrove BD, Ho AT (2015) The central role of muscle stem cells in regenerative failure with aging. Nat Med 21(8):854–862. doi:10.1038/nm.3918
8. Brack AS, Rando TA (2007) Intrinsic changes and extrinsic influences of myogenic stem cell function during aging. Stem Cell Rev 3(3): 226–237
9. Collins CA, Zammit PS, Ruiz AP, Morgan JE, Partridge TA (2007) A population of myogenic stem cells that survives skeletal muscle aging. Stem Cells 25(4):885–894. doi:10.1634/stemcells.2006-0372
10. Gopinath SD, Rando TA (2008) Stem cell review series: aging of the skeletal muscle stem cell niche. Aging Cell 7(4):590–598. doi:10.1111/j.1474-9726.2008.00399.x
11. Grounds MD (1998) Age-associated changes in the response of skeletal muscle cells to exercise and regeneration. Ann N Y Acad Sci 854: 78–91
12. Peault B, Rudnicki M, Torrente Y, Cossu G, Tremblay JP, Partridge T, Gussoni E, Kunkel LM, Huard J (2007) Stem and progenitor cells in skeletal muscle development, maintenance, and therapy. Mol Ther 15(5):867–877. doi:10.1038/mt.sj.6300145
13. Rosenblatt JD, Lunt AI, Parry DJ, Partridge TA (1995) Culturing satellite cells from living single muscle fiber explants. In Vitro Cell Dev Biol Anim 31(10):773–779. doi:10.1007/BF02634119
14. Sherwood RI, Christensen JL, Conboy IM, Conboy MJ, Rando TA, Weissman IL, Wagers AJ (2004) Isolation of adult mouse myogenic progenitors: functional heterogeneity of cells within and engrafting skeletal muscle. Cell 119(4):543–554. doi:10.1016/j.cell.2004.10.021, S0092867404010360 [pii]
15. Motohashi N, Asakura Y, Asakura A (2014) Isolation, culture, and transplantation of muscle satellite cells. J Vis Exp (86). doi:10.3791/50846
16. Cosgrove BD, Gilbert PM, Porpiglia E, Mourkioti F, Lee SP, Corbel SY, Llewellyn ME, Delp SL, Blau HM (2014) Rejuvenation of the muscle stem cell population restores strength to injured aged muscles. Nat Med 20(3):255–264. doi:10.1038/nm.3464
17. Badr CE (2014) Bioluminescence imaging: basics and practical limitations. Methods Mol Biol 1098:1–18. doi:10.1007/978-1-62703-718-1_1
18. Spring KR (2013) Cameras for digital microscopy. Methods Cell Biol 114:163–178. doi:10.1016/B978-0-12-407761-4.00008-7
19. Gilbert PM, Havenstrite KL, Magnusson KE, Sacco A, Leonardi NA, Kraft P, Nguyen NK, Thrun S, Lutolf MP, Blau HM (2010) Substrate elasticity regulates skeletal muscle stem cell self-renewal in culture. Science 329(5995):1078–1081. doi:10.1126/science.1191035

Chapter 14

Myogenic Progenitors from Mouse Pluripotent Stem Cells for Muscle Regeneration

Alessandro Magli, Tania Incitti, and Rita C.R. Perlingeiro

Abstract

Muscle homeostasis is maintained by resident stem cells which, in both pathologic and non-pathologic conditions, are able to repair or generate new muscle fibers. Although muscle stem cells have tremendous regenerative potential, their application in cell therapy protocols is prevented by several restrictions, including the limited ability to grow ex vivo. Since pluripotent stem cells have the unique potential to both self-renew and expand almost indefinitely, they have become an attractive source of progenitors for regenerative medicine studies. Our lab has demonstrated that embryonic stem cell (ES)-derived myogenic progenitors retain the ability to repair existing muscle fibers and contribute to the pool of resident stem cells. Because of their relevance in both cell therapy and disease modeling, in this chapter we describe the protocol to derive myogenic progenitors from murine ES cells followed by their intramuscular delivery in a murine muscular dystrophy model.

Key words Pluripotent, ES and iPS cells, Skeletal muscle, Pax3, Pax7, Myogenic progenitors, Muscle differentiation, Cell therapy, Transplantation

1 Introduction

Muscular dystrophies, cachexia, and sarcopenia are diseases characterized by progressive impairment of skeletal muscle function. Importantly, although several advances have been made toward the understanding of the pathogenesis of these diseases, an effective treatment to restore muscle homeostasis in these patients is still missing. In the specific case of muscular dystrophies, because of their genetic nature, the ideal treatment for successfully restoring muscle physiology would require the correction of the specific mutation using either gene or cell therapy. Although both approaches hold great potential, there are still several limitations that need to be overcome in order to successfully apply them to these patients. In the specific case of cell therapy, the cell population that will be transplanted has to fulfill several requirements: (1) ability to produce robust engraftment in the host muscle; (2) ability to contribute to substantial improvement of muscle

Michael Kyba (ed.), *Skeletal Muscle Regeneration in the Mouse: Methods and Protocols*, Methods in Molecular Biology, vol. 1460, DOI 10.1007/978-1-4939-3810-0_14, © Springer Science+Business Media New York 2016

functions; (3) contribution to the stem cell pool responsible for long-term muscle homeostasis upon delivery; (4) high potential of in vitro expansion; (5) amenability to genome editing or transgene integration followed by in vitro selection.

Studies from different research groups have suggested the feasibility of cell therapy for the treatment of muscular dystrophies in different animal models using both adult- [1–5] or pluripotent-derived progenitors [6–9]. Notably, the latest category has seen an increasing interest in the last few years due to the development of induced pluripotent stem (iPS) cell technology [10, 11]. In particular, ES/iPS cells harbor the potential to generate all the different cell types that constitute the adult organism, which overcome the need of alternative sources of tissue-specific progenitors. However, the main limitation encountered so far, which cannot be ignored, is the necessity of an appropriate differentiation protocol for the generation of a given lineage from ES/iPS cells. Taking advantage of the extensive knowledge from developmental biology studies, several investigators are defining and improving specific differentiation protocols for multiple cells types [12].

Specification of the skeletal myogenic lineage during development occurs in the somites, epithelial aggregates of cells adjacent to the neural tube and the notochord of the developing embryo (reviewed by [13]). These structures are paraxial mesodermal derivatives generated in the tail bud region through a mechanism involving controlled oscillation in gene expression in the presomitic mesodermal cells (also called clock wavefront model). After formation, signals from the nearby tissues (neural plate, neural tube and notochord) instruct cells within the somites to further differentiate into one of the three somitic derivatives: myotome, dermatome, and sclerotome. The first two derivatives are initially specified as dermomyotome, which represents a fundamental structure for the commitment of the skeletal myogenic progenitors that will form the long-range migration skeletal muscles (limbs, diaphragm, and tongue) and the axial muscles. Activation of this program involves first the expression of the Paired-box transcription factor Pax3, followed by expression of its close homolog Pax7. Double positive Pax3/7 cells located in the central domain of the dermomyotome represent the myogenic progenitors that will form all the skeletal muscles of the body [14], except for the facial muscles which have a non-somitic origin. Importantly, these transcription factors regulate the myogenic commitment through activation of the myogenic regulatory factor Myf5. The latter belongs to a family of four transcription factors (called muscle regulatory factors—MRFs) whose members play a fundamental role in the expression of the skeletal muscle-specific genes.

As briefly mentioned above, specification of the skeletal myogenic lineage during development requires the dynamic interplay between signals from different parts of the embryo which ultimately culminate in the activation of the regulatory network responsible for the skeletal myogenic-specific gene expression. Therefore it is not surprising that, although pluripotent cells have the virtual potential to generate all the cell types of the adult body, skeletal myogenic commitment is not perfectly recapitulated during ES/iPS cell differentiation due to the absence of the signals from the adjacent structures (e.g., neural tube and notochord), which do not form properly during the in vitro differentiation of pluripotent stem cells.

Such limitations, however, can be overcome by combining the extensive knowledge acquired from developmental biology studies, and with the increasing improvement in ES/iPS cell culture and their genetic manipulation, including the generation of doxycycline-inducible ES cells which allow the controlled expression of the gene of interest [15, 16]. In this regard, forced expression of the transcription factors Pax3 or Pax7 in differentiating ES cells has been proved to be sufficient in activating the gene expression network responsible for the specification of the skeletal myogenic lineage [7, 17]. Our laboratory pioneered this field by demonstrating that both mouse and human ES/iPS cells can be committed toward the myogenic lineage using the doxycycline-inducible system to express Pax3/7 proteins [6, 7]. In the specific case of murine ES cell differentiation, we showed that Pax3/7 expression during three-dimensional differentiation (from here on referred as embryoid body—EB) results in the increased frequency of the PDGFRα+FLK1− cell fraction. During development, differential expression of these two cell surface markers are associated with the mesodermal patterning toward the hemato-endothelial, cardiac and paraxial lineages in differentiating EBs [18]. Importantly, expression of Pax3 during this process is involved in the acquisition of presomitic/somitic markers and specific repression of the cardiac gene expression program in the PDGFRα+ fraction [19, 20]. Upon FACS-mediated isolation from day 5 EBs, PDGFRα+FLK1− cells from Pax3 (or Pax7)-induced cultures are able to proliferate and differentiate to skeletal myocytes upon doxycycline withdrawal. Importantly, local or systemic transplantation of proliferating ES-derived skeletal myogenic progenitors results in the participation of these cells in muscle regeneration. Since the ES/iPS system is able to recapitulate the earlier stages of mouse development leading to the specification of the skeletal myogenic lineage and to produce a cell population able to engraft in the skeletal muscle of dystrophic mouse models, it represents an important tool to both study molecular mechanisms involved in muscle development/homeostasis and also to test and develop new therapies for muscle diseases.

In this chapter, we describe in detail the protocol to derive myogenic progenitors from Pax3-inducible mouse ES cells followed by their intramuscular delivery in *mdx* mice, an established model for Duchenne Muscular Dystrophy.

2 Materials

2.1 Cell Culture Media

Media components can be purchased from different companies. The most critical factor is the ES-qualified fetal bovine serum (FBS), which we suggest to test before using for both maintenance and differentiation.

1. ES medium: Knock-out™ DMEM, optimized for ES cells (Gibco) supplemented with 15 % ES-qualified FBS, 1 % penicillin–streptomycin, 2 mM GlutaMAX (Gibco), 0.1 mM nonessential amino acids, 0.1 mM β-mercaptoethanol, and 1000 U/ml LIF.
2. EB medium: IMDM supplemented with 15 % ES-qualified FBS, 1 % penicillin–streptomycin, 2 mM GlutaMAX (Gibco), 50 μg/ml ascorbic acid (stock 50 mg/ml), and 4.5 mM monothioglycerol.
3. MEFs medium: DMEM supplemented with 10% FBS, 1% penicillin–streptomycin, and 2 mM GlutaMAX (Gibco).

Additional supplies for tissue culture:

1. A2lox Pax3 mES cells.
2. Primary mouse embryonic fibroblasts (MEFs), irradiated.
3. 0.1 % gelatin from porcine skin (w/v) in H_2O (sterilized by autoclaving).
4. Dulbecco's phosphate buffer saline (D-PBS—sterile).
5. 1 mg/ml doxycycline hyclate in sterile D-PBS (vortex well before using).
6. Trypsin–EDTA 0.25 % (sterile).
7. Murine basic fibroblast growth factor (bFGF, *see* **Note 1**).
8. Trypan blue.
9. 15 and 50 ml sterile centrifuge tubes.
10. 6-well tissue culture plate.
11. T25 and T75 tissue culture flasks.
12. Petri dish, low adherent (100 mm and 150 mm × 15 mm).
13. Centrifuge.
14. Inverted microscope.
15. 37 °C, 5 % CO_2 incubator for cell culture.
16. Laminar flow cell culture cabinet.
17. Orbital shaker for cell culture.
18. Hemacytometer.

2.2 FACS-Mediated Cell Sorting

FACS Staining medium: D-PBS supplemented with 10 % FBS and 1 % Penicillin–streptomycin.

Additional supplies for FACS staining:

1. Fc-block (purified anti-Mouse CD16/CD32—sterile).
2. PE-conjugated anti-mouse CD140a (PDGFRα—sterile).
3. APC-conjugated anti-mouse CD309 (FLK1—sterile).
4. Propidium iodide solution (50 μg/ml in PBS—sterile).
5. 70 μm strain—sterile.
6. FACS machine.

2.3 Tibialis Anterior (TA) Injury and Transplantation of Myogenic Progenitors

1. Ketamine–xylazine solution at the final concentration of 80 mg/kg and 1 ml disposable syringes.
2. X-RAD 320 Biological Irradiator (Precision X-Ray) or any similar device.
3. Surgical pad.
4. Betadine swabs.
5. Sterile stainless steel dissection scissors.
6. Sterile stainless steel tweezers.
7. Hamilton® syringe 702 RN 25 μl (Hamilton®) and/or U-100 insulin disposable 500 μl syringes.
8. Cardiotoxin from *Naja mossambica mossambica* venom at the final concentration of 10 μM (Sigma).
9. Non-absorbable black braided silk suture 18″.
10. Cell dissociation buffer, enzyme-free, PBS-based (Gibco).

2.4 Collection of TA Muscles from Recipient Mice

1. 70% ethanol.
2. Stainless steel forceps, dissection scissors and tweezers.
3. Tissue freezing medium and Peel-A-Way embedding mold square S22 (Sigma).
4. Liquid nitrogen in Styrofoam box and isopentane in stainless steel beaker.
5. Sterile stainless steel specimen forceps.

2.5 Analysis of Embedded TA Muscles

1. Cryostat and tissue freezing medium.
2. Superfrost Plus microscope slides (Thermo Scientific).
3. Non-sterile Dulbecco's phosphate buffer saline 1×.
4. Triton X-100 solution in D-PBS at the final concentration of 0.3% and bovine serum albumin (BSA) Fraction V dissolved in D-PBS at the final concentration of 3%.
5. Primary antibody solution: blocking solution supplemented with mouse monoclonal anti-dystrophin antibody NCL-DYS2 (Leica) at the final dilution of 1:20, rabbit polyclonal anti-dystrophin antibody (Abcam) at the final dilution of 1:250.

6. Secondary antibody solution: blocking solution supplemented with goat anti-mouse or goat anti-rabbit Alexa Fluor 555 (Life Technologies) at the final dilution of 1:500.
7. ProLong Gold antifade mountant with DAPI (Life Technologies).
8. Coverslips.

3 Methods

3.1 mES Maintenance

Murine ES/iPS cell propagation requires specific culture condition to prevent their spontaneous differentiation and ensure their self-renewal. In our culture settings, this is achieved by coculture ES cells with mitotically inactivated mouse embryonic fibroblasts (MEFs—Fig. 1a), which can be obtained from different sources. In addition, all the cell culture procedures should be performed in a laminar-flow cabinet and in compliance with the biohazard handling procedures of each specific institution. It is also worth to mention that ES/iPS cell cultures require daily care.

1. Prepare gelatin-coated T25 flasks by adding 4 ml of 0.1 % gelatin solution and incubate for at least 20 min at 37 °C (*see* **Note 2**).
2. Thaw feeder layer cells (MEFs) by placing the vial in a thermostat. Collect the content, transfer in a 15 ml tube, dilute with MEFs medium and centrifuge 5 min at 300 × *g* RT. Aspirate supernatant, resuspend the cell pellet with 5 ml of MEFs medium and plate in a gelatin-coated T25 flask. Incubate at 37 °C, 5 % CO_2. MEFs require several hours to completely adhere to the plate.
3. The next day, thaw ES cells (as described above) and resuspend cell pellet in ES medium. Aspirate medium from the flask containing the feeder layer, wash once with PBS and plate the ES cells suspension. Incubate at 37 °C, 5 % CO_2.
4. After 24 h, aspirate medium, wash once with 4 ml PBS and add 5 ml of fresh ES medium and place the flask back in the incubator. A good mES cell culture should have colonies of homogenous

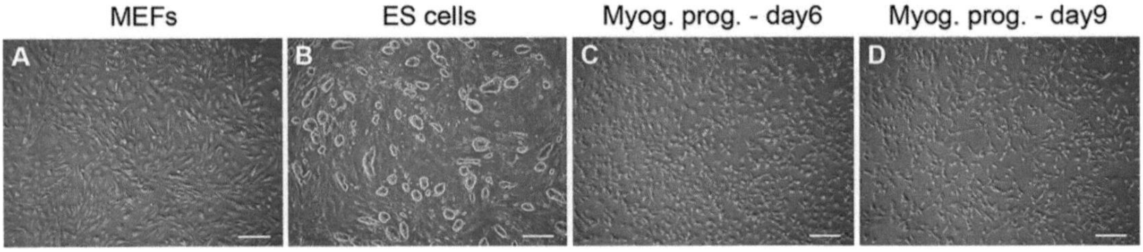

Fig. 1 Derivation of myogenic progenitors from mES cells. (**a–d**) Representative phase-contrast pictures of MEFs (**a**), mES cells growing on MEFs (**b**), PDGFRα+FLK1− cells at day 6 (**c**) and myogenic progenitors at day 9 (**d**). Scale bar 200 μm

size, characterized by a shiny border (Fig. 1b). mES cells are usually subcultured every 2 days but, in case of heterogeneous colony size, it is advisable to proceed directly to the next step without waiting for the next day.

5. After 2 days in culture, to prevent spontaneous differentiation, it is necessary to subculture mES cells. Wash cells twice with PBS then add 1.5 ml 0.25 % trypsin–EDTA and incubate at 37 °C, 5 % CO_2 for 5 min.
6. Add 5 ml MEFs medium to inactivate trypsin and resuspend eight to ten times. Check cells using an inverted microscope. If many cell clumps are still present, resuspend eight to ten times more. A single ES cell suspension is fundamental in order to properly propagate the culture.
7. Transfer cell suspension in a 15 ml tube and centrifuge 5 min at 300 × *g*.
8. Resuspend the pellet in 5 ml ES medium and place into a gelatin-coated T25 flask (remember to aspirate gelatin completely). Incubate at 37 °C for 25 min, 5 % CO_2.
9. Remove the flask from incubator carefully. Pick up supernatant (which contains cells that did not adhere) slowly and transfer into a 15 ml tube. (This procedure is used to remove MEFs from the cellular suspension thus enriching for the ES cells). Take a 10 μl aliquot.
10. Count cell number and plate 3–4 × 10^5 cells in a T25 with new MEFs using ES medium. Place in incubator.
11. Change medium after 24 h.

Following the above described procedure, mES cells can be subcultured for several passages while still maintaining their pluripotency. If cells are not going to be used for differentiation for several passages, it is advisable to freeze the cell pellet and restart the culture when needed.

3.2 mES Differentiation

Differentiation is achieved by exposing ES cells to culture conditions which force them to exit the pluripotent state and undergo commitment toward one of the three embryonic lineages. Importantly, this process is very dynamic and specific subpopulations can be identified only at given time points. To unequivocally identify the differentiation stage, we will use a progressive number for each day starting from 0 (ES cell isolation). Moreover, the culture conditions described in this protocol favor mesoderm formation and are not ideal for non-mesodermal cell types, which would require different media and procedures. As mentioned in the introduction, this protocol involves the formation of embryoid bodies (EBs), which mimics the early stages of mouse embryogenesis. EB formation can be achieved in two different ways, hanging drop and reaggregation, which are described in the following section.

3.2.1 Isolation of mES Cells

The first part of the differentiation protocol requires the isolation of mES cells. Since MEFs do not proliferate, the starting number of ES cell colonies and their size are key parameters in order to obtain a cell suspension enriched for mES cells.

1. Use ES cells cultured for 2 days.
2. Aspirate medium and wash twice with 5 ml PBS.
3. Add 1.5 ml trypsin–EDTA solution and incubate for 5 min at 37 °C, 5 % CO_2.
4. Add 5 ml of MEFs medium to inactivate trypsin and resuspend eight to ten times.
5. Transfer cell suspension in a 15 ml tube and centrifuge 5 min at 300 × *g*.
6. Resuspend the pellet in 5 ml EB medium and place into a gelatin-coated T25 flask (Remember to aspirate gelatin completely). Incubate for 25 min at 37 °C, 5 % CO_2.
7. Remove the flask from incubator carefully. Pick up supernatant (which contains cells that did not adhere) slowly and transfer into a 15 ml tube. (This procedure is used to remove MEFs from the cellular suspension thus enriching for the ES cells). Take a 10 μl aliquot.
8. Count cell number using trypan blue to exclude dead cells.

3.2.2 EB Formation by Hanging Drop Method

1. Dilute cells in EB medium at the final concentration of 10^4 cells/ml.
2. Using a multichannel pipette dispense 10 μl drops of cell suspension in a 150 mm petri dish (about 400-drops/plate). To avoid any evaporation, add 5 ml of PBS in the lid of the petri dish.
3. Incubate at 37 °C, 5 % CO_2 for 2 days (put the plate upside down—the drops will hang from the top of the plate).
4. After 2 days, harvest the EBs using 5 ml of PBS and transfer in a 50 ml tube. Let the EBs to sit down for 5 min (if you have multiple plates pull them together).
5. Aspirate carefully the PBS (the pellet of EBs should be visible) and gently resuspend the EBs with fresh EB medium. If starting from 4 × 150 mm plates, resuspend EBs in 20 ml of EB media and split in 2 × 100 mm petri dishes.
6. Place the petri dishes on an orbital shaker at 60 RPM and culture at 37 °C, 5 % CO_2.
7. After 24 h add Doxycycline to the culture medium (final concentration 1 μg/ml). Mesoderm formation happens around day 3 and the differentiation can be evaluated by FACS staining using PDGFRα and FLK1.

8. At day 4, swirl the dishes to collect the EBs in the center, aspirate half medium using a 10 ml pipette and add 12.5 ml of fresh medium supplemented with doxycycline.
9. At day 5, EBs are ready for isolation of myogenic precursors (proceed to Subheading 3.4).

3.2.3 EB Formation by Reaggreagation Method

1. Dilute cells into EB medium ($0.5–1 \times 10^6$ cells in 25 ml for a 150 mm petri dish), place the dish on an orbital shaker at 60 RPM and culture at 37 °C, 5 % CO_2.
2. After 48 h, collect all EBs with a 5 or 10 ml pipette and transfer in a 50 ml tube. Let the EBs to sit down for 1–2 min (if you have multiple plates pull them together). Remove carefully the medium with the vacuum (the pellet of EBs should be visible) and gently resuspend the EB pellet in fresh EB medium. Replate the EBs in the same petri dish (remove the old medium before) and put back the dish on the shaker at 37 °C, 5 % CO_2.
3. After 24 h add doxycycline to the culture medium (final concentration 1 μg/ml). Mesoderm formation happens around day 3 and the differentiation can be evaluated by FACS staining using PDGFRα and FLK1.
4. At day 4, swirl the dishes to collect the EBs in the center, aspirate half medium using a 10 ml pipette and add 12.5 ml of fresh medium supplemented with doxycycline.
5. At day 5, EBs are ready for isolation of myogenic precursors (proceed to Subheading 3.4).

3.3 FACS-Mediated Isolation of Myogenic Precursors and Expansion

Myogenic progenitors can be isolated using fluorescence activated cell sorting (FACS). Before starting, please refer to the FACS manufacturer's instructions. All procedures, except sorting, should be performed in a laminar-flow cabinet to prevent sample contamination.

1. Harvest EBs in a 15 ml tube (EBs will decant very quickly, and therefore centrifuge is not necessary).
2. Aspirate medium and wash the pellet of EBs twice with 10 ml PBS.
3. Add trypsin–EDTA solution (estimate pellet size, usually 3 ml are sufficient for a 300 μl pellet) and place the tube a 37 °C water bath with continuous shaking (time depends on EB size—usually day 5 EBs require 1′30″).
4. Inactivate immediately trypsin by adding 4 volumes of FACS staining medium and resuspend 4/5 times (pipetting too many times will result in increased cell death and debris).
5. Spin down cells for 5 min at $300 \times g$.
6. Resuspend cells in PBS and filter through a 70 μm strain to remove cell clumps. Take a 10 μl aliquot for cell count and spin down the rest of the cells for 5 min at $300 \times g$.

7. Resuspend cells in FACS staining medium (volume depends on cell number—approximately 100 μl/10^6 cells).
8. Add Fc Block (1 μl/four million cells). Incubate for 5 min in ice.
9. Add 0.5 μl of each antibody per one million of cells and incubate for 20 min in ice.
10. Wash cells with PBS and spin down for 5 min at 300 × g (repeat this step twice).
11. Resuspend cells in FACS staining medium + propidium iodide (volume depends on cell number—approximately 200 μl/10^6 cells) and filter using a 70 μm strain.
12. Sort PDGFRα+/FLK1− cells from PI-gate using FACS machine; collect cells in a 15 ml tube containing 3 ml of EB medium.
13. After sorting, centrifuge tube for 5 min at 300 × g and plate cells on gelatin coated flasks in EB medium supplemented with 5 ng/ml murine basic-FGF (bFGF) and 1 μg/ml doxycycline. Plate 1.5×10^6 cells in a T25 flask.

Good myogenic progenitors (Fig. 1c) are highly proliferative and they are subcultured when they reach 80 % confluency (usually every 2 days). Around day 8–9, they should acquire the typical myoblast morphology (Fig. 1d). Too many big flat cells are sign of bad differentiation.

14. Myogenic progenitor expansion: wash with PBS, add trypsin–EDTA solution (1 ml for a T25), incubate for 1 min at 37 °C, stop using 4 ml of MEF medium and transfer in a 15 ml tube.
15. Centrifuge for 5 min at 300 × g and resuspend cell pellet in EB medium supplemented with 5 ng/ml murine basic-FGF (bFGF) and 1 μg/ml doxycycline. Plate cells in new gelatin-coated flasks (1:3–1:6 ratio) and incubate at 37 °C 5 % CO_2.

3.4 Myogenic Progenitor Transplantation

For proper handling and experiments with mice, please strictly follow your institution's guidelines for animal care and use. When animals are under anesthesia, always use an ophthalmic ointment in order to preserve their eyes, and provide a pain reliever any time they undergo surgical procedures. Moreover, once in cage after surgery, we recommend to place them on a heating pad for the duration of recovery.

3.4.1 Irradiation of Mice Hind Limbs

Irradiation of hind limbs in recipient mice can be used to enhance the extent of engraftment, since it inhibits the regenerative potential of endogenous satellite cells [21, 22], creating a less competitive environment for injected myogenic progenitors to home and proliferate. Before setting up an irradiation experiment, follow the irradiator manufacturer's instructions.

1. Anesthetize the mice with intraperitoneal (IP) injection of ketamine–xylazine solution at the final concentration of 80 mg/kg.
2. Once the mice are fully anesthetized, place them with a supine position on the exposure shelf, positioning their hind limbs at the desired distance from the shelf center, according to the irradiator type. Place a lead shield on top of the mice, leaving their hind limbs exposed (*see* **Note 3** for details).
3. Set the irradiator voltage, amperage, and time in order to obtain the suitable radiation dosage, e.g., 1200 rad [23], and run the experiment. The dosage must be adjusted according to mice size (*see* **Note 4**).
4. After irradiation, move the mice in their cage and monitor until fully recovered.

3.4.2 Intramuscular Cardiotoxin Injection in TA Muscle

After 24 h from irradiation, the recipient muscles have to be injured in order to induce the regenerative stimuli and create the proper niche for the pluripotent stem cells-derived myogenic progenitors to engraft [7, 17]. Cardiotoxin has extensively been used to induce injury in mice [24–26]. Please always wear proper PPE and refer to your institution's guidelines for correct use and disposal.

1. Anesthetize mice with IP injection of 80 mg/kg ketamine–xylazine solution and wait until fully unconscious.
2. Place the mice supine on a surgery pad and shave their hind limbs with hair clippers or similar device. Remove hair from the pad and keep away from the hind limbs.
3. Rub the hind limbs with Betadine swabs.
4. Grab the skin in the lower part of the limb with thin sterile tweezers and make a small cut with scissors. Enlarge the cut towards the knee in a vertical motion with dissection scissors, until at least 1/3 of the TA muscle is exposed (*see* **Note 5**).
5. Draw 15 μl of cardiotoxin 10 μM with a glass Hamilton syringe or with a disposable insulin syringe, and inject into the exposed TA muscle (Fig. 2a—*see* **Note 6** for details).
6. Suture the skin with non-absorbable braided silk suture, making at least two loose knots, and move the mice back to their cage.
7. Apply ophthalmic ointment onto mice eyes and provide pain reliever medication inside the water bottle. Keep the cage on a heating pad and monitor mice until fully conscious and sternal and for at least 3 h post-surgery.

3.4.3 Collection of Cultured Mouse Myogenic Progenitors for Transplantation

After sorting of PDGFRα+FLK1− myogenic progenitors, cells are cultured in EB media supplemented with Dox and bFGF on gelatin-coated plates (*see* Subheading 3.4). Around passage 2–3 and after counting cells, part of them can be frozen for backup at the concentration of 1×10^6 cells/vial (*see* **Note 7**) and part of

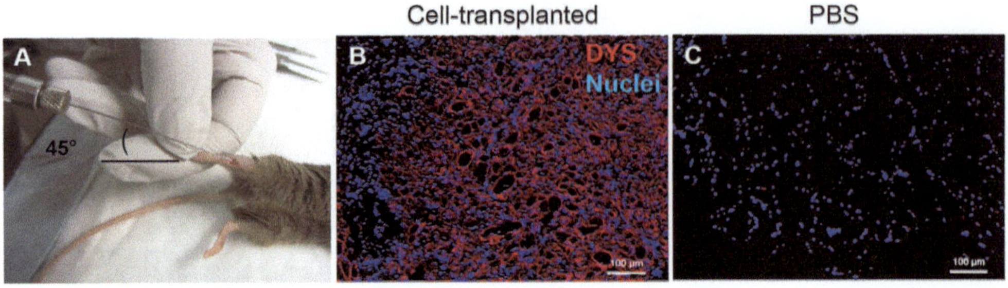

Fig. 2 Transplantation and engraftment of myogenic progenitors in recipient mice. (**a**) Representative picture showing the injection of myogenic progenitors with the Hamilton syringe in the TA of a recipient mouse, with the initial needle inclination of approximately 45°. (**b**) Representative immunostaining picture showing the Dystrophin positive staining (DMD—*red*) in a transversal section of myogenic progenitors-injected TA, indicating the successful engraftment. (**c**) Picture showing the same staining as in (**b**) in the PBS-injected contralateral TA. DAPI was used to counterstain nuclei (*Blue*). Scale bars: 100 μm

them can be plated to check the terminal myogenic differentiation capacity (*see* **Note 8**). On the day of transplantation, myogenic progenitors should not have exceeded the forth passage and 80% confluency. If myogenic progenitors become too confluent, unwanted terminal differentiation could potentially start, thus impairing the engraftment capacity.

1. Remove the media and wash once with D-PBS.
2. Add a sufficient amount of warm cell dissociation buffer, usually 10 ml/T75 flask.
3. Incubate cells for 10 min at 37 °C, checking under the microscope after 5 min in order to adjust the incubation time. Gently tap the flask in order to help cell detachment.
4. Add 10 ml of D-PBS and gently pipet cell suspension up and down until cell clumps are completely resuspended.
5. Spin down cells at 300 × *g* for 5 min and aspirate supernatant. Resuspend the cells in 2–4 ml of D-PBS, according to the pellet size. Filter the suspension with 70 μm nylon cell strainer, and count using trypan blue.
6. Transfer the desired amount of cells (e.g., 0.3–1 × 10^6 cells/TA/injection, *see* **Note 9**) in a new 15 ml conical tube and spin down again at 300 × *g* for 5 min. The rest of cell suspension can be seeded for terminal differentiation checking (*see* **Note 8**) and also frozen as backup (*see* **Note 7**).
7. Resuspend the cells in D-PBS at the final concentration of 30–50 × 10^6 cells/ml (0.3–0.5 × 10^6 cells/10 μl, *see* **Note 10**) and transfer them into a 1.5 ml tube. Place cells on ice until ready for transplantation (*see* **Note 11**).

3.4.4 Intramuscular Injection of Myogenic Progenitors into TA Muscle

After 24 h from the cardiotoxin-mediated injury, tibialis anterior muscles are ready to be injected with pluripotent stem cells-derived myogenic progenitors.

1. Anesthetize mice with IP injection of ketamine–xylazine as already described in Subheading 3.1.
2. Place the anesthetized mice on the surgical pad, rub the hind limbs with Betadine swabs and gently remove the previous suture with scissors (*see* **Note 12**).
3. Draw 10 μl from the cells-containing tube and inject into TA (*see* **Note 13**). Wait at least 30 s before pulling out the needle in order to allow the complete diffusion of cell suspension inside the muscle.
4. Suture the skin with non-absorbable braided silk suture, making at least two tight knots, and move the mice back in their cage, monitoring them as indicated in Subheading 3.2.

3.4.5 Harvesting of TA Muscle from Recipient Mice

After 4 weeks from transplantation, the injury-mediated regenerative process has been fully achieved [27] and the engraftment of injected myogenic progenitors can be assessed by immunohistochemical analysis on recipient muscles. Before removing the muscles, please refer to your institution's guidelines for animal euthanasia.

1. After euthanizing mice, spray down their hind limbs with 70% ethanol.
2. Label a cryomold and pour the tissue freezing medium until it reaches the desired volume to fully embed the muscles.
3. Place mice supine on a surgical pad and firmly pull up the skin immediately above the feet with the help of forceps and/or scissors.
4. After peeling the skin towards the knee, gently remove the fascia surrounding the TA muscle with thin tweezers.
5. Gently place the thin tweezers below the TA tendon and separate it from the underlying muscle, moving the tweezers towards the knee.
6. Keeping the tweezers between the TA and the underlying muscles, insert the scissors and make a cut at the tendon end. Gently cut the other end at the knee.
7. With the thin tweezers, grab the TA muscle from the tendon and place it inside the tissue freezing medium-containing mold. Choose the most convenient orientation for the following cryosections (*see* **Note 14**).
8. Pour a sufficient amount of isopentane inside a stainless steel beaker in order to create a thin layer at the bottom, then place the beaker inside a liquid nitrogen-containing Styrofoam box.
9. When the isopentane is completely solidified, place the cryomolds on top of it with the help of specimen forceps and wait 5 min for solidification.
10. Remove the cryomolds from the isopentane-containing beaker and store at −80 °C until ready to use.

3.4.6 Cryosections of TA Muscle from Recipient Mice

Before proceeding with cryosectioning, please refer to the cryostat manufacturer's instructions. Properly label a sufficient number of Superfrost slides and place them on a support nearby the cryostat chamber.

1. Remove the cryomolds from the −80 °C freezer and place them inside the cryostat chamber for about 30 min in order to equilibrate them.
2. Pour a small amount of tissue freezing medium on a sample stub and immediately place the embedded muscles on top. Wait until the sample is fused with the medium layer.
3. Place the sample stub inside the cryostat chuck and adjust the position until the sample reaches the knife holder.
4. Set the slide thickness at 10–12 μm, place the anti-roll plate on the knife holder and then carefully start rotating the handle to cut the embedded sample.
5. Carefully remove the anti-roll plate and place a Superfrost slide on the section, making it adhere. Repeat until a sufficient number of sections have been placed on the slide, then switch to another slide. Proceed until the desired number of sections has been cut (*see* **Note 15**).
6. Place the slides in a suitable box and store at −80 °C or immediately proceed to the immunohistochemical analysis.

3.4.7 Immunostaining of TA Muscle's Cryosections

The following protocol has been optimized for the staining of WT mouse ES/iPS-derived myogenic progenitors injected in dystrophic recipient mice.

1. Remove slides from −80 °C and let them dry at least for 30 min at RT. If immunofluorescence is performed immediately after sectioning, this step can be avoided.
2. Circumscribe the tissue sections with Super PAP pen (Life Technologies).
3. Rehydrate slides with D-PBS 1× for 5 min.
4. Remove the D-PBS and incubate for 20 min at RT with 0.3% Triton X-100 solution. This step allows the permeabilization of sarcolemma and facilitates the access of antibodies.
5. Remove Triton solution and incubate for 1 h at RT with blocking solution, in order to saturate the aspecific bindings. While blocking, prepare primary antibodies solutions for dystrophin staining (*see* **Note 16**).
6. Remove the blocking solution and incubate overnight at 4 °C with primary antibody solution. Incubate at least one slide with blocking solution alone as staining negative control.
7. The day after, remove the primary antibody solution and wash three times with D-PBS for 5 min. In the meantime, prepare

the secondary antibody solution and protect the tube with aluminum foil in order to prevent the bleaching of fluorophores.

8. Remove the last PBS wash and incubate for 45 min at RT with secondary antibody solution, taking care of protecting slides from the direct light.
9. Wash three times with PBS for 5 min and let briefly air-dry the slides after the last wash, paying attention to not overdry the tissue sections.
10. Place one to two drops of DAPI ProLong Gold mounting medium on the slides and carefully place a coverslip on top, taking care to minimize bubbles.
11. Let dry the slides for few hours at RT, then acquire pictures using an inverted fluorescence microscope or store the slides at −20 °C for future use. Successful delivery of ES-derived myogenic progenitors will result in the detection of large number of DMD+ fibers (Fig. 2b) when compared to the PBS-injected control (Fig. 2c).

4 Notes

1. Murine bFGF can be purchased from different companies as lyophilized protein or solution. Prepare 50 ng/μl stock solution and store aliquots at −20 °C (long term) or +4 °C (short term).
2. After adding gelatin, we store flasks at 37 °C without removing the coating solution, which we aspirate right before plating the cells.
3. When preparing the mice for X-irradiation, be careful in protecting the tail underneath the lead shield, and keep the TA muscles exposed placing a piece of scotch tape on mice feet, so that their soles adhere to the exposure shelf.
4. A radiation dosage between 1200 and 1800 rad directed to hind limbs is well tolerated in mice regardless of age [21, 23]. However, mild to severe circulatory and skin problems could arise, usually after a couple of weeks from the irradiation, in females and/or in smaller mice. For this reason, we suggest to adjust the dosage according to the experimental needs if these issues occur. Moreover, if mice survival becomes a major concern, we suggest to skip irradiation, despite this will increase the endogenous satellite cells competition.
5. The TA muscle is easily recognizable from its superficial position in the hind limb when mouse is supine. However, in order to individuate the correct site of injection, we suggest exposing the lower half of the muscle, starting from its distal

end so that the corresponding tendon and thus the TA are unequivocally visible.

6. We suggest using a Hamilton glass syringe instead of disposable insulin syringe since the latter has a shorter and thinner needle. We recommend inserting the needle with a 45° angle around the most distal half of the TA (Fig. 2a), then moving it parallel to the muscle towards the proximal end (the knee) in order to obtain a wider muscle coverage.
7. We usually freeze myogenic progenitors in a freezing medium made up of EB medium, 45 % FBS and 10 % DMSO.
8. To test differentiation capacity, we recommend plating myogenic progenitors into multi-wells (e.g., 24 wells at the density of 25,000–50,000 cells/cm^2), and supplement them with EB medium without Dox and bFGF for the following 7–10 days, when myotubes become clearly visible. In order to check for terminally differentiated cells, we suggest assessing the expression of MyoD, myogenin, and myosin heavy chain by immunocytochemical analysis.
9. When the recipient muscle has been preconditioned with both X-irradiation and CTX injection, we usually inject 300,000 mouse myogenic progenitors resuspended in 10 μl of D-PBS. However, if X-irradiation dosage has to be lowered or skipped (*see* **Note 4**), a higher amount of cells can be injected, usually up to 10^6.
10. We usually inject a small volume of cell suspension in one single site, e.g., 15 μl/muscle.
11. Myogenic progenitors can be prepared up to 2 h before the injection. However, we recommend preparing the cell suspension immediately before starting the surgery on mice. Moreover, we suggest to pipet carefully the suspension before injecting in order to avoid clumps, especially if they seated on ice for several min.
12. We usually try to remove the previous suture and carefully enlarge the same cut for cells injection.
13. We always include a PBS-injected muscle, usually the contralateral tibialis, in order to have an internal reliable negative control (Fig. 2c).
14. Muscle transverse sections allow observing the fiber size and nuclei position. We usually position the muscle perpendicular to the site of cut and mark that side in order to recognize it before sectioning. Then, we place the sample inside the chuck so that that marked side is perpendicular to the knife holder.

15. We recommend cutting the entire muscle to be able to analyze the engraftment. However, part of the sample can also be saved for further sectioning and for backup.
16. Anti-dystrophin antibodies can be purchased from several companies. We suggest using Leica antibody NCL-DYS 2 because it does not require a preliminary tissue fixation and it strongly reacts against human and rodents dystrophin C-terminal domain, while it does not recognize *mdx* muscles. However, it could recognize revertant *mdx* fibers, thus PBS-injected muscle sections should always be used in immunostaining experiment as negative control (*see* also **Note 11**). Alternatively, we also use rabbit polyclonal anti-dystrophin (Abcam) taking care of fixing the slides with cold acetone for 5–10 min before proceeding with the staining.

Acknowledgements

The Perlingeiro laboratory is supported by grants from the NIH (R01 AR055299), the Muscular Dystrophy Association (MDA #238127) and Parent Project Muscular Dystrophy (PPMD #00031645). A.M. was funded by a fellowship from Regenerative Medicine Minnesota (MRM 2015 PDSCH 003).

References

1. Collins CA, Olsen I, Zammit PS, Heslop L, Petrie A, Partridge TA, Morgan JE (2005) Stem cell function, self-renewal, and behavioral heterogeneity of cells from the adult muscle satellite cell niche. Cell 122(2):289–301. doi:10.1016/j.cell.2005.05.010, S0092-8674(05)00455-1 [pii]
2. Partridge TA, Morgan JE, Coulton GR, Hoffman EP, Kunkel LM (1989) Conversion of mdx myofibres from dystrophin-negative to -positive by injection of normal myoblasts. Nature 337(6203):176–179. doi:10.1038/337176a0
3. Sampaolesi M, Blot S, D'Antona G, Granger N, Tonlorenzi R, Innocenzi A, Mognol P, Thibaud JL, Galvez BG, Barthelemy I, Perani L, Mantero S, Guttinger M, Pansarasa O, Rinaldi C, Cusella De Angelis MG, Torrente Y, Bordignon C, Bottinelli R, Cossu G (2006) Mesoangioblast stem cells ameliorate muscle function in dystrophic dogs. Nature 444(7119):574–579. doi:10.1038/nature05282, nature05282 [pii]
4. Sampaolesi M, Torrente Y, Innocenzi A, Tonlorenzi R, D'Antona G, Pellegrino MA, Barresi R, Bresolin N, De Angelis MG, Campbell KP, Bottinelli R, Cossu G (2003) Cell therapy of alpha-sarcoglycan null dystrophic mice through intra-arterial delivery of mesoangioblasts. Science 301(5632):487–492. doi:10.1126/science.1082254, 1082254 [pii]
5. Tedesco FS, Hoshiya H, D'Antona G, Gerli MF, Messina G, Antonini S, Tonlorenzi R, Benedetti S, Berghella L, Torrente Y, Kazuki Y, Bottinelli R, Oshimura M, Cossu G (2011) Stem cell-mediated transfer of a human artificial chromosome ameliorates muscular dystrophy. Sci Transl Med 3(96):96ra78. doi:10.1126/scitranslmed.3002342, 3/96/96ra78 [pii]
6. Darabi R, Arpke RW, Irion S, Dimos JT, Grskovic M, Kyba M, Perlingeiro RC (2012) Human ES- and iPS-derived myogenic progenitors restore DYSTROPHIN and improve contractility upon transplantation in dystrophic mice. Cell Stem Cell 10(5):610–619. doi:10.1016/j.stem.2012.02.015, S1934-5909(12)00074-4 [pii]
7. Darabi R, Gehlbach K, Bachoo RM, Kamath S, Osawa M, Kamm KE, Kyba M, Perlingeiro RC (2008) Functional skeletal muscle regeneration from differentiating embryonic stem cells. Nat

Med 14(2):134–143. doi:10.1038/nm1705, nm1705 [pii]
8. Darabi R, Pan W, Bosnakovski D, Baik J, Kyba M, Perlingeiro RC (2011) Functional myogenic engraftment from mouse iPS cells. Stem Cell Rev 7(4):948–957. doi:10.1007/s12015-011-9258-2
9. Tedesco FS, Gerli MF, Perani L, Benedetti S, Ungaro F, Cassano M, Antonini S, Tagliafico E, Artusi V, Longa E, Tonlorenzi R, Ragazzi M, Calderazzi G, Hoshiya H, Cappellari O, Mora M, Schoser B, Schneiderat P, Oshimura M, Bottinelli R, Sampaolesi M, Torrente Y, Broccoli V, Cossu G (2012) Transplantation of genetically corrected human iPSC-derived progenitors in mice with limb-girdle muscular dystrophy. Sci Transl Med 4(140):140ra189. doi:10.1126/scitranslmed.3003541, 4/140/140ra89 [pii]
10. Takahashi K, Tanabe K, Ohnuki M, Narita M, Ichisaka T, Tomoda K, Yamanaka S (2007) Induction of pluripotent stem cells from adult human fibroblasts by defined factors. Cell 131(5):861–872. doi:10.1016/j.cell.2007.11.019, S0092-8674(07)01471-7 [pii]
11. Takahashi K, Yamanaka S (2006) Induction of pluripotent stem cells from mouse embryonic and adult fibroblast cultures by defined factors. Cell 126(4):663–676. doi:10.1016/j.cell.2006.07.024, S0092-8674(06)00976-7 [pii]
12. Sterneckert JL, Reinhardt P, Scholer HR (2014) Investigating human disease using stem cell models. Nat Rev Genet 15(9):625–639. doi:10.1038/nrg3764, nrg3764 [pii]
13. Buckingham M, Relaix F (2007) The role of Pax genes in the development of tissues and organs: Pax3 and Pax7 regulate muscle progenitor cell functions. Annu Rev Cell Dev Biol 23:645–673. doi:10.1146/annurev.cellbio.23.090506.123438
14. Relaix F, Rocancourt D, Mansouri A, Buckingham M (2005) A Pax3/Pax7-dependent population of skeletal muscle progenitor cells. Nature 435(7044):948–953. doi:10.1038/nature03594, nature03594 [pii]
15. Iacovino M, Bosnakovski D, Fey H, Rux D, Bajwa G, Mahen E, Mitanoska A, Xu Z, Kyba M (2011) Inducible cassette exchange: a rapid and efficient system enabling conditional gene expression in embryonic stem and primary cells. Stem Cells 29(10):1580–1588
16. Kyba M, Perlingeiro RC, Daley GQ (2002) HoxB4 confers definitive lymphoid-myeloid engraftment potential on embryonic stem cell and yolk sac hematopoietic progenitors. Cell 109(1):29–37, S0092867402006803 [pii]
17. Darabi R, Santos FN, Filareto A, Pan W, Koene R, Rudnicki MA, Kyba M, Perlingeiro RC (2011) Assessment of the myogenic stem cell compartment following transplantation of Pax3/Pax7-induced embryonic stem cell-derived progenitors. Stem Cells 29(5):777–790. doi:10.1002/stem.625
18. Sakurai H, Era T, Jakt LM, Okada M, Nakai S, Nishikawa S (2006) In vitro modeling of paraxial and lateral mesoderm differentiation reveals early reversibility. Stem Cells 24(3):575–586. doi:10.1634/stemcells.2005-0256, 2005-0256 [pii]
19. Magli A, Schnettler E, Rinaldi F, Bremer P, Perlingeiro RC (2013) Functional dissection of Pax3 in paraxial mesoderm development and myogenesis. Stem Cells 31(1):59–70. doi:10.1002/stem.1254
20. Magli A, Schnettler E, Swanson SA, Borges L, Hoffman K, Stewart R, Thomson JA, Keirstead SA, Perlingeiro RC (2014) Pax3 and Tbx5 specify whether PDGFRalpha+cells assume skeletal or cardiac muscle fate in differentiating embryonic stem cells. Stem Cells 32(8):2072–2083. doi:10.1002/stem.1713
21. Quinlan JG, Lyden SP, Cambier DM, Johnson SR, Michaels SE, Denman DL (1995) Radiation inhibition of mdx mouse muscle regeneration: dose and age factors. Muscle Nerve 18(2):201–206. doi:10.1002/mus.880180209
22. Wakeford S, Watt DJ, Partridge TA (1991) X-irradiation improves mdx mouse muscle as a model of myofiber loss in DMD. Muscle Nerve 14(1):42–50. doi:10.1002/mus.880140108
23. Arpke RW, Darabi R, Mader TL, Zhang Y, Toyama A, Lonetree CL, Nash N, Lowe DA, Perlingeiro RC, Kyba M (2013) A new immuno-, dystrophin-deficient model, the NSG-mdx(4Cv) mouse, provides evidence for functional improvement following allogeneic satellite cell transplantation. Stem Cells 31(8):1611–1620. doi:10.1002/stem.1402
24. Couteaux R, Mira JC, d'Albis A (1988) Regeneration of muscles after cardiotoxin injury. I. Cytological aspects. Biol Cell 62(2):171–182, 0248-4900(88)90034-2 [pii]
25. Mahdy MA, Lei HY, Wakamatsu J, Hosaka YZ, Nishimura T (2015) Comparative study of muscle regeneration following cardiotoxin and glycerol injury. Ann Anat 202:18–27. doi:10.1016/j.aanat.2015.07.002, S0940-9602(15)00103-X [pii]
26. Harris JB (2003) Myotoxic phospholipases A2 and the regeneration of skeletal muscles. Toxicon 42(8):933–945. doi:10.1016/j.toxicon.2003.11.011, S0041010103003313 [pii]
27. Charge SB, Rudnicki MA (2004) Cellular and molecular regulation of muscle regeneration. Physiol Rev 84(1):209–238. doi:10.1152/physrev.00019.2003, 84/1/209 [pii]

Chapter 15

Assaying Human Myogenic Progenitor Cell Activity by Reconstitution of Muscle Fibers and Satellite Cells in Immunodeficient Mice

Maura H. Parker

Abstract

Comparing the functional myogenic potential of various human cell populations is an important step in the preclinical evaluation of cell transplantation as a means to treat human muscle disease and degeneration. Culture systems allow one to gage the potential of cell populations to proliferate and undergo myogenic differentiation under specific conditions. An in vivo assay evaluates the ability of cells to differentiate and generate muscle fibers within a natural environment, and importantly, evaluates the potential of donor cells to reconstitute the satellite cell niche. In this chapter, we describe a technique for isolating mononuclear cells from human muscle samples, and a method of xenotransplantation for assessing functional myogenic potential in vivo. Briefly, cell populations are injected into the pre-irradiated and regenerating muscle of immunodeficient mice. The injected muscle is frozen at specific time points after injection and cryosections analyzed by immunostaining. The number of human dystrophin-expressing fibers and the number of Pax7$^+$ human lamin A/C$^+$ nuclei are determined, which provides a quantitative method of comparing the in vivo functional potential of cell populations.

Key words Xenotransplant, Muscle progenitors, Regeneration, Immunodeficient mice, Muscle fibers, Satellite cells, Human muscle

1 Introduction

Post-natal skeletal muscle growth and adult muscle regeneration require satellite cells, the presumptive stem cell of skeletal muscle. Satellite cells are positioned on the outside of muscle fibers, underneath the basal lamina [1]. The basal lamina provides a scaffold to orient cells during regeneration, and maintain the association of the satellite cells with the muscle fiber [2]. Breaking the interaction of satellite cells with the basal lamina and the muscle fiber is pivotal for generating mononuclear cell preparations for transplant. The first enzymatic isolations of skeletal muscle stem cells were performed more than 50 years ago to study skeletal muscle differentiation in culture [3–5]. A detailed description of the isolation

Michael Kyba (ed.), *Skeletal Muscle Regeneration in the Mouse: Methods and Protocols*, Methods in Molecular Biology, vol. 1460, DOI 10.1007/978-1-4939-3810-0_15, © Springer Science+Business Media New York 2016

method from chick embryo muscle is outlined by Konigsberg, and describes the basis of our method, as described in Subheading 3.2, including the preference for crude collagenase [5, 6].

Specifically, the muscle is minced using scalpel blades, followed by enzymatic digestion using crude collagenase. Sequential filtration steps removes muscle fiber fragments and results in a mixed population of mononuclear cells that include satellite cells, fibroblasts, blood-derived cells, and other as-yet unidentified cells. Once isolated from the muscle, this population of cells can be analyzed in a variety of ways. Flow cytometry can separate the mixed population of cells into subpopulations and identify cell surface markers of satellite cells. Culturing the cells can assess proliferation and differentiation potential, and demonstrate lineage progression. However, cultured cells are not within their natural environment, and results from cultured satellite cell-derived myoblasts may not accurately predict in vivo functional potential.

Assessing lineage progression and functional potential is best studied in vivo. The ideal system would be to analyze cellular function within a human system; yet, for obvious reasons, this is not possible. Therefore, we use the skeletal muscle of mice as a model of the human in vivo skeletal muscle environment. Successful mouse-to-mouse transplant of minced skeletal muscle and mononuclear cell preparations provides the platform for establishing xenotransplant to assess functional myogenic capacity [7–10]. Indeed, xenotransplant of human and canine cells has been effectively used to quantitate differences in in vivo function of different cell populations [11–17].

These studies have provided the basis for the transplantation regimen described below in Subheading 3.3, including pre-irradiation of host murine muscle before cell transplant, as described in Subheading 3.1. Pre-irradiation of the host muscle with 18 Gy 3–4 days before cell injection significantly enhances donor cell contribution by preventing host contribution to regeneration [18]. However, in our protocol, the host muscle is pre-irradiated with 12 Gy, since we use NOD/SCID immunodeficient mice, and we irradiate only 1 day before cell transplant. Moreover, we inject barium chloride to induce degeneration, which is not required in *mdx* muscle since dystrophic muscle is already in a state of degeneration.

Assessing in vivo functional capacity is achieved by measuring the contribution of cells to muscle fibers and to the host satellite cell niche. As described in Subheadings 3.4 and 3.5, we can distinguish mouse and human muscle fibers using an antibody specific for human dystrophin, and distinguish mouse and human nuclei using an antibody specific for human lamin A/C [16]. The number of dystrophin-expressing muscle fibers indicates the differentiation potential of the donor human cells, whereas the number of nuclei co-expressing Pax7 and human lamin A/C is a measure of the ability of the cells to reconstitute the satellite cell niche.

Further evolution of these protocols is the beginning step towards developing muscle cell transplantation as a means to treat human muscle disease and degeneration. Determining the best human cell population for transplant requires an assay to measure functional potential. Culture systems allow one to ask about the potential of cell populations to undergo myogenic differentiation under specific conditions. An in vivo assay evaluates the ability of cells to differentiate and generate muscle fibers within a natural environment, and importantly, evaluates the potential of donor cells to reconstitute the satellite cell niche.

2 Materials

1. NOD/SCID mice (7–10 weeks of age).
2. Cesium-137 source for irradiation.
3. Holder for mice—to keep mice in one position and only expose hindlimbs to radiation.
4. Ketamine–xylazine, or other injectable anesthetics.
5. 70% ethanol.
6. 1.2% barium chloride.
7. 0.5 cc insulin syringes (or other small volume syringe with ½″ 27-G needle).
8. Dulbecco's phosphate buffered saline (D-PBS)—Ca^{2+} and Mg^{2+} free.
9. Sterilized 10-cm glass petri dishes.
10. Sterilized paper towel or Kimwipes.
11. Scalpel blade holders with #11 blades.
12. Dulbecco's modified eagle's medium (DMEM)—high glucose.
13. Collagenase, Type 4 (Worthington Biochemical is preferred—*see* **Note 4**).
14. 15-ml conical polystyrene tubes.
15. Parafilm®.
16. Water bath set at 37 °C; preferably a shaking water bath.
17. Plastic disposable Pasteur pipettes.
18. Pipette-aid.
19. 5-ml plastic pipettes.
20. 20-ml syringes.
21. 16½ and 18½ G needles.
22. Cell strainers—75 and 40 μm.

23. 50-ml conical polystyrene tubes.
24. Centrifuge that can spin 50-ml and 15-ml tubes.
25. Inverted phase-contrast microscope.
26. P200 pipettor and appropriate tips.
27. Sterile microfuge tubes.
28. Hemocytometer or other cell counting device.
29. Cryomolds (10 mm × 10 mm × 5 mm).
30. OCT compound.
31. Aluminum block (8″ × 4″ × 4″).
32. Styrofoam box to hold aluminum block.
33. Liquid nitrogen.
34. −80 °C freezer.
35. Cryostat.
36. Superfrost Plus slides.
37. Acetone, precooled to −20 °C.
38. 4 % paraformaldehyde.
39. Glass staining jars (e.g., Coplin jars).
40. Staining chamber.
41. Hydrophobic marker.
42. Paper towels.
43. Blocking buffer: 2 % goat serum, 1 % BSA, 0.1 % cold fish skin gelatin, 0.05 % sodium azide, 1× D-PBS.
44. 1° antibody dilution buffer: 1 % BSA, 0.1 % cold fish skin gelatin, 0.05 % sodium azide, 1× D-PBS.
45. TBS: 20 mM Tris-HCl (pH 7.6), 150 mM NaCl.
46. TBS-Tw: 20 mM Tris-HCl (pH 7.6), 150 mM NaCl, 0.05 % Tween 20.
47. TBS-Tr: 20 mM Tris-HCl (pH 7.6), 150 mM NaCl, 0.25 % Triton X-100.
48. Human-specific lamin A/C antibody.
49. Human-specific dystrophin antibody (MANDYS102 or MANDYS107; Developmental Studies Hybridoma Bank).
50. Pan dystrophin antibody (MANEX1A; Developmental Studies Hybridoma Bank).
51. Anti-Pax7 antibody (PAX7; Developmental Studies Hybridoma Bank).
52. Anti-laminin antibody.
53. Secondary antibodies conjugated with fluorescent dyes.
54. 10 mg/ml Hoechst 33342.

55. ProLong Gold Antifade with API.
56. Glass coverslips.
57. Fluorescence microscope.

3 Methods

3.1 Irradiate Mice—Day 0

1. Anesthetize mice with an appropriate dose of injectable anesthesia, such as ketamine, xylazine (*see* **Note 1**). Once mice do not respond to stimulation (e.g., foot pinching), place the mice in a specially designed holder that permits exposure of one hindlimb to the beam of irradiation, while keeping the remaining body outside of the beam. Place the holder in an irradiator, such as the Markl ^{137}Cs irradiator from Shepherd and Associates.
2. Set the irradiator to provide a dose of 12 Gy of irradiation, at a rate of 0.7 Gy/min (*see* **Note 2**). Ensure that the irradiator has a source of fresh air for the mice as the duration of the irradiation is approximately 17 min.
3. When the irradiation is complete, remove the mice from the holder, and spray the irradiated lower hind limb with 70% ethanol. Inject the tibialis anterior muscle with 50 μl of 1.2% barium chloride using an insulin syringe, or syringe with an attached ½″ 27-G needle (*see* **Note 3**). Specifically, insert the needle into the skin near the foot, and parallel to the length of the tibialis anterior muscle, and push the needle gently into the muscle, so that the tip is close to the distal portion of the muscle. Inject slowly as you pull out the needle.
4. Monitor mice until fully recovered from anesthesia, and mobile.

3.2 Mononuclear Cell Isolation—Day 1

1. Place the skeletal muscle sample in sterile D-PBS and keep on wet ice until cell isolation. All remaining steps should be performed in a laminar flow hood, taking precautions to maintain sterility.
2. Remove the muscle from D-PBS and place in a 10-cm sterilized glass dish. Remove as much connective tissue and fat as possible. These will appear white against the pink or red of the muscle fibers. Cut the muscle along the length of the fibers, such that new pieces of muscle are approximately 0.2–0.5 cm wide. Place muscle pieces in a new 10-cm glass dish containing enough D-PBS to cover the muscle, and gently swirl.
3. Remove the muscle pieces and dab dry on sterilized paper towel or Kimipes. Transfer pieces to a sterile 10-cm glass dish. Using two scalpels with #11 blades attached, chop the muscle using opposing strokes, and continue until the muscle adopts

a thick paste-like appearance, which generally requires at least 5 min of continuous action.

4. Prepare two 15-ml conical polystyrene tubes containing 400 U/ml of collagenase in DMEM in a total volume of 10 ml. Shake the tubes vigorously, and wrap the caps securely with Parafilm® (*see* **Notes 4** and **5**). Incubate tubes in a 37° water bath for 60 min. Vigorously shake the tubes every 10 min (*see* **Note 6**).
5. Using a plastic Pasteur pipette, triturate the muscle by pipetting up and down 20 times. Repeat using a 5-ml plastic pipette, being careful not to breach the cotton plug barrier (*see* **Note 7**).
6. Transfer the contents of the two 15-ml conical tubes to a single 50-ml conical polystyrene tube. Triturate the sample by passing the sample 20 times through a 16½ G needle attached to a 20-ml syringe, and repeat using an 18½ G needle attached to a 20-ml syringe.
7. Place a 70-μm cell strainer in a 50-ml conical polystyrene tube and wet the membrane with 5 ml of D-PBS. Add 10 ml of D-PBS to the muscle slurry and apply the mixture to the membrane using a 5-ml pipette or a plastic Pasteur pipette. After the sample has passed through completely, wash the membrane with at least 5 ml of D-PBS (*see* **Note 8**).
8. Place a 40-μm cell strainer in a 50-ml conical polystyrene tube and wet the membrane with 5 ml of D-PBS. Apply the flow-through from **step 7** to the membrane. After sample has passed through completely, wash the membrane with at least 5 ml of D-PBS.
9. Remove the filter, and add enough D-PBS to bring the total volume to 50 ml. Cap the tube and centrifuge the sample at $478 \times g$ for 10 min at room temperature.
10. Carefully remove the supernatant by aspiration, and flick the tube vigorously to dislodge the cell pellet. Resuspend the cell pellet in 10 ml of D-PBS and transfer to a 15-ml conical polystyrene tube (*see* **Note 9**).
11. Centrifuge at $212 \times g$ for 5 min at room temperature. Remove the supernatant by aspiration, and flick the tube to dislodge the cell pellet. Resuspend the pellet in 1–2 ml of D-PBS, depending on the size of the pellet.
12. Count the cells using a hemocytometer and determine the concentration of cells in the preparation. Prepare sterile microfuge tubes with 25–50 μl of cells and keep on ice until transplant. Cryopreserve the remaining cells in FBS + 10% DMSO.

3.3 Cell Transplant into Mice—Day 1

1. Anesthetize mice. We normally use isoflurane for this step, since the mice recover most quickly with this form of anesthesia. Using an insulin syringe, or similar syringe with a ½″ 27-G needle, pull the 50 μl volume of cells into a single syringe.
2. Spray lower hindlimb with 70% ethanol, and mat down fur to expose skin. Insert the needle into the skin near the foot, and parallel to the length of the tibialis anterior muscle, and push the needle gently into the muscle, so that the tip is close to the distal portion of the muscle. Inject the cells slowly as you pull out the needle.
3. Monitor mice until fully recovered from anesthesia, and mobile.
4. Monitor mice daily for a minimum of 4 weeks, and preferably for at least 8 weeks. At the desired time point, euthanize the mice using an appropriate and approved technique, such as CO_2 overdose. Remove the injected muscle, and freeze the muscle in OCT in plastic molds, using an aluminum block submerged in liquid nitrogen. Store the muscle at −80 °C.

3.4 Analysis of Donor Muscle Fiber Generation

1. Warm frozen muscles for 30 min in a cryostat set to −18 °C. Cut cryosections from the distal to the proximal end of the muscle and adhere four serial sections to each Superfrost slide. Discard 16 sections between each slide, so that the corresponding sections between slides will represent a separation of approximately 200 μm. A typical mouse TA muscle will yield 25 slides using this technique. The discarded sections can be placed in a precooled microfuge tube for isolation of DNA, RNA, and/or protein.
2. Fix sections from even numbered slides in acetone in a glass staining jar at −20 °C for 10 min (*see* **Note 10**). Remove the slides from the acetone and lay flat in a staining chamber. Dry the slides at room temperature for 30 min. Encircle each section with a hydrophobic marker, and dry for 10 min. Rehydrate tissue sections by submerging slides in D-PBS for 10 min using a glass staining jar. Replace the D-PBS and incubate for an additional 10 min.
3. Remove the slides from D-PBS, and lay flat in a staining chamber. Carefully add blocking buffer to cover each section, and incubate for 1 h at room temperature.
4. Remove blocking buffer by tapping the long edge of the slides on paper towels on the benchtop. Carefully add primary antibodies diluted in 1° antibody dilution buffer to cover each section, and incubate for 1 h at room temperature. Primary antibodies include anti-human dystrophin, anti-human lamin A/C, anti-dystrophin (pan antibody), and one primary antibody is used per section. The fourth section is a negative control, and is incubated with a nonspecific primary antibody.

5. Remove primary antibodies by tapping the long edge of the slides on paper towels on the benchtop. Wash tissue sections by submerging slides in D-PBS for 5 min using a glass staining jar. Replace the D-PBS and incubate for an additional 5 min. Repeat for a total of three 5-min washes.
6. Remove the slides from the D-PBS, and lay flat in a staining chamber. Carefully add secondary antibody diluted in D-PBS to cover each section, and incubate for 1 h at room temperature.
7. Remove secondary antibodies by tapping the long edge of the slides on paper towels on the benchtop. Wash tissue sections by submerging slides in D-PBS for 5 min using a glass staining jar. Replace the D-PBS and incubate for an additional 5 min. Repeat.
8. Remove the slides from the D-PBS, and lay flat in a staining chamber. Carefully add Hoechst 33342 or DAPI diluted in D-PBS and incubate for 10 min at room temperature. Remove the nuclear counterstain by tapping the long edge of the slides on paper towels on the benchtop. Carefully use an aspirator to remove as much liquid as possible without disturbing the sections. Mount and coverslip sections with ProLong Gold Antifade (*see* **Note 11**).
9. Visualize staining with a fluorescence microscope. Engraftment of cells will be quantitatively assessed in two ways: (1) determining the average number of nuclei expressing human lamin A/C per cross section, indicative of overall donor cell survival, and (2) the average number of myofibers expressing human dystrophin per cross section, indicative of donor cell differentiation. The average for each injected muscle will be determined using a minimum of three cross sections from each injected muscle, from within the region of highest engraftment and covering a distance of approximately 800–1200 μm. The averages from all of the muscles injected will be used to calculate the overall average and standard deviation, and will be used in the *t*-test to calculate *p*-values (*see* **Note 12**).

3.5 Analysis of Donor Engraftment to the Satellite Cell Niche

1. Using the odd-numbered slides, encircle each section with a hydrophobic marker, and dry for 10 min. Fix sections in 4% PFA in a glass staining jar at room temperature for 15 min. Remove the slides from the PFA and wash tissue sections by submerging slides in TBS-Tw for 5 min using a glass staining jar. Replace the TBS-Tw and incubate for an additional 5 min. Repeat for a total of three 5-min washes.
2. Permeabilize by incubating sections in TBS-Tr in a glass staining jar for 10 min. Replace the TBS-Tr and incubate for an additional 10 min.

3. Remove the slides from the TBS-Tr and wash tissue sections by submerging slides in TBS-Tw for 5 min using a glass staining jar.
4. Follow **steps 3–8** of Subheading 3.4, except that the primary antibodies for each section will be as follows: (1) anti-Pax7 + anti-laminin, (2) anti-lamin A/C + anti-laminin, (3) anti-Pax7 + anti-lamin A/C + anti-laminin, and (4) negative control.
5. Visualize staining with a fluorescence microscope. Engraftment of human cells to the satellite cell niche will be quantitatively assessed by determining the average number of nuclei co-expressing human lamin A/C and Pax7 that reside underneath laminin per cross section. The average for each injected muscle will be determined using a minimum of three cross sections from each injected muscle, from within the region of highest engraftment and covering a distance of approximately 800–1200 μm. The averages from all of the muscles injected will be used to calculate the overall average and standard deviation, and will be used in the *t*-test to calculate *p*-values.

4 Notes

1. The mice will need to be immobile for at least 20 min, so use an appropriate anesthetic and dose of anesthesia. If mice awaken during the procedure, there is a risk of total body exposure, which is lethal.
2. The muscle of immune competent mice requires 18 Gy of focal irradiation to prevent host contribution to regeneration; however, NOD/SCID mouse muscle requires only 12 Gy. Importantly, the most effective rate of delivery of irradiation is 0.7 Gy/min [19]. It is important to determine the lowest dose of irradiation required for your particular strain of mice. This can be accomplished by irradiating mice with doses ranging from 8 to 20 Gy, using 2 Gy increments. After irradiation, inject the muscles with barium chloride as outlined in **step 3** of Subheading 3.1. Remove the irradiated muscle 7 days after injection, and freeze in OCT for cryosectioning. Stain cryosections from the middle of the muscle for expression of developmental myosin heavy chain (devMyHC), a marker of regeneration in adult skeletal muscle. Expression of devMyHC indicates that endogenous host cells were able to participate in regeneration, and that the dose of irradiation was inadequate. The lowest dose that prevents expression of devMyHC is the dose that should be used for transplantation of donor cells.
3. The tibialis anterior (TA) is used for transplant because it is easily accessible and injection with cells does not require exposing the muscle with an incision. When the mouse is resting on its back, the TA muscle is positioned lateral to the tibia. By

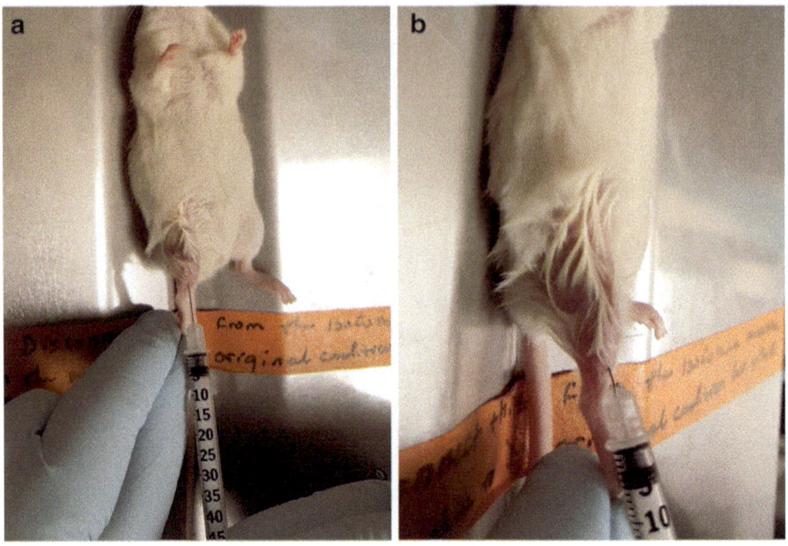

Fig. 1 Injection technique to maximize the amount of mouse tibialis anterior muscle reached. (**a**) The tip of the needle of an insulin syringe is placed to the left of the tibia of the right hindlimb. Specifically, the needle is inserted into the skin near the foot, and parallel to the length of the tibialis anterior muscle. (**b**) The needle is inserted as shown, remaining parallel to the muscle, until approximately ¾ of the needle is within the muscle. The needle is pulled out slowly as the material is injected, so as that the entire length of the muscle is injected

feeling for the tibia, you can easily locate the adjacent muscle. Injecting along the length of the tibialis anterior (TA) muscle is necessary to properly distribute donor cells. *See* Fig. 1 for an example of how to position the needle. Moreover, having a reproducible technique will be important for using the method to quantitatively compare engraftment potential. Mice will favor the injected leg for a short period of time after recovery from anesthesia; however, this normally resolves within an hour.

4. The collagenase is one of the most critical parameters for successful isolation of mononuclear cells from skeletal muscle. We use a partially purified preparation of collagenase from *Clostridium histolyticum*, which exhibits lot-to-lot variation. Type 4 collagenase from Worthington Biochemical is low in tryptic activity, which reduces cleavage of important cell surface receptors and enhances muscle cell survival; however, the level of clostripain is generally higher, as this enzyme digests the non-collagen components of the basal lamina, releasing the satellite cells. It is essential that lots of collagenase are tested and compared to choose the best lot. Once the best lots are identified, you can purchase specific lots based on enzyme activity.

5. A typical biopsy obtained from an open surgical procedure is normally 1 cm × 2 cm × 0.5 cm, or 1 cm^3. This volume of muscle is suitable for dividing between two 15-ml tubes. Adjust the number of tubes to the amount of muscle. Attempting to digest a large amount of muscle in a single 15-ml tube will result in a low cell yield.
6. The tubes can be placed upright in the water bath, but the digestion is more efficient if the tubes are laid on their side. Alternatively, the tubes can be placed upright in a shaking water bath. To prevent contamination, ensure the tube caps are securely fastened and tightly wrapped in Parafilm®. Check the muscle every 10 min, and vigorously shake the tube. The digest is complete when the muscle pieces are very small and the medium looks cloudy.
7. If it is difficult to triturate the muscle with the 5 ml pipette, return the muscle to the 37° water bath for an additional 10–15 min, and repeat the trituration process.
8. If the muscle biopsy was large or fibrous, it can be difficult to filter the slurry. A plastic Pasteur pipette can be used to pipette the slurry up and down on top of the membrane to encourage filtration. Alternatively, the samples can be split between two separate cell strainers and the filtrates combined at the end of **step 10**.
9. The cell pellet can be loosely packed, and care should be taken while aspirating the supernatant so as not to lose any of the cells. Flicking the tube to dislodge the pellet is a gentle means of facilitating cell suspension in D-PBS.
10. Stain the four sections on each slide as follows: section #1—isotype control; section #2—pan anti-dystrophin; section #3—anti-human lamin A/C; section #4—anti-human dystrophin. The section stained for anti-human dystrophin can be co-stained with anti-developmental myosin heavy chain (devMyHC) to highlight fibers in regeneration. Initially stain slide # 6, 12, and 18. Evaluate staining using a fluorescence microscope. Subsequently, stain even numbered slides surrounding the slide with the highest engraftment. Use cryosections from murine and human muscle as negative and positive controls, respectively. Lamin A/C and devMyHC antibodies are normally used at 1:100–1:250 and dystrophin antibodies (hybridoma supernatants) at 1:40. *See* Fig. 2 for an example of staining.
11. Follow the manufacturer's protocol for using ProLong Gold Antifade. Any liquid needs to be completely removed before mounting with ProLong Gold. Use an aspirator with a glass Pasteur pipette and pipette tip attached to carefully remove the D-PBS without disturbing the tissue section.

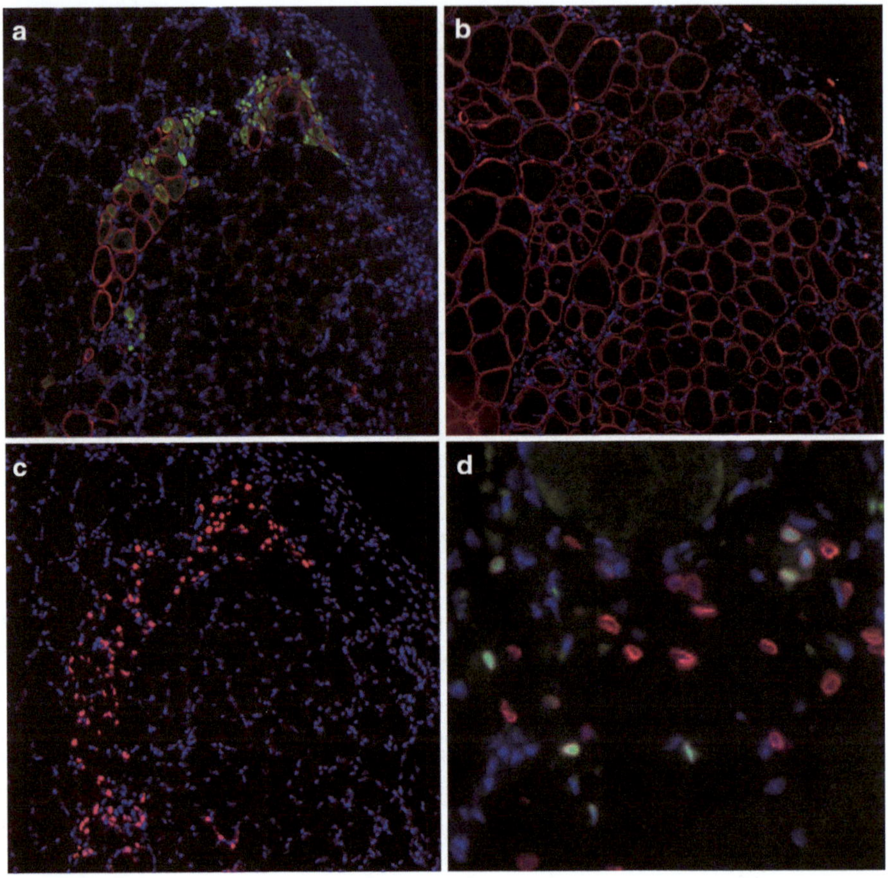

Fig. 2 Representative images of immunostaining cross sections of mouse skeletal muscle injected with 5×10^4 freshly isolated human muscle-derived cells. Cross sections were stained with anti-human dystrophin (*red*) and anti-developmental myosin heavy chain (*green*) (**a**), pan anti-dystrophin (*red*) (**b**), anti-human lamin A/C (*red*) (**c**), anti-Pax7 (*green*) and anti-human lamin A/C (*red*) (**d**), and Hoechst 33342 (*blue*) to stain nuclei. The sections shown in (**a**), (**b**), and (**c**) are from serial cross sections. The section shown in (**d**) was photographed at a higher magnification

12. Injection of 5×10^4 human cells should result in muscle fibers expressing human dystrophin and nuclei expressing canine lamin A/C in recipient mouse muscle 4–8 weeks after cell injection. An absence of staining suggests that the cell dose was not adequate, the irradiation was insufficient, or the injection technique was poor. It is important to note that the level of engraftment may vary between donor cell populations, and each cell population should be titrated by varying the cell dose, and measuring engraftment at each cell dose.

References

1. Bischoff R (1986) Proliferation of muscle satellite cells on intact myofibers in culture. Dev Biol 115(1):129–139
2. Sanes JR (2003) The basement membrane/basal lamina of skeletal muscle. J Biol Chem 278(15):12601–12604
3. Konigsberg IR, McElvain N, Tootle M, Herrmann H (1960) The dissociability of deoxyribonucleic acid synthesis from the development of multinuclearity of muscle cells in culture. J Biophys Biochem Cytol 8:333–343
4. Konigsberg IR (1960) The differentiation of cross-striated myofibrils in short term cell culture. Exp Cell Res 21:414–420
5. Rinaldini LM (1959) An improved method for the isolation and quantitative cultivation of embryonic cells. Exp Cell Res 16(3):477–505
6. Konigsberg IR (1979) Skeletal myoblasts in culture. Methods Enzymol 58:511–527
7. Carlson BM (1968) Regeneration of the completely excised gastrocnemius muscle in the frog and rat from minced muscle fragments. J Morphol 125(4):447–472
8. Partridge TA, Sloper JC (1977) A host contribution to the regeneration of muscle grafts. J Neurol Sci 33(3):425–435
9. Partridge TA, Grounds M, Sloper JC (1978) Evidence of fusion between host and donor myoblasts in skeletal muscle grafts. Nature 273(5660):306–308
10. Partridge TA, Morgan JE, Coulton GR, Hoffman EP, Kunkel LM (1989) Conversion of mdx myofibres from dystrophin-negative to -positive by injection of normal myoblasts. Nature 337(6203):176–179
11. Dellavalle A, Sampaolesi M, Tonlorenzi R, Tagliafico E, Sacchetti B, Perani L, Innocenzi A, Galvez BG, Messina G, Morosetti R et al (2007) Pericytes of human skeletal muscle are myogenic precursors distinct from satellite cells. Nat Cell Biol 9(3):255–267
12. Meng J, Adkin CF, Xu SW, Muntoni F, Morgan JE (2011) Contribution of human muscle-derived cells to skeletal muscle regeneration in dystrophic host mice. PLoS One 6(3):e17454
13. Meng J, Chun S, Asfahani R, Lochmuller H, Muntoni F, Morgan J (2014) Human skeletal muscle-derived CD133(+) cells form functional satellite cells after intramuscular transplantation in immunodeficient host mice. Mol Ther 22(5):1008–1017
14. Negroni E, Riederer I, Chaouch S, Belicchi M, Razini P, Di SJ, Torrente Y, Butler-Browne GS, Mouly V (2009) In vivo myogenic potential of human CD133+ muscle-derived stem cells: a quantitative study. Mol Ther 17(10):1771–1778
15. Parker MH, Loretz C, Tyler AE, Duddy WJ, Hall JK, Olwin BB, Bernstein ID, Storb R, Tapscott SJ (2012) Activation of notch signaling during ex vivo expansion maintains donor muscle cell engraftment. Stem Cells 30(10):2212–2220
16. Parker MH, Loretz C, Tyler AE, Snider L, Storb R, Tapscott SJ (2012) Inhibition of CD26/DPP-IV enhances donor muscle cell engraftment and stimulates sustained donor cell proliferation. Skelet Muscle 2(1):4
17. Zhu CH, Mouly V, Cooper RN, Mamchaoui K, Bigot A, Shay JW, Di Santo JP, Butler-Browne GS, Wright WE (2007) Cellular senescence in human myoblasts is overcome by human telomerase reverse transcriptase and cyclin-dependent kinase 4: consequences in aging muscle and therapeutic strategies for muscular dystrophies. Aging Cell 6(4):515–523
18. Morgan JE, Hoffman EP, Partridge TA (1990) Normal myogenic cells from newborn mice restore normal histology to degenerating muscles of the mdx mouse. J Cell Biol 111(6 Pt 1):2437–2449
19. Gross JG, Bou-Gharios G, Morgan JE (1999) Potentiation of myoblast transplantation by host muscle irradiation is dependent on the rate of radiation delivery. Cell Tissue Res 298(2):371–375

Chapter 16

Methods for Mitochondria and Mitophagy Flux Analyses in Stem Cells of Resting and Regenerating Skeletal Muscle

Laura García-Prat, Marta Martínez-Vicente, and Pura Muñoz-Cánoves

Abstract

Mitochondria generate most of the cell's supply of ATP as a source of energy. They are also implicated in the control of cell's growth and death. Because of these critical functions, mitochondrial fitness is key for cellular homeostasis. Often, however, mitochondria become defective following damage or stress. To prevent accumulation of damaged mitochondria, the cells clear them through mitophagy, which is defined as the selective degradation of mitochondria by autophagy (the process for degradation of long-lived proteins and damaged organelles in lysosomes). Recently, constitutive mitophagic activity has been reported in quiescent muscle stem cells (satellite cells), which sustain regeneration of skeletal muscle. In response to muscle damage, these cells activate, expand, and differentiate to repair damaged myofibers. Mitophagy was shown to be required for maintenance of satellite cells in their healthy quiescent state. Conversely, damaged mitochondria accumulated in satellite cells with aging and this was attributed to defective mitophagy. This caused increased levels of reactive oxygen species (ROS) and loss of muscle stem cell regenerative capacity at old age. In this chapter, we describe different experimental strategies to evaluate mitochondria status and mitophagy in muscle stem cells from mice. They should improve our ability to study muscle stem homeostasis in adult life, and their loss of function in aging and disease.

Key words Skeletal muscle regeneration, Stem cell, Satellite cell, Mitochondria, Mitophagy, Autophagy, Mitocondria membrane potential, Reactive oxygen species (ROS)

1 Introduction

Mitochondria are the main source of energy for the cell and key modulators of cell death. For these reasons, cells possess multiple regulatory and protective mechanisms to maintain them in a healthy state at molecular and network levels. A major mechanism by which cells maintain fit mitochondria is through clearance of damaged ones via a targeted process known as selective mitochondrial autophagy, or mitophagy [1, 2]. Autophagy is a process whereby cytoplasmic cell components are degraded by the lysosome [3–7]. Mitophagy is the only known pathway through which complete, damaged and dysfunctional mitochondria are eliminated [8] and was first described in the 1950s [9].

Michael Kyba (ed.), *Skeletal Muscle Regeneration in the Mouse: Methods and Protocols*, Methods in Molecular Biology, vol. 1460, DOI 10.1007/978-1-4939-3810-0_16,

Mitochondria exist as a highly dynamic intracellular pool that undergo constant fission (splitting) and fusion (joining) to control the mitochondrial network shape. Additionally, energy production in the cell is controlled by biogenesis of de novo mitochondria and elimination of the unwanted mitochondria by mitophagy. The accurate and timely degradation of dysfunctional mitochondria by mitophagy and the maintenance of the appropriate mitochondrial pool is essential for their quality and quantity control and consequently for cellular homeostasis and survival.

Mitophagy is responsible for basal mitochondrial turnover, and can also be induced in certain pathophysiological conditions. Indeed, mitophagy is activated during the development and differentiation of certain cell types (like maturation of erythrocytes and fertilized oocytes) but also functions as a stress-response mechanism (like in response to depolarization of the mitochondrial membrane or hypoxia).

In recent years, different molecular mechanisms for mitophagy have been described and it appears that different mechanisms involving numerous proteins are activated in response to different conditions or stressors [10, 11]. However, all these mechanisms follow a similar pattern and involve adapter molecules that act as a bridge between the mitochondria to be eliminated and LC3 protein present in the autophagosome membrane. Accordingly, all mitophagy adapters contain an LC3-interacting region (LIR) motif to bind LC3 in the autophagosome membrane, which during the mitophagy process grows around mitochondria, engulfing them within the autophagosome. Subsequent fusion with lysosomes eliminates the targeted mitochondria [12] (Fig. 1).

Some mitophagy adapters are outer mitochondrial membrane (OMM) proteins like Nix [13], BNPI3 [14, 15], and FUNDC1 [16] and act as elimination signals on the mitochondria surface. Furthermore, the adapter role can also be carried out by cytosolic proteins like p62 [17], NBR1 [18], NDP52 [19], and optineurin [19, 20], which contain the LIR motif together with a ubiquitin-associated (UBA) domain to bind polyubiquitinated proteins (mostly mediated by Parkin-ubiquitin activity) on the mitochondrial membrane [21–23].

During development of particular cell types—like oocyte fertilization or red blood cell maturation—elimination of mitochondria is necessary and the mitochondrial protein NIX is the adapter protein mediating this particular type of selective autophagy [24]. Under hypoxia conditions, mitophagy is induced and BNIP3 and FUNDC1 act as the adapters between the mitochondria that are to be eliminated and the autophagosome [10, 25]. However, induction of PINK1/Parkin-mediated mitophagy upon depolarization of the mitochondrial membrane is by far the best characterized mechanism of mitophagy, probably favored by the simplicity of the experimental manipulation and monitoring with pharmacological uncouplers of

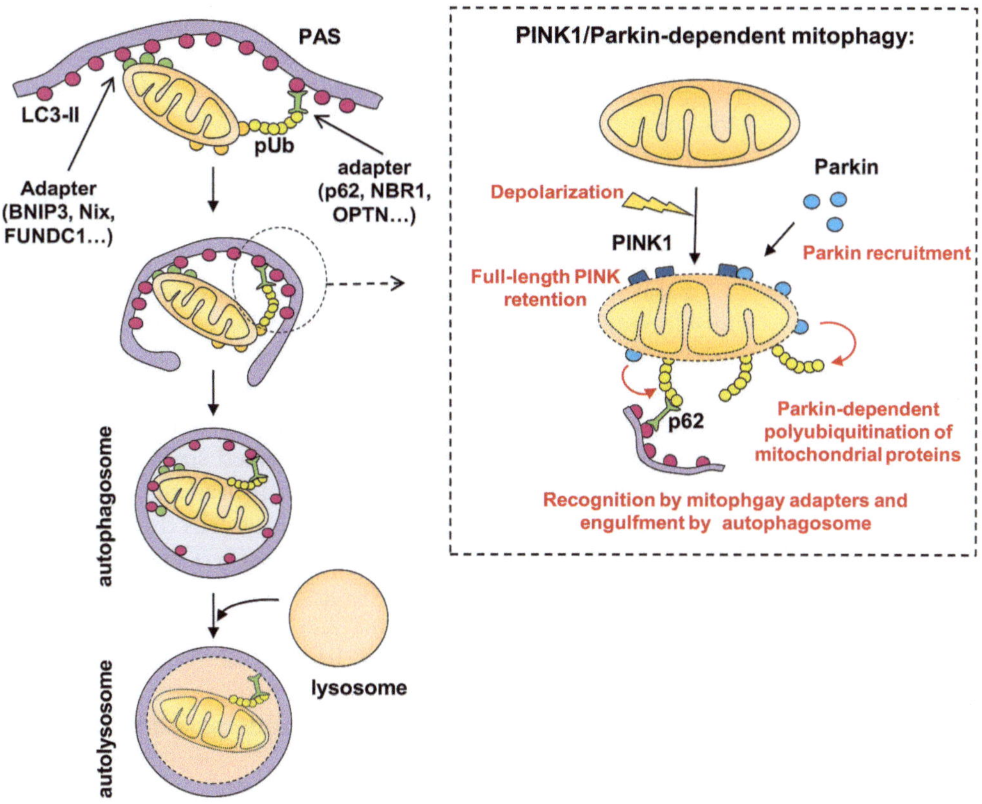

Fig. 1 The process of mitophagy

the mitochondrial membrane like CCCP [12]. A cascade of events occurs following loss of mitochondrial membrane potential. In brief, under resting conditions, the PTEN-induced putative kinase 1 (PINK1) is rapidly degraded after import into healthy mitochondria. Upon loss of mitochondrial membrane potential, the proteolysis of PINK1 is impaired, thus causing accumulation of PINK1 on the surface of damaged mitochondria. PINK1 promotes the recruitment of the E3-ligase PARK1/Parkin via phosphorylation of Parkin by PINK1. Accumulated and active Parkin on the mitochondria surface triggers the polyubiquitination of several proteins in the mitochondria membrane that act as an elimination signal. Different mitophagy adapters such as p62, NBR1, NDP52, and optineurin can recognize and bind these polyubiquitinated chains and simultaneously bind to LC3 to recruit the autophagic limiting membrane to engulf the targeted mitochondria [12].

Other alternative Parkin-independent pathways have been described very recently. Translocation of the phospholipid cardiolipin from the inner to the outer membrane of mitochondria can also act as an adapter for selective mitophagy and recruit LC3 to damaged mitochondria in neurons [26]. Additional Parkin-independent pathways have also been proposed, thus adapters

NDP52 and optineurin can be recruited to mitochondria directly by PINK1 (and independently of Parkin) where they recruit the autophagic proteins [19]. Other proteins present in the mitochondrial membrane (VDAC [17] and the TOMM complex [27]) are involved in the ubiquitin machinery (SMURF1 [28], Gp78 [29]) and the autophagic pathway (AMBRA1 [30], RHEB [31]), and can also participate in selective mitophagy under various conditions.

Basal mitophagy has been recently described for first time in muscle stem cells (also known as satellite cells). These normally quiescent cells activate and expand in response to muscle damage to sustain regeneration [32]. Mitophagy was shown to be necessary for proper maintenance of satellite cell homeostasis in the resting quiescent state [33]. In the referred study, damaged mitochondria with lower mitochondrial membrane potential accumulated in aged compared to young satellite cells due to an impaired autophagic flux, similar to what was observed in mice with genetically impaired autophagy in young satellite cells by specific deletion of the Atg7 gene. This accumulation of dysfunctional mitochondria caused increased ROS levels and subsequent cell senescence in aged and autophagy-deficient satellite cells. In this chapter, we review a variety of methods to study mitophagy in muscle stem cells.

2 Materials

2.1 Satellite Cell Isolation by Fluorescence-Activated Cell Sorting (FACS)

1. DMEM (Dulbecco's Modified Eagle Medium).
2. Penicillin/streptomycin.
3. Fetal bovine serum (FBS).
4. Filters 100, 70, and 40 μm.
5. Lysis buffer (BD Pharm Lyse, 555899).
6. FACS buffer: PBS 1×, 5 % goat serum.

Antibodies:

7. PE/Cy7 anti-mouse/human Ly-6A/E (Sca-1) (Biolegend, 108114).
8. PE/Cy7 anti-mouse/human CD11b (Biolegend, 101216).
9. α-7 integrin R-phycoerythrin (AbLab, 53-0010-05).
10. Alexa Fluor 647 rat anti-mouse CD34 (BD Pharmingen, 560230).

Digestion mix (four limb muscles of one mouse require 100 mL of digestion mix):

11. Collagenase D, 0.8 % final concentration.

12. Trypsin 2.5 %, 0.125 % final concentration.
13. DMEM, 1 % P/S.
    DAPI, stock solution 1 mg/mL, final concentration 1 μg/mL.

2.2 Mitochondrial Mass Quantification in Satellite Cells

2.2.1 Flow Cytometry Analysis of Mitochondria Content

1. DAPI, stock solution 1 mg/mL, final concentration 1 μg/mL.
2. FACS buffer: PBS 1×, 5 % Goat Serum.
3. MitoTracker Red CMXRos (Invitrogen, M7512).
4. Fetal bovine serum (FBS).

2.2.2 Fluorescence Microscopy Analysis of Mitochondria Content

1. MitoTracker Red CMXRos (Invitrogen, M7512).
2. FACS buffer: PBS 1×, 5 % goat serum.
3. DAPI, stock solution 1 mg/mL, final concentration 1 μg/mL.
4. 8-well glass slides.
5. Poly L-lysine.
6. PBS 1×, 0.02 % azide.
7. Fixation: 4 % PFA.
8. Mowiol.

2.2.3 Western Blot Analysis of Mitochondria Markers

Immunoprecipitation (IP) Buffer:

1. 50 mM Tris–HCl.
2. 150 mM NaCl.
3. 1 % NP-40.
4. 5 mM EGTA.
5. 5 mM EDTA.
6. 20 mM NaF.
   IP Working Solution (for 10 mL):
7. One tablet of protease inhibitor (CompleteMini, Roche, 11836153001).
8. 50 μL phosphatase inhibitor Cockatil 1 (Sigma, P2850).
9. 50 μL phosphatase inhibitor Cocktail 2 (Sigma, P5726).
10. NaVanadate, 0.1 mM final concentration.
11. PMSF, 1 mM final concentration.
12. Glycerophosphate, 10 mM final concentration.
    Resolving gel 12 % (two gels):
13. 8.6 mL ddH_2O.
14. 6 mL 40 % acrylamide.
15. 3 mL 1.5 M Tris–HCl, pH 8.8.
16. 200 μL 10 % SDS.

17. 200 μL 10% APS.
18. 8 μL TEMED.
    Stacking gel 5% (two gels):
19. 5.9 mL ddH_2O.
20. 1 mL 40% acrylamide.
21. 1 mL 1.0 M Tris–HCl, pH 6.8.
22. 80 μL 10% SDS.
23. 80 μL 10% APS.
24. 8 μL TEMED.
    Running Buffer (pH 8.3):
25. 0.025 M Tris–HCl.
26. 0.192 M Glycine.
27. 0.1% SDS.
    Transfer Buffer:
28. 0.025 M Tris–HCl.
29. 0.192 M glycine.
30. 20% methanol.
31. Blocking solution: 5% milk, TBS-T.
32. Antibody solution: 5% BSA, TBS-T.
33. PVDF membrane (Amersham Hybon PVDF, GE Healthcare LifeScience, 10600023).
34. Tris–HCl buffered saline (TBS, 10×): 1.5 M NaCl, 0.1 M Tris–HCl, pH 7.4.
35. TBS-T: TBS 1×, 0.05% Tween.
    5× Laemmli Buffer:
36. 100 mM Tris–HCl pH 6.8.
37. 32% glycerol.
38. 2% SDS.
39. 0.05% β-mercaptoethanol.
40. 0.1% bromopehnol blue.
41. ddH_2O.
    Antibodies:
42. Anti-TOM20 (Abcam, ab56783).
43. Anti-tubulin (Sigma T-6199).
44. ECL reagent (Amersham ECL Western Blotting Detection Reagents, GE Healthcare LifeScience, RPN2209).
45. X-ray film (Amersham Hyperfilm ECL, GE Healthcare LifeScience, 28906837).

2.3 Mitophagy Quantification: Colocalization of Mitochondrial and Autophagosomal/Lysosomal Markers by Immunostaining

1. Blocking solution: 10% goat serum, 10% FBS in PBS 1×.
2. Washes: PBS 1×, 0.01% Tween.
3. Permeabilization: PBS 1×, 0.5% Triton.
4. DAPI, stock solution 1 mg/mL, final concentration 1 μg/mL.
5. 8-well glass slides.
6. Poly L-lysine.
7. FACS buffer: PBS 1×, 5% goat serum.
8. Bafilomycin A1, 10 nM final concentration (Sigma, B1793).
9. DMSO (dimethyl sulfoxide).

Antibodies:

10. Anti-TOM20 (Abcam, ab56783).
11. LAMP-1 (Santa Cruz Biotechnology, sc-19992).
12. Mouse monoclonal antibody to LC3 (NanoTools, 5F10).
13. Rabbit anti-Parkin (Abcam ab15954).

2.4 Determination of Mitochondria Quality and Fitness by Flow Cytometry

1. DAPI, stock solution 1 mg/mL, final concentration 1 μg/mL.
2. FACS buffer: PBS 1×, 5% goat serum.
3. TMRM (Tetramethylrhodamine methyl ester perchlorate).
4. MitoTracker Green FM (Invitrogen, M7514).

2.5 Mitophagy Flux Determination by Flow Cytometry

2.5.1 Assay 1

1. CCCP (Carbonyl cyanide 3-chlorophenylhydrazone) (Sigma, C2759).
2. DAPI, stock solution 1 mg/mL, final concentration 1 μg/mL.
3. FACS buffer: PBS 1×, 5% goat serum.
4. TMRM.
5. MitoTracker Green FM (Invitrogen, M7514).
6. Growth medium: HAM'S F-12, 20% FBS, 1% L-glutamine, 1% P/S.
7. Coating solution: 0.1 mg/mL of collagen (Becton Dickinson, 354236), 0.02 N acetic acid.
8. DMSO.
9. Trypsin 2.5%, 0.25% final concentration.

2.5.2 Assay 2

1. Growth medium: HAMS F-12, 20% FBS, 1% L-glutamine, 1% P/S.
2. Coating solution: 0.1 mg/mL of collagen, 0.02 N acetic acid.
3. Bafilomycin A1, 10 nM final concentration (Sigma, B1793).
4. DMSO.
5. Trypsin 2.5%, 0.25% final concentration.
6. FACS buffer: PBS 1×, 5% goat serum.
7. MitoTracker Red CMXRos (Invitrogen, M7512).

3 Methods

3.1 Satellite Cell Isolation by FACS

1. Quiescent satellite cells are isolated from resting muscles of mice. To isolate satellite cells from regenerating muscles, mouse muscles are previously injected 50 μL of cardiotoxin to induce damage and regeneration. Activated/proliferating satellite cells are usually isolated 24–72 h after injury.
2. Collect muscles from fore and hind limbs in cold DMEM 1 % P/S into 50 mL falcon tubes.
3. Decant all the muscles collected in a petri dish and remove DMEM 1 % P/S completely.
4. Mince muscles with scissors.
5. Mince muscles further with razor blades.
6. Collect minced muscles into a 50 mL falcon tube and add cold DMEM 1 % P/S. Leave muscle sediment and remove DMEM 1 % P/S, discarding floating fat pieces. Repeat this step to further clean the sample from non-muscle pieces.
7. In the last wash, remove DMEM 1 % P/S as much as possible and split the minced muscle into two 50 mL falcon tubes.
8. Add 10 mL of the prepared digestion mix (collagenase/trypsin) to each tube.
9. Incubate for 25 min at 37 °C in a shaking water bath.
10. At the end of the digestion, leave the tube for 5 min on ice to let the sample sediment.
11. Transfer the digestion supernatant to a new 50 mL falcon tube on ice.
12. Add 10 mL of the digestion mix again and incubate for 25 min at 37 °C.
13. Collect the new digestion supernatant and pool it with the supernatant from the first digestion round.
14. **Steps 6–12** are repeated two more times.
15. If some pieces of muscle still remain, decant the sample on a petri dish for a new mincing process until no more are seen, and repeat one round of digestion. At the end of the fourth round of digestion, all muscle tissue should be digested.
16. Add 5 mL of fetal bovine serum (FBS) to each 50 mL falcon tube to block the digestion.
17. Filter the supernatant with 100 and 70 μm filters.
18. Centrifuge the supernatant 10 min at 50 × *g* and at 4 °C.
19. Collect the supernatant and discard the pellet (optional: the pellet can be washed and the supernatant collected and pooled with the previous supernatants).

20. Centrifuge at 600 × *g* for 15 min at 4 °C, repeat three times. The supernatant is discarded at each round and the pellet is resuspended gently in cold DMEM 1 % P/S.
21. After three centrifugations at 600 × *g*, pellets from both tubes are pooled together in cold DMEM 1% P/S, and passed through a 40 μm filter.
22. Centrifuge at 600 × *g* for 15 min at 4 °C.
23. Resuspend the pellet in 3 mL of 1× lysis buffer and incubate at 4 °C for 10 min (protect from light). Before centrifugation add cold DMEM 1 % P/S up to 50 mL.
24. Centrifuge at 600 × *g* for 15 min at 4 °C.
25. Discard the supernatant and resuspend the pellet in 1 mL of cold DMEM 1 % P/S.
26. Count the number of cells for each sample.
27. Centrifuge at 600 × *g* for 15 min at 4 °C and resuspend the pellet at 1×10^4 cells/μL (1×10^6 cells in 100 μL) in FACS Buffer.
28. Staining with antibodies:
    (a) Negative selection: Sca1-PECy7 and CD11b-PECy7 (0.5 μL/100 μL FACS buffer)
    (b) Positive selection: α-7 integrin-PE (1 μL/100 μL FACS buffer) and CD34-APC (1.5 μL/100 μL FACS buffer). CD34 is only used to stain quiescent satellite cells.
    (c) Controls: Single staining and FMO controls are required to set up the gates.
29. Incubate the cells with antibodies for 20 min at 4 °C, protected from light.
30. Centrifuge at 600 × *g* for 15 min at 4 °C.
31. Discard the supernatant and resuspend the cell bulk in 1 mL of FACS buffer for sample sorting.
32. Add DAPI (final concentration 1 μg/mL) 5 min prior FACS to detect and exclude dead cells.
33. Collect $Sca1^-/CD11b^-/CD34^+/\alpha7\text{-integrin}^+$ satellite cells in Eppendorf tubes with 500 μL of FACS buffer at 4 °C.

3.2 Mitochondrial Mass Quantification

3.2.1 Flow Cytometry Analysis of Mitochondria Content in Satellite Cells (Fig. 2a)

1. After **step 30** of the protocol for satellite cell isolation from muscles of WT mice (*see* **Note 1**), discard the supernatant, resuspend the cell pellet in 500 μL of FACS buffer, and transfer it into an Eppendorf tube.
2. Add MitoTracker Red (*see* **Note 2**) CMXRos (100 nM) and incubate for 30 min at 37 °C in the incubator (open lid). Samples without staining and with single staining must be used as negative controls.

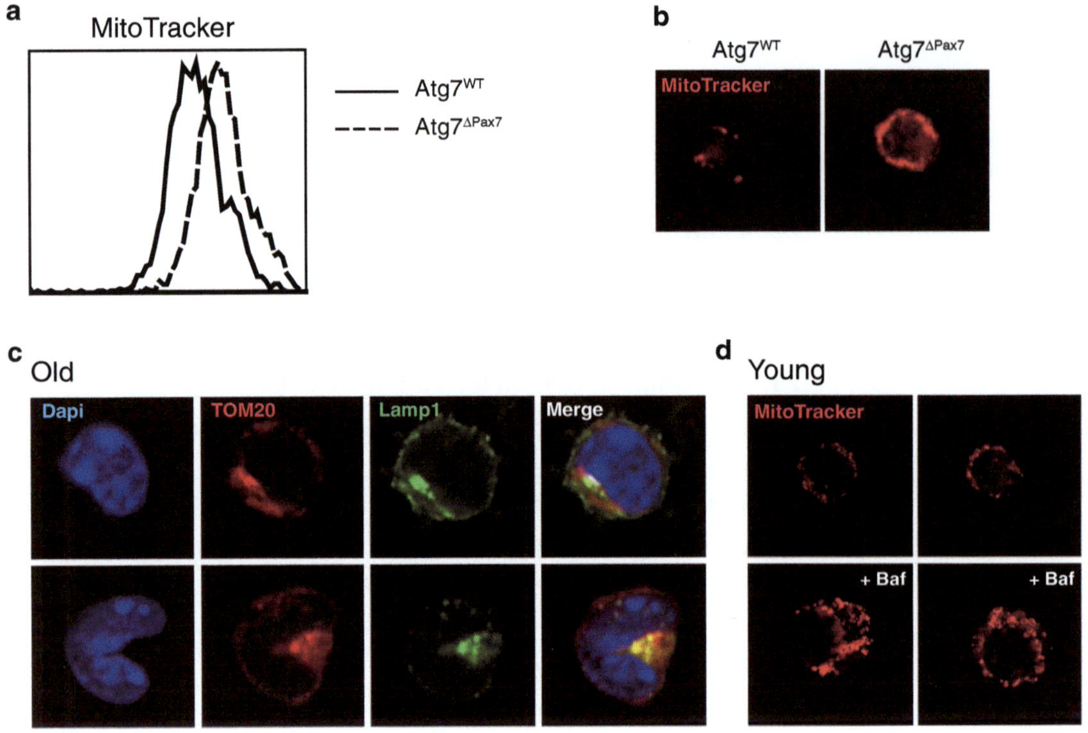

Fig. 2 (**a**) Mitochondrial content quantified by MitoTracker staining of quiescent satellite cells isolated from $Atg7^{WT}$ and $Atg7^{\Delta Pax7}$ mice by flow cytometry. Representative images are shown. (**b**) Representative images from (**a**). (**c**) TOM20 and Lamp1 immunostaining in quiescent satellite cells freshly isolated from old mice. (**d**) MitoTracker staining of young satellite cells treated with bafilomycin A1 or vehicle for 24 h. Representative images are shown

3. Centrifuge Eppendorf tubes at 900×*g* for 10 min at 4 °C.
4. Discard the supernatant and add 1 mL of FACS buffer with 10% FBS.
5. Add DAPI (final concentration 1 μg/mL) 5 min prior FACS to detect and exclude dead cells.
6. Analyze PE fluorescence in FACS LSR Fortesa (Becton Dickinson) of 10,000 satellite cells ($Sca1^-$/$CD11b^-$/$\alpha7$-integrin$^+$) for each sample.
7. MitoTracker Red CMXRos fluorescence signal is achieved by determining the mean fluorescence intensity (MFI) of the whole histogram signal for satellite cells and compared to the corresponding negative control sample (no stained) in an overlaid histogram [34].

3.2.2 Fluorescence Microscopy Analysis of Mitochondria Content (Fig. 2b)

After satellite cell isolation by FACS:

1. Add MitoTracker Red CMXRos (100 nM) for 30 min at 37 °C in the incubator (open lid).
2. Centrifuge Eppendorf tubes at 900×*g* for 10 min at 4 °C.
3. Resuspend the cell pellets in FACS buffer.

4. Coat 8-well glass slides with poly L-lysine for 30 min at RT.
5. Remove poly L-lysine (can be reused).
6. Add cell suspension (10,000–15,000 cells/well). Ensure that cell suspension is distributed evenly in the well; if necessary, add PBS 1×.
7. Spin the slides 5 min at 50 × *g*.
8. Remove supernatant from each well.
9. Add 4 % PFA (200 μL) in each well. Incubate for 10 min at RT.
10. Remove 4 % PFA and perform two washes with PBS 1×. At this point, slides can be stored at 4 °C with PBS 1× 0.02 % azide.
11. Mount slides with mowiol.
12. Analyze fluorescence in images obtained with Leica SPE confocal laser scanning microscope and Superresolution microscopy using stimulated emission depletion (STED), Leica TCS SP5 STED.
13. Quantification of mitochondrial mass (Red fluorescence) can be determined on digital images with Fiji and the cell image analysis software CellProfiler.
14. This protocol is compatible with an immunostaining for different antibodies (*see* **Note 3**) for further analysis.

3.2.3 Western Blot Analysis of Mitochondria Markers

After satellite cell isolation by FACS:

1. Centrifuge at 14,000 × *g* for 5 min at 4 °C.
2. Remove the supernatant as much as possible. At this step, samples can be stored at −80 °C.
3. Add 100 μL of IP working solution in each Eppendorf tube for cell lysis.
4. Ensure complete cell lysis by incubating suspensions for 45 min at 4 °C with shaking.
5. Remove remaining cell debris by centrifuging at 14,000 × *g* for 15 min at 4 °C. The supernatant contains the proteins and can be quantified using the standard Bradford Assay. However, since quiescent satellite cells contain low amounts of proteins, and since the same number of satellite cells is used for each sample, the protein quantification step can be omitted.
6. Prepare samples for gel loading: 40 μL of each sample with 5× Laemmli buffer (10 μL).
7. Incubate samples for 5 min at 95 °C.
8. Prepare the stacking gel at 5 % and the resolving gel at 12 %, use 1.5 mm gel and 10-well comb to be able to load 50 μL of sample.
9. Load the maximum sample volume (around 50 μL).

10. Run the samples at 100 V with running buffer.
11. Transfer at 100 V for 1 h, using PVDF membrane and transfer buffer.
12. Block membrane with blocking solution for 1 h at RT. The membrane can be divided in different parts by cutting it horizontally at different levels, to allow Western blotting of proteins with different molecular weights, by using different antibodies.
13. Incubate O/N with the primary antibody (TOM20) in antibody solution.
14. Wash 3× with TBS-T.
15. Incubate with the secondary HRP antibody for 2 h at RT, protected from light.
16. Wash 3× with TBS-T.
17. Detect HRP using ECL reagent and X-ray film. Ensure that the exposures are in the linear range.
18. Wash 3× with TBS-T.
19. Incubate O/N with the primary antibody (Tubulin) used as loading control.
20. Wash 3× with TBS-T.
21. Incubate with the secondary HRP antibody for 2 h at RT, protected from light.
22. Wash 3× with TBS-T.
23. Detect HRP using ECL reagent and X-ray film. Ensure that the exposures are in the linear range.
24. Quantify the bands using ImageJ (*see* **Notes 4** and **5**).

3.3 Mitophagy Quantification: Colocalization of Mitochondrial and Autophagosomal/Lysosomal Markers by Immunostaining (Fig. 2c)

After satellite cell isolation by FACS (Subheading 3.1, above), mitophagy can be assessed by colocalization of TOM20 (mitochondrial marker) and GFP-LC3 (or endogenous LC3) as marker of autophagosomes and LAMP-1 as marker of lysosomes. Satellite cells might be treated with lysosomal inhibitors (i.e., bafilomycin A1) for 4 h at 37 °C prior fixation to block clearance of mitochondria inside the autolysosomes and detect higher colocalization of mitochondria and autophagosomes/lysosomes.

1. Coat 8-well glass slide with poly L-lysine for 30 min at room temperature.
2. Remove poly L-lysine (it can be reused).
3. Add cell suspension (10,000–15,000 cells/well). Ensure that cell suspension is evenly distributed in the well; if necessary, add PBS 1×.
4. Spin the slides for 5 min at 50 × *g*.

5. Remove supernatant from each well.
6. Add 4 % PFA (200 μL) in each well. Incubate for 10 min at RT.
7. Remove 4 % PFA and perform two washes with PBS 1×. At this point, slides can be stored at 4 °C with PBS 1× 0.02 % azide.
8. Permeabilize cells with PBS 1× 0.5 % Triton for 15 min at RT.
9. Wash 3× with PBS 1×.
10. Add blocking solution for 30 min at RT.
11. Incubate with primary antibody in blocking solution O/N at 4 °C.
12. Wash 3× with PBS 1×, 0.01 % Tween.
13. Incubate with secondary antibody in PBS 1×, 0.01 % Tween 1 h 30 min at RT (seat in dark).
14. Wash 3× PBS 1×, 0.01 % Tween.
15. Add DAPI (final concentration 1 μg/mL) for 5 min at RT.
16. Wash 3× PBS 1×, 0.01 % Tween.
17. Mount with mowiol.
18. Analyze fluorescence in confocal images taken with Leica SPE confocal laser scanning microscope system and Superresolution microscopy using stimulated emission depletion (STED), Leica TCS SP5 STED.
19. Quantification of colocalization can be determined on digital images with Fiji and the cell image analysis software CellProfiler (*see* **Notes 6–8**).

3.4 Determination of Mitochondria Quality and Fitness by Flow Cytometry Analysis

TMRM is a mitochondrial dye dependent on mitochondrial membrane potential sequestered only by active mitochondria, while MitoTracker Green FM stains all mitochondria regardless of their membrane potential. The ratio between MFI of these two dyes indicates the proportion of active/healthy mitochondria from total mitochondria content of the cell.

1. After **step 30** of the protocol for satellite cell isolation from muscles of WT mice, discard the supernatant, resuspend the cell pellet in 500 μL of FACS buffer, and transfer it into an Eppendorf tube.
2. Add MitoTracker Green FM (500 nM final concentration) and TMRM (1 μM final concentration) and incubate for 30 min at 37 °C in the incubator (open lid). Samples without staining and with single staining must be used as negative controls.
3. Centrifuge Eppendorf tubes at 900 ×*g* for 10 min at 4 °C.
4. Discard the supernatant and add 1 mL of FACS buffer with 10 % FBS.

5. Add DAPI (final concentration 1 μg/mL) 5 min prior FACS to detect and exclude dead cells.
6. Analyze FITC and PE fluorescence in FACS LSR Fortesa (Becton Dickinson) of 10,000 satellite cells (Sca1⁻/CD11b⁻/α7-integrin⁺) for each sample.
7. Calculate membrane potential of each sample as TMRM MFI and MitoTracker Green MFI ratio [35] (*see* **Note 9**).

3.5 Mitophagy Flux Determination by Flow Cytometry Analysis

3.5.1 Assay 1

To detect mitophagy flux, cells are treated with CCCP (carbonyl cyanide 3-chlorophenylhydrazone) which causes mitochondria uncoupling and consequently induces selective autophagy of CCCP-damaged mitochondria. After 1 h, CCCP-treatment, a huge reduction of TMRM/MitoTracker Green FM MFI ratio should be seen (dyes are respectively dependent and independent of mitochondrial membrane potential) while at 24 h this ratio in young satellite cells must be recovered by mitophagy, to levels similar of those of non-treated cells.

1. After **step 32** of the protocol for satellite cell isolation, satellite cells (Sca1⁻/CD11b⁻/CD34⁺/α7-integrin⁺) are collected by FACS in Eppendorf tubes with 500 μL of Growth Medium (GM).
2. Coat petri dishes for cell culture with coating solution (0.1 mg/mL collagen in sterile water and 0.02 N acetic acid).
3. Plate sorted satellite cells in coated petri dishes and culture them in GM.
4. After 4 days in culture, treat cells with CCCP (10 μM final concentration) or DMSO for 1 h (one set of cells are not treated at that time point).
5. Change the media and culture cells for 24 h.
6. Treat control cells with CCCP (10 μM final concentration) for 1 h.
7. After 1 h, trypsinize (0.25 % Trypsin) all sets of cells.
8. Add MitoTracker Green FM (500 nM final concentration) and TMRM (1 μM final concentration) incubate for 30 min at 37 °C in the incubator (open lid). Samples without staining and single staining must be done as controls.
9. Centrifuge Eppendorf tubes at 900 × *g* for 10 min at 4 °C.
10. Discard the supernatant and add 1 mL of FACS buffer with 10 % FBS.
11. Add DAPI (final concentration 1 μg/mL) 5 min prior FACS to detect and exclude dead cells.
12. Analyze FITC and PE fluorescence in FACS LSR Fortesa (Becton Dickinson) of 10,000 satellite cells for each sample.
13. Calculate membrane potential of each sample as TMRM MFI and MitoTracker Green FM MFI ratio.

3.5.2 Assay 2 (Fig. 2d)

The quantification of mitochondria under control conditions and after treatment with lysosomal inhibitors (like bafilomycin A1) can indicate the amount of mitochondria degraded by lysosomes during mitochondrial turnover. Differences in MFI of MitoTracker Red CMXRos samples treated with or without bafilomycin A1 correspond to accumulated mitochondria.

1. After **step 32** of the protocol for quiescent satellite cell isolation, satellite cells ($Sca1^-$/$CD11b^-$/$CD34^+$/$\alpha7$-integrin$^+$) are collected by FACS in Eppendorf tubes with 500 μL of Growth Medium (GM).
2. Coat petri dishes for cell culture with coating solution (0.1 mg/mL of collagen in sterile water and 0.02 N acetic acid).
3. Plate sorted satellite cells in coated petri dishes and culture them in GM.
4. Add bafilomycin A1 (10 nM) or DMSO in cell media for 24 h.
5. Trypsinize (0.25 % trypsin) all sets of cells.
6. Incubate with MitoTracker Red CMXRos (100 nM) for 30 min at 37 °C in the incubator (open lid) to stain mitochondria. Samples without staining and with single staining must be used as controls.
7. Centrifuge Eppendorf tubes at 900 × *g* for 10 min at 4 °C.
8. Resuspend the cell pellet with 1 mL of FACS buffer.
9. Analyze FITC and PE fluorescence in FACS LSR Fortesa (Becton Dickinson) of 10,000 satellite cells for each sample (*see* **Note 10**).

4 Notes

1. For this type of analysis Sca1-PECy7, CD11b-PECy7 (negative selection) and α7-integrin7-647, CD34-FITC (positive selection) staining is needed for identification of satellite cells.
2. Other mitochondrial probes could also be used in this protocol. In our experience, MitoTracker Red CMXRos stains all mitochondria regardless their membrane potential, in contrast to what is described in the manufacturer's protocol.
3. Other mitochondrial fluorescent probes can be used but it is important to use a mitochondria dye that is retained after cell fixation as MitoTracker Red CMXRos. Similarly, antibodies against mitochondrial proteins can be used. The use of lysosomal inhibitors (i.e., bafilomycin A1) are useful to quantify mitochondrial turnover by mitophagy.

4. Other markers for mitochondria can be used as mitochondrial mass markers: hsp60, VDAC, subunits of the Complex IV or other member of the TOMM complex. Alternatively, Parkin recruitment to mitochondria (measured in the mitochondrial fraction, not in total lysate) can be used as markers of mitophagy initiation.
5. In addition, p62 levels can be assessed by Western blotting as an indicator of autophagic flux.
6. Quiescent satellite cells have low cytoplasm content, and autophagosomes and mitochondria are hard to discern in such small cytoplasm. For this reason, high objectives with good resolution are necessary. In addition, doing several z-sections can improve their visualization.
7. Alternatively, to detect the initiation of the mitophagy process, colocalization of mitochondria (by fluorescent probes or by antibody) with Parkin (endogenous Parkin or preferably transfected EGFP-Parkin) which is recruited to depolarized mitochondria, can be used.
8. Co-localizations can be determined on digital images Fiji, according to ref. [36], with respect to the total cellular area. The Pearson's coefficient (r) is used to analyze the correlation of the intensity values of green and red pixels in dual-channels images. This coefficient measures the strength of the linear relationship between the intensities in two images calculated by linear regression and ranges from 1 to −1, with 1 standing for complete positive correlation.
9. For this type of analysis Sca1-PECy7, CD11b-PECy7 (negative selection) and α7-integrin7-647 staining are needed for identification of satellite cells.
10. Alternatively, this protocol could be used to detect mitophagic flux by fluorescent microscopy if cells are fixed with 4 % PFA.

Acknowledgments

Work in the authors' laboratories has been supported by: MINECO, Spain SAF2012-38547, AFM, E-Rare/Eranet, Fundació Marató-TV3, MDA, EU-FP7 (Myoage, Optistem and Endostem) and DuchennePP-NL, to PM-C; and ISCIII, Spain (FIS-PS09/01267, FIS-PI13/02512, CP09/00184, PI14/01529) and CIBERNED to MM-V. L.G.-P. was supported by a Predoctoral Fellowship from Programa de Formación de Personal Investigador (Spain).

References

1. Mauro-Lizcano M et al (2015) New method to assess mitophagy flux by flow cytometry. Autophagy 11(5):833–843
2. Dolman NJ et al (2013) Tools and techniques to measure mitophagy using fluorescence microscopy. Autophagy 9(11):1653–1662
3. Cuervo AM (2004) Autophagy: many paths to the same end. Mol Cell Biochem 263(1–2): 55–72
4. Levine B, Klionsky DJ (2004) Development by self-digestion: molecular mechanisms and biological functions of autophagy. Dev Cell 6(4):463–477
5. Shintani T, Klionsky DJ (2004) Autophagy in health and disease: a double-edged sword. Science 306(5698):990–995
6. Klionsky DJ (2005) Autophagy. Curr Biol 15(8):R282–R283
7. Kroemer G, Marino G, Levine B (2010) Autophagy and the integrated stress response. Mol Cell 40(2):280–293
8. Ashrafi G, Schwarz TL (2013) The pathways of mitophagy for quality control and clearance of mitochondria. Cell Death Differ 20(1):31–42
9. Clark SL Jr (1957) Cellular differentiation in the kidneys of newborn mice studies with the electron microscope. J Biophys Biochem Cytol 3(3):349–362
10. Wei H, Liu L, Chen Q (2015) Selective removal of mitochondria via mitophagy: distinct pathways for different mitochondrial stresses. Biochim Biophys Acta 1853(10 Pt B):2784–2790
11. Yoshii SR, Mizushima N (2015) Autophagy machinery in the context of mammalian mitophagy. Biochim Biophys Acta 1853(10 Pt B):2797–2801
12. Youle RJ, Narendra DP (2011) Mechanisms of mitophagy. Nat Rev Mol Cell Biol 12(1):9–14
13. Novak I et al (2010) Nix is a selective autophagy receptor for mitochondrial clearance. EMBO Rep 11(1):45–51
14. Rikka S et al (2011) Bnip3 impairs mitochondrial bioenergetics and stimulates mitochondrial turnover. Cell Death Differ 18(4):721–731
15. Hanna RA et al (2012) Microtubule-associated protein 1 light chain 3 (LC3) interacts with Bnip3 protein to selectively remove endoplasmic reticulum and mitochondria via autophagy. J Biol Chem 287(23):19094–19104
16. Liu L et al (2012) Mitochondrial outer-membrane protein FUNDC1 mediates hypoxia-induced mitophagy in mammalian cells. Nat Cell Biol 14(2):177–185
17. Geisler S et al (2010) PINK1/Parkin-mediated mitophagy is dependent on VDAC1 and p62/SQSTM1. Nat Cell Biol 12(2):119–131
18. Kirkin V et al (2009) A role for NBR1 in autophagosomal degradation of ubiquitinated substrates. Mol Cell 33(4):505–516
19. Lazarou M et al (2015) The ubiquitin kinase PINK1 recruits autophagy receptors to induce mitophagy. Nature 524(7565):309–314
20. Ying H, Yue BY (2016) Optineurin: the autophagy connection. Exp Eye Res 144:73–80
21. Johansen T, Lamark T (2011) Selective autophagy mediated by autophagic adapter proteins. Autophagy 7(3):279–296
22. Kraft C, Peter M, Hofmann K (2010) Selective autophagy: ubiquitin-mediated recognition and beyond. Nat Cell Biol 12(9):836–841
23. Wild P, McEwan DG, Dikic I (2014) The LC3 interactome at a glance. J Cell Sci 127(Pt 1): 3–9
24. Sandoval H et al (2008) Essential role for Nix in autophagic maturation of erythroid cells. Nature 454(7201):232–235
25. Zhang J, Ney PA (2009) Role of BNIP3 and NIX in cell death, autophagy, and mitophagy. Cell Death Differ 16(7):939–946
26. Chu CT et al (2013) Cardiolipin externalization to the outer mitochondrial membrane acts as an elimination signal for mitophagy in neuronal cells. Nat Cell Biol 15(10):1197–1205
27. Bertolin G et al (2013) The TOMM machinery is a molecular switch in PINK1 and PARK2/PARKIN-dependent mitochondrial clearance. Autophagy 9(11):1801–1817
28. Orvedahl A et al (2011) Image-based genome-wide siRNA screen identifies selective autophagy factors. Nature 480(7375):113–117
29. Fu M et al (2013) Regulation of mitophagy by the Gp78 E3 ubiquitin ligase. Mol Biol Cell 24(8):1153–1162
30. Van Humbeeck C et al (2011) Parkin interacts with Ambra1 to induce mitophagy. J Neurosci 31(28):10249–10261
31. Melser S et al (2013) Rheb regulates mitophagy induced by mitochondrial energetic status. Cell Metab 17(5):719–730
32. Garcia-Prat L, Sousa-Victor P, Munoz-Canoves P (2013) Functional dysregulation of stem cells during aging: a focus on skeletal muscle stem cells. FEBS J 280(17):4051–4062
33. Garcia-Prat L et al (2016) Autophagy maintains stemness by preventing senescence. Nature 529(7584): 37–42

34. Warnes G (2015) Flow cytometric assays for the study of autophagy. Methods 82:21–28
35. Cottet-Rousselle C et al (2011) Cytometric assessment of mitochondria using fluorescent probes. Cytometry A 79(6):405–425
36. Bolte S, Cordelieres FP (2006) A guided tour into subcellular colocalization analysis in light microscopy. J Microsc 224(Pt 3):213–232

Chapter 17

Identification, Isolation, and Characterization of Mesenchymal Progenitors in Mouse and Human Skeletal Muscle

Akiyoshi Uezumi, Takehiro Kasai, and Kunihiro Tsuchida

Abstract

Mesenchymal progenitors residing in the muscle interstitial space contribute to pathogeneses such as fat infiltration and fibrosis. Because fat infiltration and fibrosis are hallmarks of diseased muscle, it is important to establish an accurate and reproducible method for isolating mesenchymal progenitors for research on muscle diseases. In this chapter, we describe methods based on fluorescence-activated cell sorting (FACS) to purify mesenchymal progenitors from mouse and human skeletal muscle using the most reliable marker for mesenchymal progenitors, PDGFRα. These methods allow concurrent isolation of the muscle stem cells called satellite cells. The quality of isolated mesenchymal progenitors is confirmed by their remarkable adipogenic potential without myogenic capacity, while purified satellite cells possess robust myogenic activity with no adipogenic potential. Simultaneous isolation of both mesenchymal progenitors and satellite cells from mouse and human tissues provides a powerful platform for studying skeletal muscle regeneration and diseases.

Key words Skeletal muscle, Mesenchymal progenitors, Satellite cells, Cell isolation, FACS

1 Introduction

Skeletal muscle has remarkable regenerative capacity that depends on a stem cell population called satellite cells [1]. Satellite cells are mononucleated cells residing between the plasma membrane of myofibers and the basal lamina [2]. Their high myogenic potential is essential to adult muscle regeneration and cannot be replaced by other cell types [3–5]. On the other hand, mesenchymal progenitors reside in the muscle interstitium, and therefore represent a cell population distinct from satellite cells [6]. These cells lack myogenic capacity but possess high potential to differentiate into mesenchymal lineages such as adipose and osteogenic cells [6, 7]. Mesenchymal progenitors are pathologically important because they directly contribute to fat infiltration and fibrosis [6, 8–10], which are commonly observed in diseased muscle. Because skeletal muscle cells dissociated by enzymatic digestion

Michael Kyba (ed.), *Skeletal Muscle Regeneration in the Mouse: Methods and Protocols*, Methods in Molecular Biology, vol. 1460, DOI 10.1007/978-1-4939-3810-0_17, © Springer Science+Business Media New York 2016

contain both mesenchymal progenitors and satellite cells, it is important to distinguish mesenchymal progenitors from satellite cells. Identifying target cells by specific markers and purifying them by fluorescence-activated cell sorting (FACS) represent the most effective and reliable method to achieve this. As a marker of mesenchymal progenitors, PDGFRα is considered the best because it is highly specific and can be used for both mouse and human tissues [11].

In the following sections, we describe detailed methods for the purification of mesenchymal progenitors from mouse and human skeletal muscle using PDGFRα as a positive marker. These methods enable simultaneous isolation of both mesenchymal progenitors and satellite cells. Methods to assess the quality of isolated cells by inducing adipogenic differentiation of mesenchymal progenitors or myogenic differentiation of satellite cells are also included.

2 Materials

2.1 Dissociating Cells from Skeletal Muscle

1. Mice: 8- to 10-week-old female mice (*see* **Note 1**).
2. Human muscle tissues (*see* **Note 2**).
3. Forceps and scissors. For trimming and mincing muscle tissues, fine-tipped forceps and curved scissors are recommended.
4. Sterile 60 mm tissue culture dishes.
5. Sterile 5, 10, and 25 ml pipettes.
6. PBS without Ca^{2+} and Mg^{2+} (sterile).
7. Dissection microscope.
8. Hank's balanced saline solution (HBSS) (sterile).
9. Collagenase Type II (Worthington, cat#: CLSS2).
10. Sterile 5 or 10 ml syringe.
11. Sterile 0.22-μm PVDF membrane syringe-driven filter unit.
12. A 20 ml beaker.
13. Magnetic stirrer and stir bar.
14. Tissue culture incubator (humidified, 5 % CO_2, ambient O_2, 37 °C).
15. An 18-gauge needle.
16. Sterile 40 μm cell strainers.
17. Sterile 100 μm cell strainers.
18. Sterile 15 and 50 ml conical tubes.
19. Benchtop centrifuge.
20. Hypotonic solution: prepare 0.83 % NH_4Cl and 0.17 M Tris HCl (pH 7.65) separately, and autoclave them separately. Mix at the ratio of 9:1.

21. Sterile 200 and 1000 μl pipette tips.
22. Hemocytometer.
23. Washing buffer: PBS supplemented with 2.5 % fetal bovine serum (FBS).
24. Growth medium (GM): high glucose Dulbecco's modified Eagle's medium (DMEM) supplemented with 20 % FBS, 2 mM L-glutamine, 1 % penicillin–streptomycin, and 2.5 ng/ml basic fibroblast growth factor (bFGF).
25. Sterile 60 mm dishes coated with collagen I.
26. Tissue culture incubator (humidified, 5 % CO_2, 3 % O_2, 37 °C).

2.2 Purification of Mesenchymal Progenitors and Satellite Cells from Dissociated Cells by FACS

1. Sterile phosphate-buffered saline (PBS) without Ca2+ and Mg2+.
2. Sterile 0.05 % trypsin–EDTA.
3. Trypsin stop solution: PBS supplemented with 10 % FBS.
4. Washing buffer, as above.
5. Sterile 1.5 ml microcentrifuge tubes.
6. Benchtop centrifuge.
7. Sterile 10, 200, and 1000 μl pipette tips.
8. Antibodies and secondary reagents: Alexa Fluor 488-conjugated rat anti-mouse CD31 (BioLegend, clone: 390), Alexa Fluor 488-conjugated rat anti-mouse CD45 (BioLegend, clone: 30-F11), goat polyclonal anti-mouse PDGFRα (R&D, cat#: AF1062), biotinylated rat anti-mouse satellite cells ([12], clone: SM/C-2.6), biotinylated goat polyclonal anti-human PDGFRα (R&D, cat#: BAF322), PE-conjugated mouse anti-human CD56 (Miltenyi Biotec, clone: AF12-7H3), PE-conjugated donkey polyclonal anti-goat IgG (Jackson ImmunoResearch, cat#: 705-116-147), streptavidin-PE/Cy5 (BioLegend), Alexa Fluor 488-conjugated rat isotype control (BioLegend), and PE-conjugated mouse isotype control (Miltenyi Biotec).
9. Sterile 40 μm cell strainers.
10. Sterile 50 ml conical tubes.
11. 5 ml round-bottom FACS tubes.
12. Propidium iodide (PI) solution.
13. Cell sorter.

2.3 Culture and Differentiation of Mesenchymal Progenitors and Satellite Cells

1. GM, as above.
2. Adipogenic induction medium (AdipoIM): DMEM supplemented with 10 % FBS, 0.5 mM IBMX, 0.25 μM dexamethasone, 10 μg/ml insulin, 2 mM L-glutamine, and 1 % penicillin–streptomycin.

3. Adipogenic maintenance medium (AdipoMM): DMEM supplemented with 10% FBS, 10 μg/ml insulin, 2 mM L-glutamine, and 1% penicillin–streptomycin.
4. Differentiation medium (DM): DMEM supplemented with 5% horse serum, 2 mM L-glutamine, and 1% penicillin–streptomycin.
5. Benchtop centrifuge.
6. Sterile 200 and 1000 μl pipette tips.
7. Matrigel diluted to 1 mg/ml with DMEM.
8. 48-well cell culture plates.
9. Tissue culture incubator (humidified, 5% CO_2, 3% O_2, 37 °C; or humidified, 5% CO_2, ambient O_2, 37 °C).
10. Inverted phase-contrast microscope.

3 Methods

3.1 Dissociating Cells from Skeletal Muscle

The procedures for dissociating cells from mouse and human skeletal muscle are essentially identical. Therefore, we describe procedures for mouse unless otherwise specified.

1. Before processing muscle tissue, autoclave forceps and scissors. Place a stir bar in a 20 ml beaker, cover beaker with aluminum foil, and then autoclave them.
2. Weigh a 60 mm tissue culture dish containing PBS.
3. Excise hind limb muscles of mouse. Transfer excised mouse muscles (or human muscle tissues) to the 60 mm tissue culture dish containing PBS (*see* Fig. 1a) and then weigh dish.
4. Calculate tissue weight by subtracting weight in **step 2** from weight in **step 3**.
5. Carefully remove remaining nerves, blood vessels, tendons, and fat using fine-tipped forceps under a dissection microscope.

The following steps should be performed in a sterile laminar flow hood.

6. Transfer trimmed muscles into a new 60 mm tissue dish and mince them thoroughly using the curved scissors (*see* Fig. 1b).
7. Dissolve collagenase type II in HBSS to make 0.2% collagenase solution. 4 ml collagenase solution per g of tissue is required for digestion (*see* **Note 3**). Sterilize collagenase solution by forcing it through 0.22 μm syringe-driven filter unit into the autoclaved beaker.
8. Transfer minced muscles into the beaker containing collagenase solution with the stir bar. Cover the beaker with aluminum foil.
9. Place the beaker and magnetic stirrer in tissue culture incubator, and then digest muscles for 50 min (30 min for human)

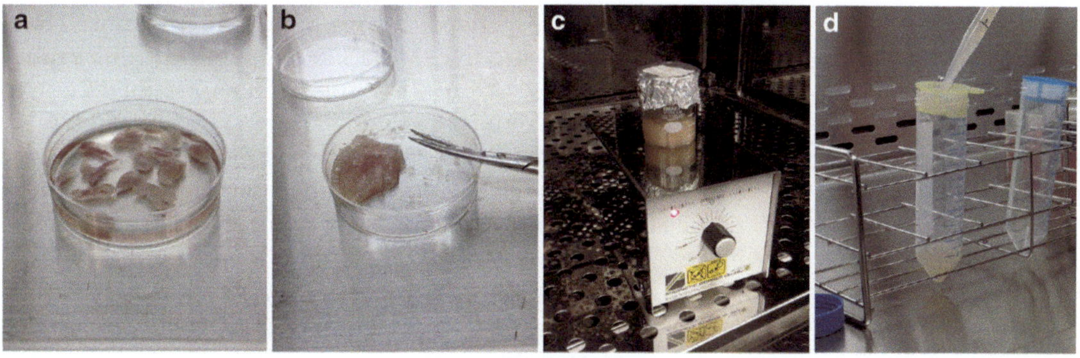

Fig. 1 Dissociating cells from skeletal muscle. (**a**) Excised muscles in PBS. (**b**) Minced muscle. (**c**) Digestion of muscle using magnetic stirrer. (**d**) Filtration of digested muscle slurry

at 37 °C with stirring by the magnetic stirrer (*see* Fig. 1c, *see* **Note 4**).

10. Pass digested muscles through the 18 gauge needle five to seven times using a sterile syringe.
11. Continue digestion for 20 min (15 min for human tissue) at 37 °C with stirring by the magnetic stirrer (*see* **Note 4**).
12. Add 10 ml PBS to digested slurry and mix thoroughly using 10 ml pipette.
13. Filter digested slurry through 100 μm cell strainer over a 50 ml conical tube (*see* Fig. 1d). Dilute digested slurry by washing cell strainer with PBS and adjust total volume to 25 ml/g of tissue (*see* **Note 5**).
14. Filter the slurry filtered in **step 13** through a 40 μm cell strainer on a new 50 ml conical tube. Dilute the slurry by washing cell strainer with PBS and adjust total volume to 50 ml/g of tissue (*see* **Note 5**).
15. Centrifuge cells for 5 min at $760 \times g$.
16. Resuspend the pellet in hypotonic solution. Use 2 ml hypotonic solution per gram of tissue. Transfer cells to a 15 ml conical tube and incubate for 1 min at room temperature to eliminate erythrocytes. Add at least 1 volume of PBS. Count cells using a hemocytometer.
17. Centrifuge cells for 5 min at $760 \times g$.
18. For mouse tissues, resuspend the pellet in washing buffer and adjust cell concentration to 1×10^7 cells/ml. Proceed to Subheading 3.2.
19. For human tissues, resuspend the pellet in 5 ml GM and culture on 60 mm collagen I-coated dish at 37 °C in 5% CO_2 and 3% O_2 (*see* **Note 6**). Maintain culture by changing GM every 3–4 days until cells reach 70–80% confluence.

3.2 Purification of Mesenchymal Progenitors and Satellite Cells from Dissociated Cells by FACS

3.2.1 Antibody Staining of Mouse Cells for FACS

1. Divide cells resuspended in washing buffer (**step 18** in Subheading 3.1) into five sterile 1.5 ml microcentrifuge tubes. Label these tubes "A" to "E": tube A for unstained control, tube B for Alexa Fluor 488 single-stained control, tube C for PE single-stained control, tube D for PE/Cy5 single-stained control, and tube E for triple-stained sample. Unstained or single-stained controls are used for compensation settings (*see* **Note 7**).
2. Add Alexa Fluor 488-conjugated rat isotype control (1:250) to tube A. Add Alexa Fluor 488-conjugated rat anti-mouse CD31 (1:250) and Alexa Fluor 488-conjugated rat anti-mouse CD45 (1:250) to tube B; add goat polyclonal anti-mouse PDGFRα (final conc. 2.5 μg/ml) to tube C; add biotinylated rat SM/C-2.6 (1:250) to tube D; and add Alexa Fluor 488-conjugated rat anti-mouse CD31, Alexa Fluor 488-conjugated rat anti-mouse CD45, goat polyclonal anti-mouse PDGFRα, and biotinylated rat SM/C-2.6 to tube E.
3. Incubate samples at 4 °C for 30 min in the dark.
4. Fill the tubes with washing buffer, and centrifuge cells for 5 min at 760 × *g*.
5. Resuspend the pellet in washing buffer and adjust cell concentration to 1×10^7 cells/ml.
6. Add PE-conjugated donkey polyclonal anti-goat IgG (1:200) and streptavidin-PE/Cy5 (1:200) to tube A. Add PE-conjugated donkey polyclonal anti-goat IgG to tube C, streptavidin-PE/Cy5 to tube D, and PE-conjugated donkey polyclonal anti-goat IgG and streptavidin-PE/Cy5 to tube E.
7. Incubate samples at 4 °C for 30 min in the dark.
8. Fill the tubes with washing buffer and centrifuge cells for 5 min at 760 × *g*.
9. Resuspend the pellet in washing buffer and adjust volume to 1 ml/tube.
10. Filter control and stained samples through 40-μm cell strainers placed on 50 ml conical tubes. Transfer filtered control and stained samples to 5 ml FACS round-bottom tubes.

3.2.2 Antibody Staining of Human Cells for FACS

1. Human cells at 70–80 % confluence (**step 19** in Subheading 3.1) are ready for FACS. Aspirate GM and wash cells with PBS.
2. Add 1 ml 0.05 % trypsin–EDTA per 60 mm dish and incubate cells for 3–5 min at 37 °C.
3. Add 1 ml trypsin stop solution and detach cells by gentle pipetting using a 1000 μl pipette tip. Transfer cells to a 15 ml conical tube. Count cells using a hemocytometer.
4. Centrifuge cells for 5 min at 430 × *g*.

5. Resuspend the pellet in washing buffer and adjust cell concentration to 5×10^6 cells/ml.
6. Divide cells resuspended in washing buffer into four sterile 1.5 ml microcentrifuge tubes. Label these tubes "A" to "D": tube A for the unstained control, tube B for PE single-stained control, tube C for PE/Cy5 single-stained control, and tube D for double-stained sample. Unstained and single-stained controls are used for compensation settings (*see* **Note 8**).
7. Add PE-conjugated mouse isotype control (1:20) to tube A. Add PE-conjugated mouse anti-human CD56 (1:20) to tube B, biotinylated goat polyclonal anti-human PDGFRα (final conc. 2.5 μg/ml) to tube C, and PE-conjugated mouse anti-human CD56 and biotinylated goat polyclonal anti-human PDGFRα to tube D.
8. Incubate samples for 30 min at 4 °C in the dark.
9. Fill up the tubes with washing buffer and centrifuge cells for 5 min at $430 \times g$.
10. Resuspend the pellet in washing buffer and adjust cell concentration to 5×10^6 cells/ml.
11. Add streptavidin-PE/Cy5 (1:200) to tubes A, C, and D.
12. Incubate samples for 30 min at 4 °C in the dark.
13. Fill the tubes with washing buffer and centrifuge cells for 5 min at $430 \times g$.
14. Resuspend the pellet in washing buffer and adjust volume to 1 ml/tube.
15. Filter control and stained samples through 40-μm cell strainer caps on 5 ml FACS round-bottom tubes.

3.2.3 Purification of Mesenchymal Progenitors and Satellite Cells from Mouse by FACS

Alexa Fluor 488, PE, and PE/Cy5 are excited by a 488 nm laser and detected by an appropriate filter set. Because the emission wavelength of PI excited by a 488 nm laser overlaps with that of PE/Cy5, PI should be excited by a UV laser and detected by a 620/29 bandpass filter. General settings of the cell sorter should follow the manufacturer's instructions. We describe here detailed procedures specific to purification of mesenchymal progenitors and satellite cells from skeletal muscle briefly.

1. Exclude debris on a forward scatter (FSC) vs. side scatter (SSC) graph.
2. Adjust detector voltage and compensation to optimal levels to clearly visualize negative and positive populations by analyzing unstained or single-stained controls (tubes A–D in Subheading 3.2.1).
3. Add 10 μl PI solution per 1×10^6 cells and mix gently. Exclude dead cells (PI-positive cells) on a PI vs. FSC or SSC graph.

4. Analyze triple-stained sample (tube E in Subheading 3.2.1). Gate for $CD31^-CD45^-$ cells as shown in Fig. 2a.
5. Display data of $CD31^-CD45^-$ cells on a PDGFRα-PE vs. SM/C-2.6-PE/Cy5 graph. Gate for $PDGFR\alpha^+SM/C\text{-}2.6^-$ cells (red gate) and $PDGFR\alpha^-SM/C\text{-}2.6^+$ cells (green gate) as shown in Fig. 2b.
6. Prepare collection tubes by adding 1 ml washing buffer to sterile 1.5 ml microcentrifuge tubes.
7. Sort $CD31^-CD45^-PDGFR\alpha^+SM/C\text{-}2.6^-$ cells as mesenchymal progenitors and $CD31^-CD45^-PDGFR\alpha^-SM/C\text{-}2.6^+$ cells as satellite cells (*see* **Note 9**).

3.2.4 Purification of Mesenchymal Progenitors and Satellite Cells from Human Tissue by FACS

1. Exclude debris on a forward scatter (FSC) vs. side scatter (SSC) graph.
2. Adjust detector voltage and compensation to optimal levels to clearly visualize negative and positive populations by analyzing unstained or single-stained controls (tubes A–C in Subheading 3.2.2).
3. Add 10 μl PI solution per 1×10^6 cells and mix gently. Exclude dead cells (PI-positive cells) on a PI vs. FSC or SSC graph.
4. Analyze double-stained sample (tube D in Subheading 3.2.2). Display data on a PDGFRα-PE/Cy5 vs. CD56-PE graph. Gate for $PDGFR\alpha^+CD56^-$ cells (red gate) and $PDGFR\alpha^-CD56^+$ cells (green gate) as shown in Fig. 3a.
5. Prepare collection tubes by adding 1 ml washing buffer to sterile 1.5 ml microcentrifuge tubes.
6. Sort $PDGFR\alpha^+CD56^-$ cells as mesenchymal progenitors and $PDGFR\alpha^-CD56^+$ cells as satellite cells (*see* **Note 10**).

3.3 Culture and Differentiation of Mesenchymal Progenitors and Satellite Cells

1. Add Matrigel to 48-well cell culture plates to coat the surface of the wells and then remove excess. Incubate the plates for 15 min at 37 °C. In a sterile laminar flow hood, uncover the plates and air dry.
2. Centrifuge sorted cells for 5 min at $760 \times g$ ($430 \times g$ for human cells) and resuspend the pellet in GM.
3. Add 1×10^4 cells to each well with 350 μl/well of the medium.
4. Incubate cells at 37 °C in 5 % CO_2 and 3 % O_2 for 3–4 days (*see* **Note 11**). After this growth period, mesenchymal progenitors and satellite cells reach ~70 % confluence and are ready for induction of differentiation (*see* Figs. 2c, d, and 3b, c, *see* **Note 12**).
5. To induce adipogenic cells from mouse mesenchymal progenitors, remove GM and add adipoIM to the wells, and incubate at 37 °C in 5 % CO_2 and ambient O_2 for 3 days. Then remove adipoIM and add adipoMM to the wells and incubate at 37 °C in 5 % CO_2 and ambient O_2 for 3 days (*see* **Note 13**). After

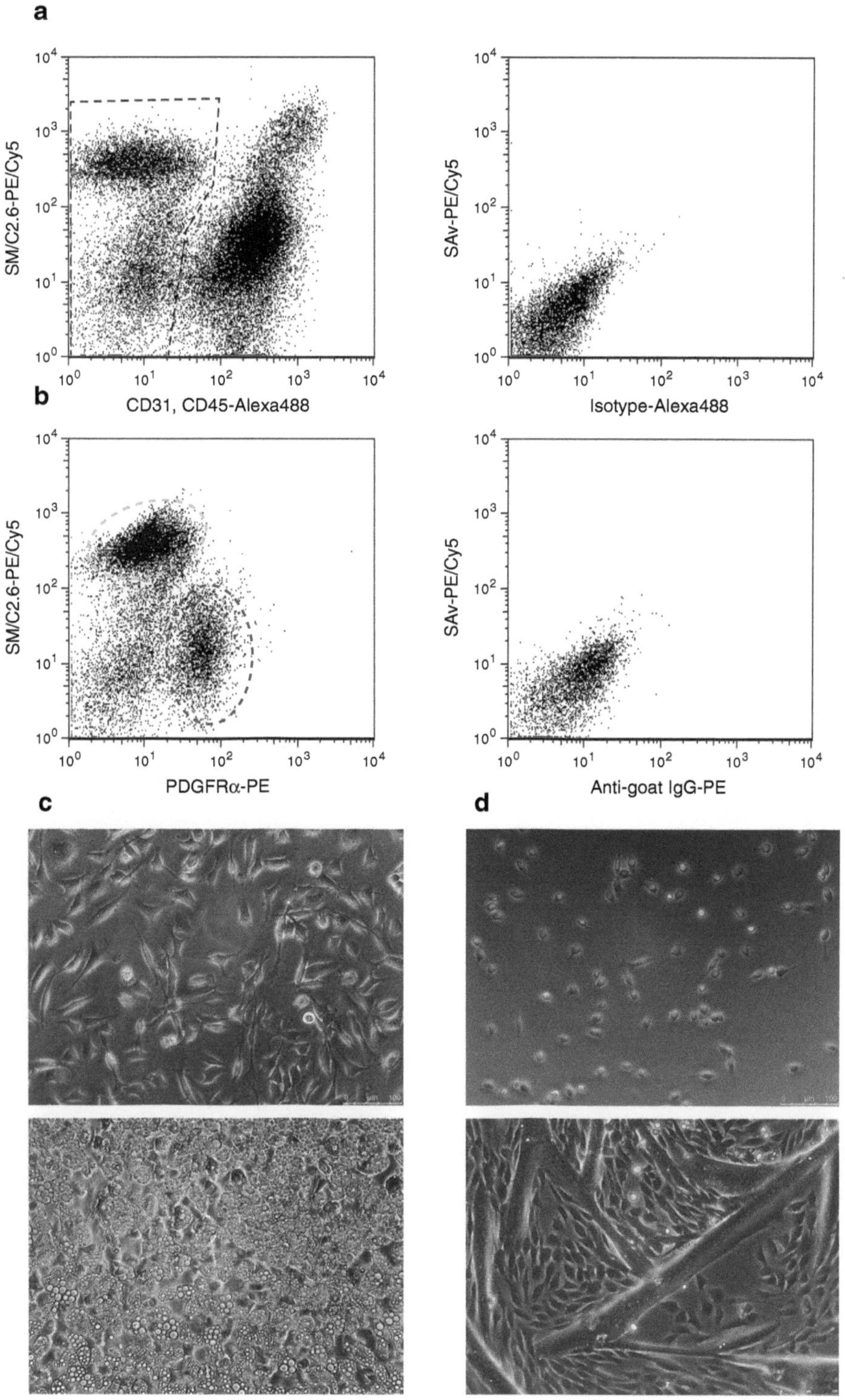

Fig. 2 Isolation and culture of mouse mesenchymal progenitors and satellite cells. (**a**) FACS dot plots showing the expression of CD31 and CD45 (*x*-axis) and SM/C-2.6 (*y*-axis). The unstained control is shown in the *right*. (**b**) FACS dot plots showing the expression of PDGFRα (*x*-axis) and SM/C-2.6 (*y*-axis) in $CD31^{-}CD45^{-}$ cells. The unstained control is shown in the *right*. (**c**) Morphology of cultured mouse mesenchymal progenitors in GM (*upper panel*) and after adipogenic differentiation (*lower panel*). (**d**) Morphology of cultured mouse satellite cells in GM (*upper panel*) and after myogenic differentiation (*lower panel*)

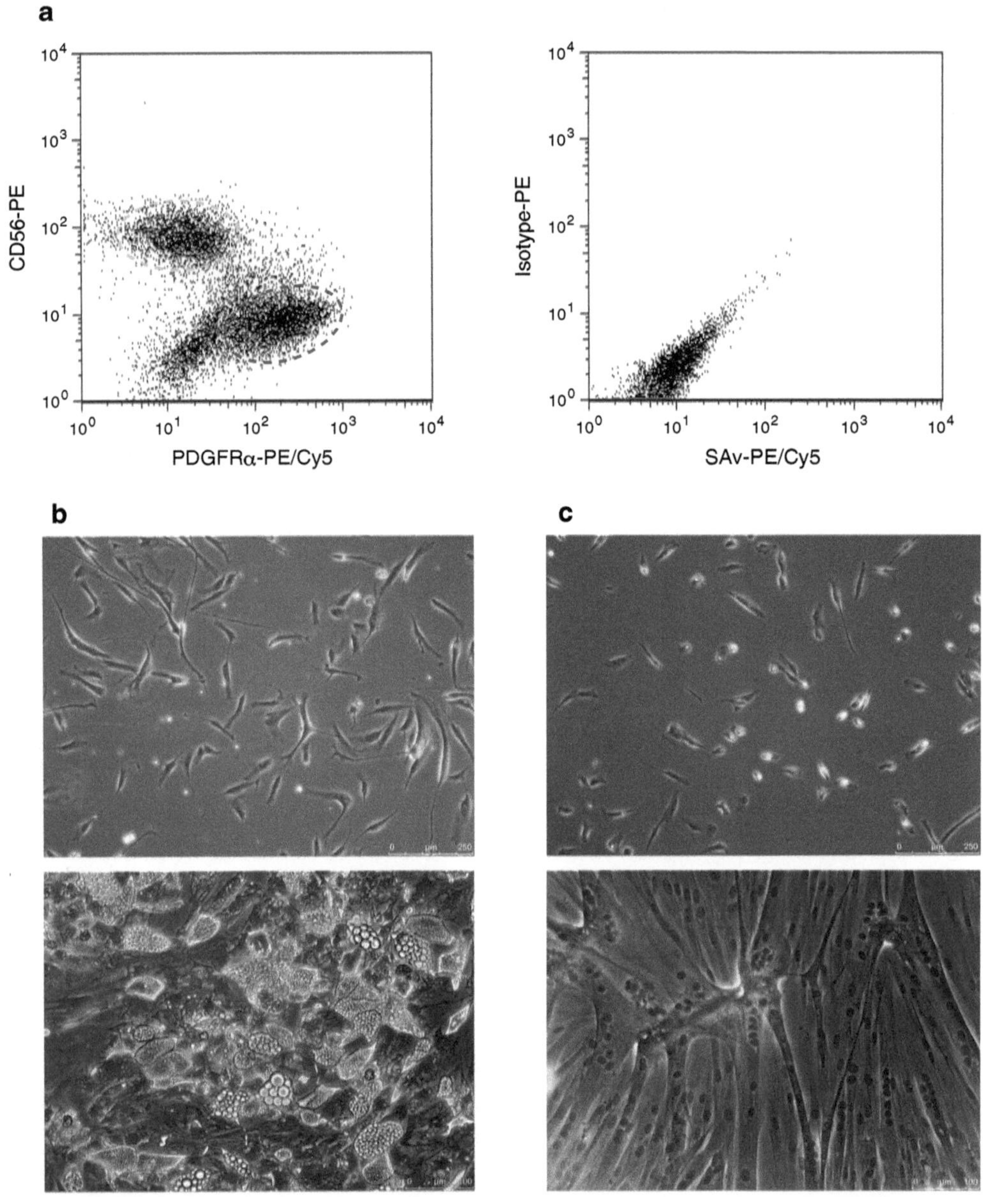

Fig. 3 Isolation and culture of human mesenchymal progenitors and satellite cells. (**a**) FACS dot plots showing the expression of PDGFRα (*x*-axis) and CD56 (*y*-axis). The unstained control is shown in the *right*. (**b**) Morphology of cultured human mesenchymal progenitors in GM (*upper panel*) and after adipogenic differentiation (*lower panel*). (**c**) Morphology of cultured human satellite cells in GM (*upper panel*) and after myogenic differentiation (*lower panel*)

adipogenic induction, mesenchymal progenitors differentiate into adipocytes containing lipid droplets that are readily revealed by an inverted phase-contrast microscope (*see* Fig. 2c).

6. To induce myogenic differentiation of mouse satellite cells, remove GM and add DM to the wells and incubate at 37 °C in

5 % CO_2 and ambient O_2 for 2–3 days (*see* **Note 13**). After myogenic differentiation, satellite cells develop into large myotubes that are readily revealed by inverted phase-contrast microscopy (*see* Fig. 2d).

7. To induce adipogenic cells from human mesenchymal progenitors, remove GM and add adipoIM to the wells and incubate at 37 °C in 5 % CO_2 and ambient O_2 for 3 days. Then remove adipoIM, add adipoMM to the wells, and incubate at 37 °C in 5 % CO_2 and ambient O_2 for 1 day (*see* **Note 13**). Repeat this induction and maintenance cycle three times. After adipogenic induction, mesenchymal progenitors differentiate into adipocytes containing lipid droplets that are readily revealed by inverted phase-contrast microscopy (*see* Fig. 3b).
8. For myogenic differentiation of human satellite cells, remove GM and add DM to the wells and incubate at 37 °C in 5 % CO_2 and ambient O_2 for 5 days (*see* **Note 13**). After myogenic differentiation, satellite cells develop into large myotubes that are readily revealed by an inverted phase-contrast microscope (*see* Fig. 3c).

4 Notes

1. Muscles of young female mice are digested more easily than those of male or older mice. We usually use C57BL/6, but we have confirmed that other strains such as BALB/c or ICR can be used as well.
2. Experiments using human tissues or cells must be approved by the institutional ethical review board.
3. Use at least 4 ml collagenase solution for efficient stirring even when tissue weight is less than 1 g.
4. Digestion time depends on sample condition. If male or older mice are used, longer digestion should be performed (e.g., we digest muscles of aged mice for 60 + 30 min).
5. Use one 50 ml conical tube with a 100-μm cell strainer and one 50 ml conical tube with a 40-μm cell strainer per gram of tissue (e.g., if starting tissue weight is 3 g, use three sets of conical tubes with cell strainers). If starting tissue weight is less than 1 g, use one conical tube with cell strainer and dilute digested slurry adjusting final volume to 50 ml.
6. Cell sorting by FACS requires a relatively large number of cells and therefore requires a large amount of starting tissue. However, a large amount of human tissue is difficult to obtain. We usually have less than 1 g of human muscle tissue. Therefore, it is necessary to expand cells before FACS in human experi-

ments. Note that a low oxygen concentration (3 % O_2) is used for cell culture. This is critical for efficient expansion of human cells.

7. In mouse experiments, at least 5×10^5 cells should be used for the compensation setting (tubes A–D). All the remaining cells are used for the triple-stained sample (tube E).
8. In human experiments, at least 3×10^5 cells should be used for compensation setting (tubes A–C). All the remaining cells are used for the double-stained sample (tube D).
9. If the cell concentration is too high to analyze by FACS, dilute the sample with washing buffer as appropriate. It is important to use bright fluorophores for clear identification of mesenchymal progenitors and satellite cells. After careful optimization, we determined the use of goat anti-PDGFRα plus PE-anti-goat IgG (for identification of mesenchymal progenitors) and biotinylated-SM/C-2.6 plus streptavidin-PE/Cy5 (for identification of satellite cells) to be the most effective method. Cell yields by these methods are about 5×10^4 cells/young female mouse for mesenchymal progenitors and $1–1.5 \times 10^5$ cells/young female mouse for satellite cells.
10. Human cells are cultured for several days prior to FACS. Endothelial cells ($CD31^+$ cells) and hematopoietic cells ($CD45^+$ cells) are eliminated during this expansion period. If the cell concentration is too high to analyze by FACS, dilute the sample with washing buffer as appropriate. Cell yields from humans are highly variable among patients, as described in our paper [10].
11. A low oxygen concentration (3 % O_2) is important for efficient growth of both mouse and human cells.
12. Mesenchymal progenitors are bigger than satellite cells and show a flattened morphology, while satellite cells show a compact morphology (*see* Figs. 2c, d, and 3b, c).
13. Although a low oxygen concentration is good for cell growth and expansion, cells differentiate better in the normoxic condition compared with the hypoxic condition. Thus, differentiation experiments should be performed under normoxic conditions.

Acknowledgements

We thank K. Ono for proofreading the paper. This work was supported by AMED Health and Labour Sciences Research Grants for Comprehensive Research on Persons with Disabilities, Japan Foundation for Aging and Health, JSPS KAKENHI Grant Number 15K12675, and the 24th General Assembly of the Japanese Association of Medical Sciences.

References

1. Bischof R (2004) Satellite and stem cells in muscle regeneration. In: Engel AG, Franzini-Armstrong C (eds) Myology, vol 1, 3rd edn. McGraw-Hill, New York, pp 66–86
2. Mauro A (1961) Satellite cell of skeletal muscle fibers. J Biophys Biochem Cytol 9:493–495
3. Lepper C, Partridge TA, Fan CM (2011) An absolute requirement for Pax7-positive satellite cells in acute injury-induced skeletal muscle regeneration. Development 138(17):3639–3646. doi:10.1242/dev.067595
4. Murphy MM, Lawson JA, Mathew SJ et al (2011) Satellite cells, connective tissue fibroblasts and their interactions are crucial for muscle regeneration. Development 138(17):3625–3637. doi:10.1242/dev.064162
5. Sambasivan R, Yao R, Kissenpfennig A et al (2011) Pax7-expressing satellite cells are indispensable for adult skeletal muscle regeneration. Development 138(17):3647–3656. doi:10.1242/dev.067587
6. Uezumi A, Fukada S, Yamamoto N et al (2010) Mesenchymal progenitors distinct from satellite cells contribute to ectopic fat cell formation in skeletal muscle. Nat Cell Biol 12(2):143–152. doi:10.1038/ncb2014
7. Wosczyna MN, Biswas AA, Cogswell CA et al (2012) Multipotent progenitors resident in the skeletal muscle interstitium exhibit robust BMP-dependent osteogenic activity and mediate heterotopic ossification. J Bone Miner Res 27(5):1004–1017. doi:10.1002/jbmr.1562
8. Joe AW, Yi L, Natarajan A et al (2010) Muscle injury activates resident fibro/adipogenic progenitors that facilitate myogenesis. Nat Cell Biol 12(2):153–163. doi:10.1038/ncb2015
9. Uezumi A, Ito T, Morikawa D et al (2011) Fibrosis and adipogenesis originate from a common mesenchymal progenitor in skeletal muscle. J Cell Sci 124(Pt 21):3654–3664. doi:10.1242/jcs.086629
10. Uezumi A, Fukada S, Yamamoto N et al (2014) Identification and characterization of PDGFRalpha+mesenchymal progenitors in human skeletal muscle. Cell Death Dis 5:e1186. doi:10.1038/cddis.2014.161
11. Uezumi A, Ikemoto-Uezumi M, Tsuchida K (2014) Roles of nonmyogenic mesenchymal progenitors in pathogenesis and regeneration of skeletal muscle. Front Physiol 5:68. doi:10.3389/fphys.2014.00068
12. Fukada S, Higuchi S, Segawa M et al (2004) Purification and cell-surface marker characterization of quiescent satellite cells from murine skeletal muscle by a novel monoclonal antibody. Exp Cell Res 296(2):245–255

Chapter 18

FACS Fractionation and Differentiation of Skeletal-Muscle Resident Multipotent Tie2+ Progenitors

Arpita A. Biswas and David J. Goldhamer

Abstract

The skeletal muscle niche is complex and heterogeneous. Over the past few decades, various groups have reported the existence of multiple adult stem cell populations within this environment. Techniques commonly used to identify and assess the differentiation capacities of these cellular fractions, oftentimes rare populations, include the use of lineage tracers, immunofluorescence and histochemistry, flow cytometry, gene expression assays, and phenotypic analysis in culture or in vivo. In 2012, our lab identified and characterized a skeletal-muscle resident Tie2+ progenitor that exhibits adipogenic, chondrogenic, and osteogenic differentiation potentials (Wosczyna et al., J Bone Miner Res 27:1004–1017, 2012). This Tie2+ progenitor also expresses the markers PDGFRα and Sca-1 which in turn label a population of muscle-resident fibro/adipogenic progenitors (FAPs) (Joe et al., Nat Cell Biol 12:153–163, 2010; Uezumi et al., Nat Cell Biol 12:143–152, 2010), suggesting similar identities or overlap in the two mesenchymal progenitor populations. Our study demonstrated that these Tie2-expressing mesenchymal progenitors contribute robustly to BMP-induced heterotopic ossification (HO) in mice, and therefore could represent a key cellular target for therapeutic intervention in HO treatment (Wosczyna et al., J Bone Miner Res 27:1004–1017, 2012). In this chapter, we provide a detailed description of our updated fluorescence-activated cell sorting (FACS) strategy and describe cell culture methods for differentiation of Tie2+ progenitors to adipogenic and osteogenic fates. This strategy is easily adaptable for the prospective isolation of other rare subpopulations resident in skeletal muscle.

Key words Flow cytometry, Skeletal muscle stem cell, Cre, Tie2, Fibro/adipogenic progenitors, Sca-1, PDGFRα, Heterotopic ossification, Adipogenesis, Osteogenesis

1 Introduction

Flow cytometry has emerged as a tremendously useful experimental tool in muscle biology, particularly to define the cellular components of the muscle microenvironment. The ability to discover new, unique progenitor/stem cell populations, as well as to assess functional heterogeneity within previously known fractions, is afforded, in large part, by this technology. FACS-isolation of progenitor populations coupled with phenotyping using cell culture assays can also be very useful to elucidate the mechanisms that

Michael Kyba (ed.), *Skeletal Muscle Regeneration in the Mouse: Methods and Protocols*, Methods in Molecular Biology, vol. 1460, DOI 10.1007/978-1-4939-3810-0_18, © Springer Science+Business Media New York 2016

impact the regulation of muscle development and homeostasis. In fact, such studies have led to the identification of potential drug targets for diseases and candidates for regenerative medicine [1–6].

In our previously published work, we reported that a population of Tie2+ mesenchymal progenitors resides within the skeletal muscle interstitium and mediates heterotopic ossification, a pathological condition in which bone develops within muscle and associated soft tissues [7–9]. In this chapter, we describe the isolation of Tie2+ progenitors both from wild type (WT) mice, as well as from mice in which Tie2+ progenitors were permanently labeled with Green Fluorescent Protein (GFP), using Tie2-Cre transgenic mice and the Cre-dependent reporter, $R26^{NG}$ [1]. The use of Cre/loxP lineage tracing was critical in our studies to identify the cell-of-origin for HO since it enabled us to track the developmental fate of target cells in vivo. However, the *Tie2* gene is not a unique marker of target cells. In fact the predominant cell types labeled within these mice are endothelial and hematopoietic cells [10], whereas target cells are defined as muscle-resident non-endothelial (CD31−), non-hematopoietic (CD45−), Tie2-labeled (GFP+) progenitors that co-express the cell surface receptors PDGFRα and Sca-1. Additionally, our analysis showed that skeletal muscle tissue also contains **GFP**−CD31−CD45−PDGFRα+Sca-1+ cells that are likely identical to their GFP+ counterparts. We demonstrated that these GFP− cells occupy the same niche within the muscle interstitium, express *Tie2*, and show similar growth and differentiation characteristics in culture [1], suggesting that inefficiency of Tie2-Cre-mediated recombination is responsible for incomplete labeling of the target cell population. As such, we use the term 'Tie2+ progenitor' when referring to CD31−CD45−PDGFRα+Sca-1+ cells within mouse skeletal muscle, regardless of whether they are lineage labeled with Tie2-Cre.

Although Tie2+ progenitors are present in multiple tissues, this chapter focuses solely on purification from skeletal muscle. We provide information on how to fractionate GFP-labeled progenitors using Tie2-Cre;$R26^{NG/+}$ mice. However, our sort gating progression is compatible with collection of CD31−CD45−PDGFRα+Sca-1+ cells from WT mice (please refer to Subheading **3.3** for details).

2 Materials

2.1 Mice

Experimental mice hemizygous for the Tie2-Cre allele (B6.Cg-Tg(Tek-cre)1Ywa/J from Jackson Laboratory Stock# 008863 [10]) and heterozygous for the $R26^{NG}$ reporter allele (made in our lab [11] and available at Jackson Laboratory FVB.Cg-*Gt(ROSA)26Sortm1(CAG-lacZ,-EGFP)Glh*/J Stock # 012429) or an alternative GFP reporter will need to be generated. We recommend using two experimental WT or Tie2-Cre;$R26^{NG/+}$ mice to increase yield (*see* Subheading 3.3.3). Mice that are heterozygous for Tie2-Cre or $R26^{NG}$ can be used as non-fluorescent control (NFC) mice We typically use mice between 2 and 3 months of age, although comparable results were obtained with mice up to 6 months old (unpublished observations).

2.2 Reagents and Equipment

2.2.1 FACS Isolation

Antibodies should be stored at 4 °C and kept protected from light. During staining, keep vials on ice and minimize light exposure. All solutions used for FACS preparation should be precooled before use. Use sterile reagents.

1. Muscle digest media: 1× DMEM, supplemented with 700–800 U/ml collagenase Type II, 0.3 U/ml dispase (required for digestion of harvested muscle, does not need precooling).
2. Cold media: 1× DMEM, supplemented with 20% fetal bovine serum (FBS).
3. Blocking media/wash media: 1× DPBS, supplemented with 10% FBS.
4. 1× DPBS, supplemented with 2% FBS.
5. Antibodies (*see* **Note 1**).
   - Anti-mouse APC-conjugated CD140A (PDGFRα receptor) antibody, 1:100 dilution.
   - Rat anti-mouse V450-conjugated Ly-6A/E (Sca-1 receptor) antibody, 1:400 dilution.
   - CD31 microbeads, mouse (Miltenyi Biotech).
   - CD45 microbeads, mouse (Miltenyi Biotech).
6. 200 μg/ml 7-AAD.
7. Dissection tools including sharp and blunt scissors and forceps.
8. Sterile plastic petri dishes.
9. Sterile 1× DPBS.
10. Sterile 50, 15, and 2 ml tubes.
11. Sterile 5 ml FACS tubes with 35 μm cell strainer.
12. Sterile microcentrifuge tubes.
13. Sterile serological pipettes.
14. 70 and 100 μm cell strainers.
15. Sterile tissue culture hood.
16. Water bath, 37 °C.
17. Refrigerated centrifuge with swinging bucket rotor.
18. QuadroMACS Separator (Miltenyi Biotech).
19. MACS Multistand (Miltenyi Biotech).
20. MACS LS separation columns (Miltenyi Biotech).

21. Pre-separation filters (Miltenyi Biotech).
22. BD FACSAria II.
23. Fluorescence stereomicroscope.

2.2.2 Cell Culture and Maintenance

It is ideal to use freshly prepared media. Warm in a water bath at 37 °C just prior to feeding cells.

1. Growth media (GM): 1× DMEM, supplemented with 20% FBS, 2.5 ng/ml human recombinant bFGF, 1% penicillin–streptomycin (Pen/Strep).
2. Osteogenic differentiation media: 1× DMEM, supplemented with 5% FBS, 300 ng/ml human recombinant BMP2 (bone morphogenetic protein 2), 1% Pen/Strep (*see* **Note 2**).
3. Matrigel growth factor reduced basement membrane matrix, phenol-red free, LDEV-free (Corning, specific handling and storage instructions, please refer to Subheading 3.1.1).
4. Rabbit anti-mouse perilipin antibody, 1:500 dilution.
5. Alkaline phosphatase detection kit.
6. Tissue culture-treated plates.
7. Tissue culture incubator (5% CO_2, 37 °C).

3 Methods

3.1 Equipment Preparation

3.1.1 Matrigel Coating of Plates

We use Matrigel-coated plates for culturing Tie2+ cells. Cells can be sorted into a sterile tube coated with blocking media, washed, spun down and plated. Alternatively, cells can be sorted directly into plates. Tie2+ cells grow well when plated in wells coated with a thin gel. As per the manufacturer's recommendations, 50 μl of Matrigel matrix per cm^2 of growth area is suitable for supporting cell attachment and proliferation (*see* **Note 3**). We typically use Matrigel diluted with DMEM to a protein concentration of 1 mg/ml for coating. Cell culture plates can be coated beforehand and stored in a refrigerator at 4 °C for up to 2 weeks, although it is ideal to use them the same day (Corning Guidelines for Use).

1. Matrigel should be aliquoted and stored at −20 °C.
2. Depending on the number and size of plates being coated, calculate Matrigel requirement. Each batch of Matrigel has a different protein concentration so final dilutions will vary based on the stock solution.
3. Thaw aliquots on ice overnight/few hours in a 4 °C refrigerator. In gel form, Matrigel is translucent but appears transparent when properly liquefied.

4. Store culture plates (that are to be coated), micropipette tips, 15 ml conical tubes and 5/10 ml serological pipettes in a −20 °C freezer overnight. It is essential for all consumables utilized for dispensing and handling the Matrigel to be chilled.
   *It is important to perform the subsequent steps on ice in a sterile culture hood.
5. Dilute the thawed Matrigel to the appropriate protein concentration, i.e., 1 mg/ml with DMEM, making sure to use chilled pipette tips and tubes. It is best to prepare additional working solution to account for any pipetting error. Also, Matrigel is a very viscous liquid and needs to be handled carefully.
6. Dispense diluted Matrigel to the center of each well of the culture plates (on ice).
7. Cover the plates and swirl gently to spread Matrigel solution evenly over the entire surface of each well.
8. Place the culture plates in the hood at room temperature for 1 h.
9. After an hour, aspirate the Matrigel and wash wells with 1× sterile DPBS. If sorting directly into plates, dispense appropriate volume of growth media into the wells, seal plates with tape and store in a 4 °C refrigerator until FACS setup is complete. If plates are not being used immediately, dispense 1× DPBS into the wells and store at 4 °C.

3.1.2 BD FACSAria II Setup

Sorting and analysis is done on a BD FACSAria II equipped with 405, 488, and 633 nm lasers. We have found that a 100 μm nozzle and threshold rate of 3000–4000 events per second provides a good combination of sorting speed, and cell purity, yield, and survival.

3.1.3 QuadroMACS Cell Separator Setup

1. The setup process can be initiated during a 15 min interval period at **step 3**, Subheading 3.2.2.
2. Place the QuadroMACS separator components in a sterile culture hood.
3. Assemble the instrument. Place LS columns in the magnetic field of the cell separator and place a filter on top of the column. Prepare as many columns as number of samples being processed.
4. Wash by running 3 ml of blocking media through each column. Do not attempt to accelerate rate of flow through by tapping or negative pressure.
5. Once the reservoir is empty, the columns are ready for use (Proceed to **step 4**, Subheading 3.2.2).

3.2 Preparation for FACS

The protocol assumes that two experimental mice (WT or Tie2-Cre;*R26*$^{NG/+}$) will be used for cell isolation. A third, non-fluorescent control mouse (NFC) will be needed only if Tie2-Cre;*R26*$^{NG/+}$ experimental mice are being used (*see* **step 2**, Subheading 3.2.3). The Tie2-Cre;*R26*$^{NG/+}$ cell suspension sample will be referred to as NG in the subsequent sections.

3.2.1 Muscle Dissociation

1. Anesthetize mice in an isoflurane chamber and euthanize by cervical dislocation (or as per institutional and government regulations).
2. For each sample, make a 10 ml aliquot of digest media in a 50 ml sterile conical tube.
3. Spray the mice with 70% ethanol, making sure that the fur is thoroughly soaked. This prevents stray hair from getting incorporated into your tissue sample.
4. Make an incision in the skin at the mouse belly and carefully retract the skin by pulling downwards to expose the hindlimbs. Assess and confirm tissue-specific recombination using a fluorescence stereomicroscope for the Tie2-Cre;*R26*$^{NG/+}$ mice (GFP+ vasculature can be easily detected at a whole mount level).
5. Harvest all skeletal muscle from both hind limbs. Remove and exclude as much overlying/intermuscular fat as possible. Process muscle from each mouse independently (*see* **Note 4**). Collect tissue in a petri dish.
6. Add two drops of sterile 1× DPBS per sample to prevent desiccation of the tissue. Maintain sterile working conditions for the subsequent steps.
7. Mince the muscle thoroughly for 7–8 min.
8. Transfer each sample of minced muscle into its respective tube containing the digest media.
9. Using a 5 ml pipette, triturate the slurry about 10–15 times. Use a separate pipet for each sample to avoid cross contamination.
10. Place the tubes in a 37 °C water bath and incubate for 1 h, triturating every 15 min.
11. Quench enzymatic digestion by adding 5 ml of cold media to each sample.
12. Run each cell suspension sample through 100 μm and 70 μm filter cell strainers respectively. After each filtration step, wash the filter with an additional 5 ml volume of cold media.
13. Centrifuge samples at 800 ×*g* at 4 °C for 5 min.

14. Aspirate supernatant and resuspend the pellet with 10 ml of cold 1× DPBS per sample.
15. Repeat **step 13**.
16. Carefully aspirate supernatant and gently resuspend pellet in 180 μl of blocking media (*see* **Note 5**).

3.2.2 Magnetic Depletion of CD31+ (Endothelial) and CD45+ (Hematopoietic) Cell Fractions

1. Add 20 μl of anti-CD31 and 20 μl of anti-CD45 magnetic bead-conjugated antibodies to each sample. Mix sample gently but thoroughly with a micropipette.
2. Incubate for 15 min on ice at 4 °C.
3. Meanwhile, set up the MACS cell separator instrument (*see* Subheading 3.1.3).
4. After completion of the 15 min incubation period, add 500 μl of blocking media to each sample.
5. Run each sample through an individual column (*see* **Note 4**). Collect cell suspension in a sterile 15 ml tube.
6. Wash each column with blocking media, 3 × 3 ml.
7. Spin at 800 × *g* for 5 min, aspirate supernatant completely and resuspend cell suspension in 1 ml of blocking media.
8. The two WT or NG experimental samples can be combined at this point to make a total volume of ~2 ml.
9. Proceed to antibody staining step.

3.2.3 Antibody Staining

1. Label nine, 2 ml microcentrifuge tubes as per Table 1.
2. First add 100 μl of the appropriate cell suspension to tubes **1–8**. Depending on the experimental mice being used, the cell suspension added to each tube will vary. When using WT mice, the same cell suspension can be used for all tubes. When using Tie2-Cre;$R26^{NG/+}$ mice, use either NG or NFC samples as indicated in Table 1.
3. Tubes 1–5 serve as single color compensation controls and 6–8 as Fluorescence-Minus-One (FMO) controls, for setting gate parameters (Fewer single-stained and FMO control tubes are needed for WT mice, *see* Subheading 3.3.2). Add antibodies Sca-1-V450 (0.25 μl) or PDGFRα-APC (1 μl) to the control tubes as per Table 1. 7-AAD is added just prior to sorting and not at this time.
4. Tube **9** is the sort sample. The bulk of the WT or NG cell suspension sample (~1600–1800 μl) is added to this tube. Add 0.25 μl of Sca-1-V450 and 1 μl PDGFRα-APC, per 100 μl of sample.
5. Mix the contents by gentle pipetting with a micropipette.
6. Incubate at 4 °C on ice for 30 min.

Table 1
Antibody staining of control and experimental cell suspensions for FACS sorting

			Antibody		
Tube #[a]	Tube name	Cell suspension[b]	7AAD	PDGFRα-APC	Sca-1–V450
1	No color	NFC	–	–	–
2	7AAD only	NFC	+	–	–
3	APC only	NFC	–	+	–
4	V450 only	NFC	–	–	+
5	GFP only	NG	–	–	–
6	GFP FMO	NFC	+	+	+
7	APC FMO	NG	+	–	+
8	V450 FMO	NG	+	+	–
9	Experimental (All markers)	NG	+	+	+

[a]Tubes 1–8 serve as controls and tube 9 is the experimental sample
[b]If using WT experimental mice, the same cell suspension can be added to all the tubes (*see* **step 2**, Subheading 3.2.3)

7. Add 200–400 μl of ice-cold sterile 1× DPBS to each tube and mix by inverting.
8. Spin samples at 800 ×*g* at 4 °C for 5 min.
9. Aspirate the supernatant and resuspend cells in 800 μl of ice-cold sterile 1× DPBS for tubes 1–8 and 1600 μl for tube 9.
10. Repeat **step 8**.
11. Resuspend cells in 400 μl of 2% FBS/1× DPBS for tubes 1–8 and 1600 μl for tube 9.
12. Transfer the contents of each tube into a 5 ml FACS tube that has a 35 μm cell strainer cap.
13. Add 3 μl of 7-AAD per 100 μl of cell suspension as indicated in Table 1.
14. Samples are now ready for cell sorting.

3.3 Cell Sorting

3.3.1 Isolation of Target Cells from Tie2-Cre; $R26^{NG/+}$ Mice

1. Run the single color controls (tubes 1–5) and adjust compensation.
2. Draw a gate on the FSC-A and SSC-A plot that excludes any debris (Scatter gate, Fig. 1a).
3. Create a gate to exclude any dead cells that have been stained with 7-AAD (Live gate, Fig. 1a).

4. Draw gates to include only intact single cells and exclude any doublets and cell aggregates (Doublet discrimination gates, Fig. 1a).
5. By comparing FMO controls (tubes 6–8) and the stained sample (tube 9), create a gate to establish boundaries that accurately discriminate between positive and negative signals for APC and V450. This gate will demarcate the PDGFRα+ Sca-1+ population (Fig. 1a).
6. Using the GFP FMO, draw a gate to collect the GFP+ fraction of the PDGFRα+Sca-1+ population (Fig. 1b). Cells can be sorted into 300 μl of 1× DPBS supplemented with 2 % FBS or directly into Matrigel-coated plates containing GM.
7. We use backgating to examine the distribution of the target population at higher levels in the gating hierarchy to verify that target cells display similar scatter properties and protein expression levels. This helps to ensure that all of our gates are positioned correctly (data not shown).

3.3.2 Isolation of Target Cells from Wild Type Mice

If using WT mice, the gating progression to be followed is almost identical to what is listed above in Subheading 3.3.1. Prepare tubes (1–4) and (7–9) as per Table 1. Use the gating strategy indicated in Fig. 1a and collect all PDGFRα+Sca-1+ cells.

3.3.3 Analysis

With this protocol, there is some degree of variability in the yield. PDGFRα+Sca-1+ cells represent 8–9 % of the total live mononucleated cells after depletion of the CD31+CD45+ (lin−) fraction

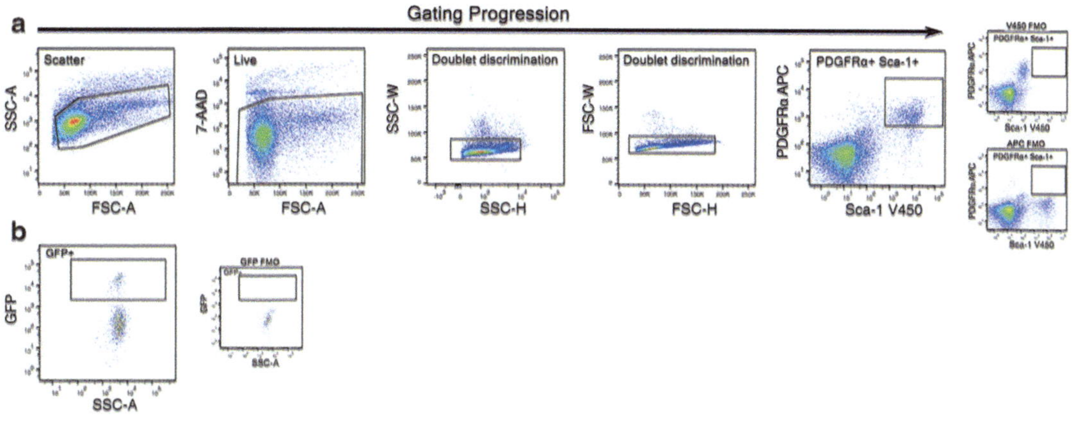

Fig. 1 FACS enrichment scheme for skeletal muscle-derived Tie2+ progenitors. (**a**) Forward and side scatter gates were used to exclude debris. A live gate (7-AAD− cells) and doublet discrimination gates were used to exclude dead cells and cell aggregates respectively. APC and V450 FMOs were used to gate out cells that may display background fluorescence, which might confound accurate identification of the PDGFRα+ Sca-1+ fraction. (**b**) The GFP+ gate was used to collect labeled cells from the PDGFRα+ Sca-1+ population. The GFP FMO is used to distinguish between GFP-labeled and unlabeled cells

and antibody staining. GFP+PDGFRα+Sca-1+ cells represent only 1.2–1.6 % of total live lin– mononucleated cells. Although the use of the Miltenyi enrichment step will reduce yield of the target population, pre-FACS cell enrichment will significantly reduce sort time, with a consequent reduction in cell stress and improved work-flow efficiency. Reduced cell yield can be offset by the use of two experimental mice.

3.4 Cell Culture and Analysis

Seeding density and plate size can be determined as per specific application. We have had success growing cells after plating at multiple seeding densities and using different well sizes. In our previous publication, we conducted clonal analysis using 96-well plates to assess heterogeneity within the Tie2+ population [1]. For assessing differentiation capacities, we typically plate 200 cells per well in 12-well plates (~50 cells per cm^2), as described below.

1. Plate sorted Tie2+ cells at a density of 200 cells per well in a 12-well plate and grow in GM at 5 % CO_2.
2. For the first 3 days after plating, allow cultures to become established.
3. Change GM 3 days after seeding and every 2 days thereafter.
4. Propagate cells until the density reaches 80–90 % confluency.
5. To promote adipogenic differentiation, maintain high serum conditions; adipocytes are typically observed at day 13 or 14.
6. To induce osteogenic differentiation, switch to osteogenic differentiation media at day 14 and continue to culture the cells for an additional 8–10 days, with media changes every 2 days.
7. Typically, we use freshly isolated cells for all of our analyses since extensive passaging may result in reduced differentiative capacity (unpublished observations). If passaging the cells is necessary, allow the cells to grow until the density reaches about 70–80 % confluence. Ensure that the density does not get too high so as to avoid excessive adipogenic differentiation.

Figure 2 shows examples of GFP+PDGFRα+Sca-1+ cells grown at high density in GM or osteogenic differentiation media (BMP-2 treated). Adipogenic differentiation, which is obvious by visual inspection under contrast microscopy, can be confirmed by staining for the adipogenic marker, Perilipin. Cells robustly undergo adipogenic differentiation by day 13–14 in culture under both GM and osteogenesis-inducing conditions (a, b and d). Osteogenic capacity can be assessed by staining for osteogenic markers, such as Osterix [1], or by Alkaline phosphatase (ALP) staining. Cells cultured in GM show relatively mild ALP staining (left column, e), which is likely due to BMPs present in the FBS (unpublished observations). Under osteogenesis-inducing conditions, cells show intense ALP activity (right column, e).

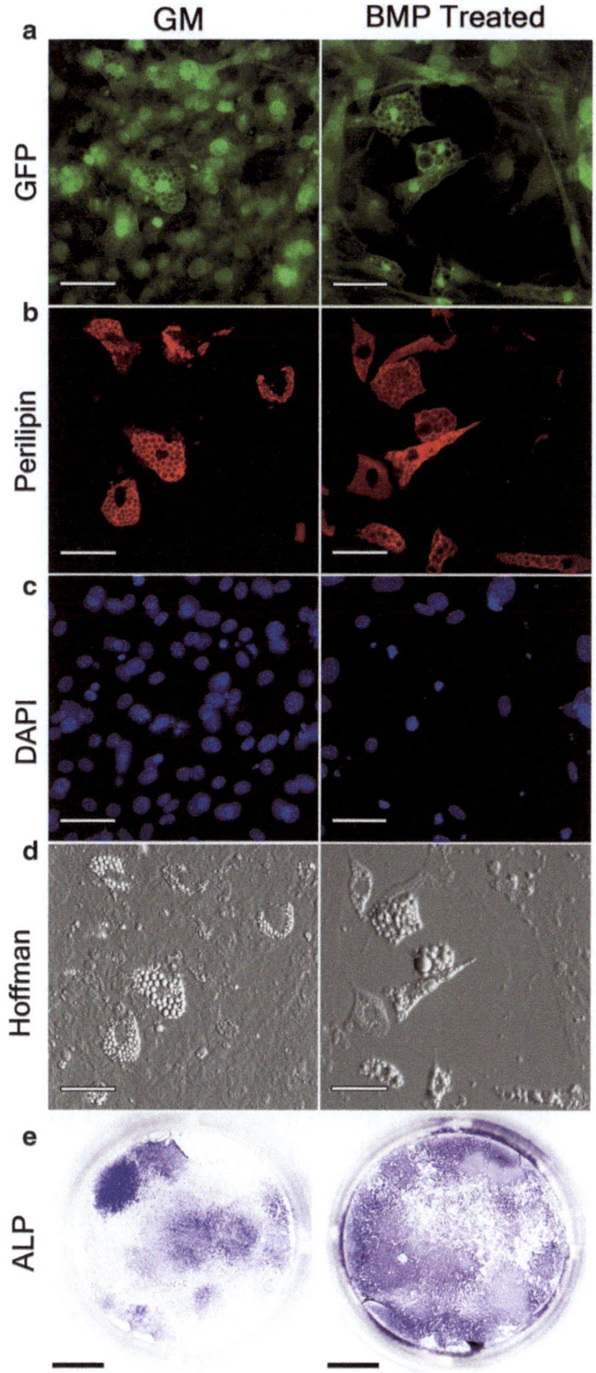

Fig. 2 Adipogenic and osteogenic differentiation of GFP+PDGFRα+Sca-1+ progenitors in culture. Cells were sorted directly into 12-well plates at a density of 200 cells per well and grown for 14 days in GM (*left column*). Parallel wells were grown for an additional 8 days in osteogenic differentiation medium (BMP2-treated; *right column*). (**a–d**) Under both culture conditions, adipogenic differentiation was evident by staining with the adipogenic marker, Perilipin (**b**), and by visual inspection using contrast microscopy (**d**). (**e**) After 14 days in GM, cells show mild ALP activity, which is likely due to endogenous BMPs present in the GM (*left column*). Cells cultured in osteogenic differentiation medium (*right column*) show strong ALP activity. Scale bars, 50 μm (*white*) and 0.5 cm (*black*)

4 Notes

1. It is important to titrate the FACS antibodies to determine the appropriate working dilution since lot-to-lot variation can affect performance. The use of isotype controls while titering antibodies is recommended.
2. We have detected strong ALP activity in Tie2 progenitors cultured in osteogenic differentiation media containing 50 ng/ml BMP-2 (unpublished observations).
3. Since Matrigel is very viscous, coating the well surface evenly with the volume recommended by the manufacturer can be challenging. Increasing the volume by 20% helps expedite the process.
4. For the purposes of isolation of mononucleated cells (Subheading 3.2.1) and depletion using the MACS separator system (Subheading 3.2.2), each tissue sample (if using multiple mice) should be processed separately. Timing for mincing and muscle dissociation has been optimized for muscle mass that is generated from one mouse (1.5–2 g per mouse). Also, combining samples will cause clogging during filtration steps and inefficient magnetic depletion, resulting in reduced yield and enrichment. Samples from mice of the same genotype can be mixed after magnetic depletion is complete (**step 8**, Subheading 3.2.2).
5. We do not recommend conducting red blood cell lysis prior to FACS, since it adversely affects cell yield.

Acknowledgements

We thank FACS Core Facility scientists Dr. Carol Norris and Robert Smith for their expert technical assistance, and Dr. Michael Wosczyna for developing the initial methodologies for isolating Tie2+ progenitors. We also thank Dr. John Lees-Shepard, Nicholas Legendre, Anthony Patelunas, and Sarah-Anne Nicholas for their valuable comments during the course of this work and on the manuscript. This work was supported by NIH grant R01AR057371.

References

1. Wosczyna MN, Biswas AA, Cogswell CA, Goldhamer DJ (2012) Multipotent progenitors resident in the skeletal muscle interstitium exhibit robust BMP-dependent osteogenic activity and mediate heterotopic ossification. J Bone Miner Res 27:1004–1017. doi:10.1002/jbmr.1562
2. Joe AWB, Yi L, Natarajan A et al (2010) Muscle injury activates resident fibro/adipogenic progenitors that facilitate myogenesis. Nat Cell Biol 12:153–163. doi:10.1038/ncb2015
3. Uezumi A, Fukada S, Yamamoto N et al (2010) Mesenchymal progenitors distinct from satellite

cells contribute to ectopic fat cell formation in skeletal muscle. Nat Cell Biol 12:143–152. doi:10.1038/ncb2014

4. Pretheeban T, Lemos DR, Paylor B et al (2012) Role of stem/progenitor cells in reparative disorders. Fibrogenesis Tissue Repair 5:20. doi:10.1186/1755-1536-5-20
5. Uezumi A, Ikemoto-Uezumi M, Tsuchida K (2014) Roles of nonmyogenic mesenchymal progenitors in pathogenesis and regeneration of skeletal muscle. Front Physiol 5:68. doi:10.3389/fphys.2014.00068
6. Yin H, Price F, Rudnicki MA (2013) Satellite cells and the muscle stem cell niche. Physiol Rev 93:23–67. doi:10.1152/physrev.00043.2011
7. Potter BK, Burns TC, Lacap AP et al (2007) Heterotopic ossification following traumatic and combat-related amputations prevalence, risk factors, and preliminary results of excision. J Bone Joint Surg Am 89:476–486
8. Cipriano CA, Pill SG, Keenan MA (2009) Heterotopic ossification following traumatic brain injury and spinal cord injury. J Am Acad Orthop Surg 17:689–697
9. Shore EM, Kaplan FS (2008) Insights from a rare genetic disorder of extra-skeletal bone formation, fibrodysplasia ossificans progressiva (FOP). Bone 43:427–433. doi:10.1016/j.bone.2008.05.013
10. Kisanuki YY, Hammer RE, Miyazaki J et al (2001) Tie2-Cre transgenic mice: a new model for endothelial cell-lineage analysis in vivo. Dev Biol 230:230–242. doi:10.1006/dbio.2000.0106
11. Yamamoto M, Shook NA, Kanisicak O et al (2009) A multifunctional reporter mouse line for Cre- and FLP-dependent lineage analysis. Genesis 47:107–114. doi:10.1002/dvg.20474

Part III

Muscle Functional Assays

Chapter 19

In Vitro Assays to Determine Skeletal Muscle Physiologic Function

Justin E. Sperringer and Robert W. Grange

Abstract

In vitro muscle contractile function assays are important to characterize the differences between different muscle types (e.g., slow vs. fast), between a diseased and non-diseased muscle, or importantly, to demonstrate the efficacy of a muscle treatment such as a drug, an overexpressed transgene, or knockout of a specific gene. Fundamental contractile properties can be assessed by twitch, tetanic, force–frequency, force–velocity, and fatigue assays. Many of these assays are conducted with the muscle at a constant length, e.g., an isometric contraction. However, to better represent the dynamic purpose of muscles in vivo (e.g., to move limbs), dynamic assays such as the force–velocity (concentric contractions) or stretch-injury (eccentric contractions) should also be obtained. Characterizing skeletal muscle function in vitro is a powerful approach to demonstrate efficacy of a treatment to rescue diseased muscle and to assess functional regeneration.

Key words Strength, Power, Rescued function, Muscle physiology

1 Introduction

In this chapter we describe the equipment, software, and protocols to obtain fundamental mouse skeletal muscle isometric and dynamic contractile properties in vitro, including twitch, tetanus, force–frequency, force–velocity (and power), and fatigue assays. The stretch injury protocol for isolated muscles is described in Chapter 1 of this volume [1].

Muscles are specialized and adaptable, largely because of the signaling pathways that respond during development or in response to a stress such as exercise to drive adaptation [2]. Mouse muscles are composed of different fiber types (e.g., I, IIa, IIx, and IIb), and each has different properties (e.g., force produced; speed of shortening) [3]. When a muscle is composed of a greater proportion of a particular fiber type(s), the muscle takes on the properties of those predominant fibers. For example, a slow soleus muscle composed primarily of type I and IIA fibers [4] produces less absolute

Michael Kyba (ed.), *Skeletal Muscle Regeneration in the Mouse: Methods and Protocols*, Methods in Molecular Biology, vol. 1460, DOI 10.1007/978-1-4939-3810-0_19, © Springer Science+Business Media New York 2016

force and shortens more slowly (see examples below), than a fast extensor digitorum (EDL) muscle composed of primarily type IIx and IIb fibers [5]. These differences are easily distinguished by the functional assays noted above. For example, to determine the effects of knocking out Sox 6, a slow phenotype repressor [6], functional analyses demonstrated that the fast EDL demonstrated properties similar to a slow muscle phenotype (e.g., decreased V_{max}; less fatiguable; Fig. 2b, c [6]).

Within the context of a diseased muscle, fundamental contractile properties can also differ from those of a non-diseased muscle (e.g., dystrophic vs. non-dystrophic; [7]. In this regard, a treatment to improve or even rescue function, can also be discriminated by these functional assays. Because of the importance of muscles such as the diaphragm to breathing, or skeletal muscles for speech, swallowing, and mobility, functional improvement in response to treatment provides powerful support for the efficacy of the treatment (*see* Fig. 3c) [8]. From the perspective of myocyte engraftment into a muscle to rescue a specific disease state, there may be evidence the gene is transcribed, the encoded protein is translated, has localized correctly and is functional, and engraftment of these muscle cells in vivo has been successful in the regenerated muscle; a final determination of treatment success is demonstrating that muscle function has improved [9].

1.1 In Vitro Muscle Preparation

In vitro is the term used to describe the isolated skeletal muscle preparation, but can also be referred to as ex vivo. Typically, the muscle of interest is removed intact from a deeply anesthetized mouse and placed in a temperature-controlled bath immersed in a salt solution (e.g., physiological salt solution) that mimics closely the environment in the animal and thereby keeps the muscle viable. The muscle is mounted in the bath fixed at one tendon, and attached by the other tendon to a transducer that can measure the force response when the muscle is activated to contract by an electrical stimulator. The electrical stimulation pulses are delivered to the muscle as a field stimulation (i.e., the electrodes are close to but do not touch the muscle) via platinum electrodes. Various in vitro assays are used to characterize the contractile force properties of the muscle.

1.2 Brief Description of Assays

A twitch is the force response to a single electrical stimulation (i.e., 1 Hz). A maximal tetanus is the force response to some maximal stimulation frequency; there is no further increase in force when frequency of activation is increased. A force–frequency relation is a plot of the force responses to activation frequencies between a twitch and the maximal tetanic frequency. A force–velocity relation is a plot of the peak speed of shortening for each associated load moved by the muscle. The load is typically expressed as a percent (or decimal fraction) of the maximal tetanic force response and would vary from ~5% (0.05) to 100% (1.00) of maximal tetanic force. From this plot, V_{max} or the maximal speed of shortening can be calculated. In addition, the power the muscle can produce can be determined by multiplying the load (force) moved by the muscle times the peak shortening speed (velocity) at that load. Fatigue is the inability of a muscle to produce a desired force. Fatigue is often induced by stimulating the muscle over time using a specific stimulation pattern (e.g., intermittent or continuous) at a specific stimulation frequency. Over the course of the stimulation pattern, the force output of the muscle decreases (i.e., fatigues).

1.3 Typical Muscles Assayed

The extensor digitorum longus (EDL), a muscle with fast contractile properties, and the soleus, a muscle with slow contractile properties are typically used because of their accessibility for dissection and their anatomical size provides good viability in the tissue bath because of suitable oxygen diffusion. Other in vitro preparations include diaphragm strips [10] and the tibialis anterior muscle [9], among others (e.g., lumbrical, [11]); these are not discussed below.

In a C57Bl/6J mouse, which is commonly used for muscle contractile studies, the fiber type distribution for EDL muscles is approximately 1% type I, 20% type IIx, 79% IIb (Leudeke et al. [5]), and for soleus muscles is approximately 40–45% type I, 55–60% type IIa, 3% IIx, IIb [4, 5]. Because of the differences in fiber type distribution, the contractile properties between these muscles are easily distinguished (i.e., EDL: type IIx, IIb—fast; soleus: type 1, IIa—slow). The EDL extends the hind limb digits, and is a contributor to dorsiflexion during mouse gait. The soleus attaches to the calcaneus via the Achilles tendon and contributes to plantarflexion during mouse gait.

2 Materials

To perform the various in vitro protocols, specialized equipment is needed. Details of dissection of the EDL are provided in Chapter 1 [1]; we have added additional details where our methods differ. Below, we also describe the dissection of the Soleus. We also provide details herein on data collection and describe some key variables and their units that can be obtained from the various assays. Our equipment and software were purchased from Aurora Scientific, Inc. (ASI) because both isometric (e.g., twitch, tetanus) and dynamic (e.g., force–velocity, eccentric) contractions can be obtained.

2.1 In Vitro Muscle System (*See* Note 1)

1. 300C, Dual-Mode Muscle Lever System.
2. 800A in vitro Muscle Apparatus (Fig. 1; *see* **Note 2**).
3. Radno*ti 1583 Seri*es Tissue Organ Bath (25 ml) (*see* **Note 3**).
4. 701C Electrical Stimulator.
5. Stimulation electrodes—Pure platinum wire [1].
6. Series 604A Analog/Digital Signal Interface (*see* **Note 4**).

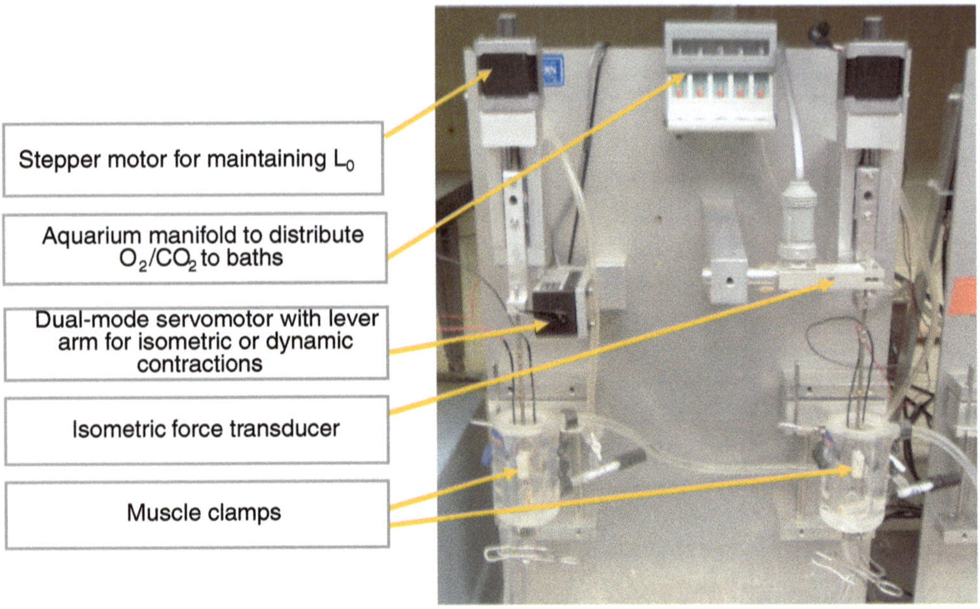

Fig. 1 Stepper motor, dual-mode servomotor, and baths used for in vitro mouse skeletal muscle contractile function assays. A transducer for only measuring isometric forces and the manifold for distributing gas to the baths are shown but not discussed in the chapter

7. PC computer with National Instruments Analog/Digitial board installed.
8. 605A Dynamic Muscle Control and Analysis Software (*see* **Note 5**).

2.2 Additional Materials for In Vitro Contractile Assays

1. Physiological Salt Solution (PSS) (Table 1, *see* **Note 6**).
2. Thermometer −20 to +50 °C in 0.1 °C gradations to confirm temperature of PSS.
3. Dissection dish for PSS (e.g., 100 mm culture dish).
4. Adroit Medical Systems HTP-1500 Heat Therapy Pump; 5/16″ inside diameter tubing to run water to tissue organ baths (*see* **Note 7**).
5. 95 % O_2/5 % CO_2 specialty gas.
6. Gas tank regulator (e.g., Concoa Series 402 Regulator); 3/16″ inside diameter tubing to run gas lines to tissue organ baths.

2.3 Dissection Equipment, Surgical Instruments, Anesthetic: As Described in Chapter 1 [1]; Additional Materials are Listed Below

1. Anesthetic: ketamine–xylazine. Make a fresh 1 ml cocktail of 0.2 ml ketamine (20 mg/ml final) + 0.1 ml (10 mg/ml final) xylazine + 0.7 ml sterile saline.
2. 4-O silk suture (e.g., Deknatel).
3. 1.0 ml polyethylene transfer pipette (e.g., Samco).
4. 10 μl plastic pipette tip.
5. Hemostat (e.g., Crile hemostat, Fine Science Tools #13004-14).

Table 1
Physiological saline solution (PSS)

Reagent	Mass (g)	□ final (mM)
Deionized water (diH_2O)	1.000 L	N/A
Sodium chloride (NaCl)	7.042 g	120.5
Potassium chloride (KCl)	0.348 g	4.7
Magnesium sulfate heptahydrate ($MgSO_4 \cdot 7H_2O$)	0.296 g	1.2
Sodium phosphate monobasic monohydrate ($NaH_2PO_4 \cdot H_2O$)	0.166 g	1.2
Sodium bicarbonate ($NaHCO_3$)	1.714 g	20.4
Calcium chloride dihydrate ($CaCl_2 \cdot 2H_2O$)	0.236 g	1.6
Dextrose ($C_6H_{12}O_6$)	1.802 g	10
Pyruvate ($C_3H_4O_3$)	0.110 g	1.0

3 Methods

3.1 Computer, Software, and In Vitro Data Collection System

1. Turn on computer.
2. Run DMC.
3. Turn on 300C-LR electronics and stimulator.
4. Set up data folders.
   (a) Create a data folder by month and year (e.g., August 2015).
   (b) Create folder for each animal (Mouse_C1_EDL; C-control animal).

3.2 Bath Setup

1. Turn on the circulating pump ~20–30 min before the start of the experiment. Set the desired temperature (we use 30 °C); confirm temperature of the PSS with the thermometer; adjust the temperature control as needed.
2. Make 1 L PSS.
3. Clamp the bath drains.
4. Turn on the gas (95 % O_2/5 % CO_2), before adding PSS to baths.
5. Fill each bath with ~25 ml PSS. When gassed with 5 % CO_2 the pH will be ~7.4; can check pH with litmus paper or take a sample and use a pH meter.
6. Fill the dissection dish with PSS and bubble with 95 % O_2/5 % CO_2.

3.3 EDL Muscle Dissection

Differences between the protocol described in Chapter 1 and ours are described below reflecting necessary variations based on slightly different equipment and experienced preferences.

1. Deliver 0.2 ml of ketamine–xylazine cocktail per 20 g mouse body mass intraperitoneal (i.p.) Inject additional ketamine–xylazine i.p. as needed to obtain and maintain a deep plane of anesthesia (e.g., no foot reflex). After muscles are excised, we euthanize the mouse by CO_2 asphyxiation.
2. To mount the EDL between clamp and arm of servomotor, prepare two sutures for each muscle prior to dissection: cut one length ~12–15 cm, and another ~6 cm.
3. Expose EDL as described [1]; keep the muscle moist by dripping oxygenated PSS from transfer pipette ~every 30–60 s.
4. Suture for distal tendon. Slip tips of Dumont forceps under the distal tendon and provide an opening just large enough to grab the suture and gently pull it through ~3–4 cm. This should leave a second long suture tail of 11–12 cm. Position the suture at the myotendinous junction and tie two very tight half hitch knots; the first perpendicular to the length of the muscle, the second in line with the muscle. When the distal suture is secure, use the Student Vannas scissors to cut the distal tendon of the EDL to obtain ~2 mm of the tendon distal to the myotendinous junction.
5. Suture for proximal tendon. With the other suture length, make a loop 2–3 mm in diameter with one loose half-hitch; trim the tails to ~3 cm long. Gently pull the EDL towards the knee and expose the proximal tendon (*see* Chapter 1 Call and Lowe [1]).
6. Hold the loop with one of the tails using forceps and very gently pass the long distal suture and then the EDL through the loop, until the loop is positioned at the proximal myotendinous junction (keep the muscle moist throughout). Tie two very tight half hitch knots as described for the distal tendon; cut the proximal tendon of the EDL.
7. The excised EDL should be moved quickly to the bath by holding the long suture, and be initially immersed in the bath PSS for 15–30 s before hanging.
8. Mounting the EDL between clamp and arm of servomotor. The proximal tendon of the EDL and the two tails of the suture are placed into the clamp at the base of the bath, and secured tightly with the knot at the myotendinous junction just showing above the clamp (Fig. 3a). Clamping the muscle should take no more than 15–30 s. Keep the muscle moist with PSS. Leave the long suture tail draped over the edge of the bath. Raise the tissue bath to immerse the muscle in PSS.
9. Gently lift the long suture (with muscle attached) up to the lever arm, ensure the muscle is not twisted, and do not stretch the muscle. Slip the suture over the hook on the lever arm and

tie two half hitches to secure the muscle to the hook. The knots should be tight, but there should be some slack in the suture between the lever arm and the muscle. Manually adjust the tension on the muscle to 1.0 g of resting tension. In our system, the clamp moves relative to the fixed servomotor (i.e., moving the clamp down tightens the suture, and increases muscle tension).

10. Let the muscle equilibrate at 1.0 g tension for 10 min.

3.4 Soleus Muscle Dissection

1. Dissection instruments and equipment are as described in Chapter 1 [1].
2. The soleus is a plantarflexor muscle located on the posterior aspect of the hind limb, between the medial and lateral heads of the gastrocnemius muscle.
3. Prepare two lengths of 4-O suture for each muscle prior to dissection: One length ~12–15 cm; another ~6 cm.
4. Deeply anesthetize the animal with ketamine–xylazine as described for the EDL above.
5. With the mouse supine (i.e., on its back), pull skin back past the knee and over the calcaneus).
6. Once the muscles of the hind limb are exposed, keep them moist with PSS every ~30–60 s.
7. Suture for distal tendon. The soleus lies beneath the medial and lateral heads of the gastrocnemius muscle, but both muscles (and the plantaris) share the Achilles tendon. Use the Dumont forceps to make a small hole through the fascia between the Achilles and the posterior portion of the tibia, ~2–3 mm proximal from the calcaneus. The Student Vannas scissors can be used in blunt dissection to open the hole if needed. Thread the short suture through the hole ~3 cm, grab with the Dumont forceps and position the suture to the proximal edge of the hole. This should position the suture close to or at the myotendinous junction for the soleus.
8. Tie two tight half hitches (perpendicular to and then in series with the soleus) around the Achilles tendon. Trim the two tails to ~2 cm.
9. Before cutting the Achilles tendon, use the Student Vannas scissors to open the fascia along the medial and lateral sides of the gastrocnemius up to the knee; do not go too deep as the soleus could be damaged. Any overlying muscles medial and lateral to the knee can be cut so the knee is well-exposed.
10. Use the Student Vannas scissors to cut the Achilles tendon distal to the suture and close to the calcaneous.
11. Elevate the foot with one hand and gently grasp the Achilles tendon with the Graefe forceps and peel back towards the knee to reveal the soleus muscle still centered between and against the heads of the gastrocnemius muscle. As needed, by blunt dissection or by cutting, trim the fascia on either side of the

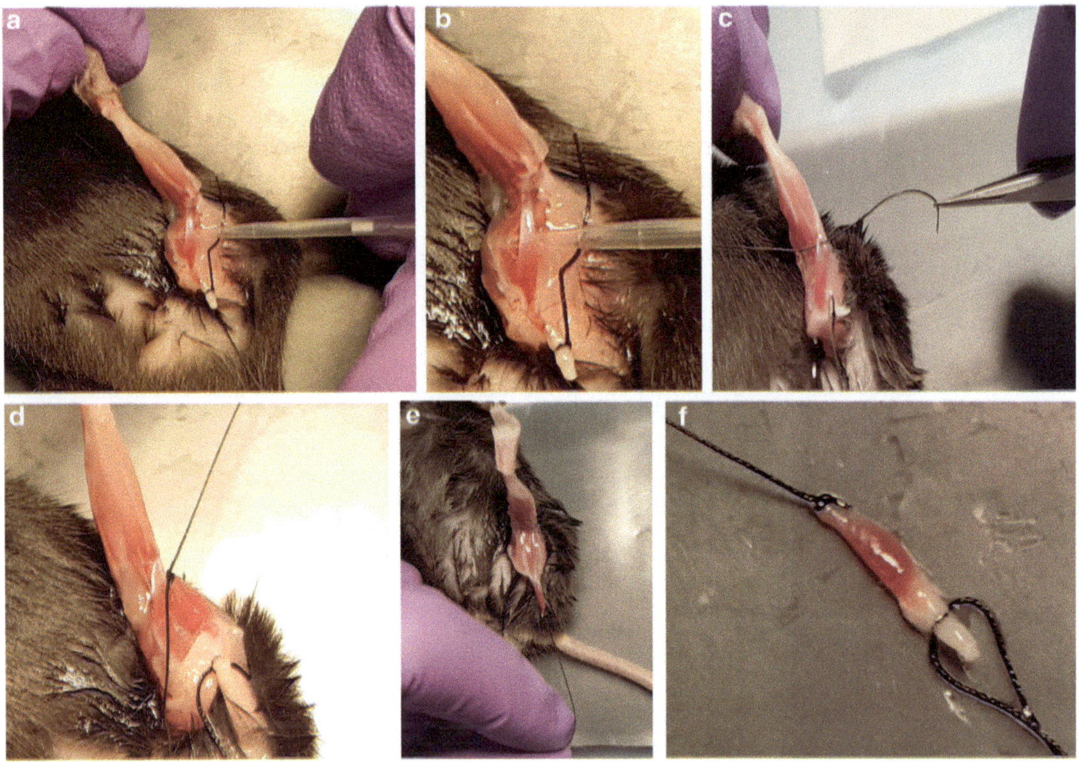

Fig. 2 Soleus dissection. (**a**) Short suture is tied at Achilles tendon; Achilles severed from calcaneus to reveal soleus; soleus muscle is separated from the gastrocnemius muscles using a 10 μl pipette tip. (**b**) Soleus muscle lifted to show proximal attachment at knee. (**c**) 4-0 suture passed under proximal tendon of soleus. (**d**) Long 4-0 suture tied at proximal tendon. (**e**) Soleus proximal tendon cut at knee; long suture used to gently pull soleus back towards Achilles; (**f**) completely dissected soleus: short suture at Achilles; long suture at proximal tendon

gastrocnemius with the Student Vannas scissors as you peel back to reveal the knee and the proximal tendon of the soleus. This tendon is flat and white, located on the lateral posterior aspect of the tibia at the knee. To keep the posterior side of the tibia exposed to perform the remaining steps, hold the foot with your fingers or with a small hemostat so the knee is against the mouse body. The foot should be away from the dissector and the Achilles close to the dissector.

12. Use the Dumont forceps to open a small gap between the proximal tendon and the gastrocnemius (close the tips, press down under the tendon and slowly move under and perpendicular to the tendon; do not touch the muscle with the forceps). Remove the forceps. When a small gap is made under the tendon, wet the 10 μl pipette tip with PSS. Gently pass a small length (1–2 mm) of the tip under the proximal tendon, and then press down and gently roll the tip towards the calcaneus (Fig. 2a, b). This action should help mostly separate the soleus from the gastrocnemius, but final separation is described below.

13. Suture for proximal tendon. Thread the long suture under the proximal tendon ~3 cm, and grab with the Dumont forceps (Fig. 2c). Tie two very tight half hitch knots at the myotendinous junction (perpendicular and in series) but be sure not to stretch the muscle (Fig. 2d); and then cut the tendon as close to its attachment on the tibia as possible to leave a 1–1.5 mm of tendon proximal to the suture.
14. Use the Graefe forceps, grab the long suture tail and gently pull the soleus back towards the Achilles (Fig. 2e). Use the Student Vannas scissors to free the fascia between the soleus and the gastrocnemius muscles carefully to avoid cutting the soleus. When the Achilles is reached, cut the Achilles from the gastrocnemius to free the soleus proximal to the Achilles suture knot.
15. The excised soleus (Fig. 2f) should be moved quickly to the bath by holding the long suture, and be initially immersed in the bath PSS for 15–30 s before hanging.
16. Mounting the soleus between clamp and arm of servomotor. The Achilles tendon and the two tails of the suture are placed into the clamp at the base of the bath, and secured tightly with the knot at the myotendinous junction just showing above the clamp (Fig. 3b). Clamping the muscle should take no more than 15–30 s. Keep the muscle moist with PSS. Leave the long suture tail draped over the edge of the bath. Raise the tissue bath to immerse the muscle in PSS.
17. Gently lift the long suture (with muscle attached) up to the lever arm, ensure the muscle is not twisted. Slip the suture over the hook on the lever arm and tie two half hitches to secure the muscle to the hook. The knots should be tight, but there should be some initial slack in the suture between the lever arm and the muscle. Manually adjust the tension on the muscle to 1.0 g of resting tension; this should take up the slack in the suture. In our system, the clamp moves relative to the fixed

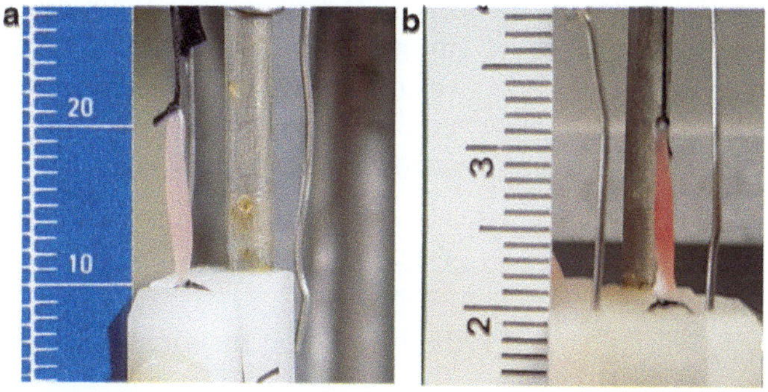

Fig. 3 (**a**) EDL muscle mounted in clamp. (**b**) Soleus muscle mounted in clamp. For both muscles, the suture knot is visible

servomotor (i.e., moving the clamp down tightens the suture, and increases muscle tension).

18. Let the muscle equilibrate at 1.0 g tension for 10 min.

3.5 Common Terms (Units) to Describe Contractile Properties

1. Active force/tension (g, mN, g/mm^2, mN/mm^2)—difference between peak force/tension and resting force/tension.
2. Concentric contraction—muscle produces force and shortens.
3. Contraction time (ms)—the time period between onset of the contractile response and the return of active muscle force to resting force.
4. Cross-sectional area (CSA; mm^2). CSA can be estimated from the equation: muscle mass (mg)/(length (mm) × muscle density (1.06 mg/mm^3) (Mendez and Keys [12]).
5. Eccentric contraction—muscle produces force and is lengthened by an external force greater than the force produced by the muscle.
6. Fatigue—typically expressed as a percentage of initial force (i.e., force at the beginning of the fatigue protocol) or as the percentage of force lost since the beginning of the protocol. Numerous fatigue protocols have been reported in the literature.
7. Force–Frequency (F-F) relation: force responses at frequencies of electrical activation between 1 Hz (twitch) and the maximal tetanic frequency (e.g., 150 Hz). Typically collected in frequency steps (e.g., 1, 30, 50, 80, 100, 120, 150 Hz) to provide data to plot the force–frequency relation.
8. Force/Tension time integral (g × s; mN × s)—area under the force profile between the beginning of the contractile response and return to resting tension (i.e., Contraction time).
9. Force–Velocity (*F-V*) relation: speed of shortening (i.e., velocity; mm/s; L_o/s; L_f/s) at a given load, typically expressed as a decimal fraction or percent of the maximal tetanic force. Simple explanation: if the load a muscle moves is a low load, shortening is fast; if the load is high, shortening is slow. Two common methods: tetanic quick release; tetanic afterload. The resulting curve of velocity vs. load can be fit with the Hill equation (see below). With this equation, a load of zero can be used to determine V_{max}).
10. Half Relaxation Time (HRT; ms)—the time for the muscle to drop from peak active force/tension to half of peak active force/tension.
11. Hill Equation for force–velocity curve fitting: $Y = \dfrac{b(1-x)}{(x+a)}$

    x is the imposed percent load (the afterload; expressed as a decimal), Y is the measured velocity (mm/s) at a given load, and a and b are mathematically derived values. Initial values of $b = 0.3$ (m/s) and $a = 0.25$ (N) can be used to initiate the curve-fitting process in a program such as Graphpad Prism. The values for a and b that result from the curve fit can be used in the equation above to predict V_{max} (i.e., when the afterload $x = 0$).

12. Isometric contraction—force/tension measured with the muscle at the same length, typically L_o.
13. L_f—fiber lengths (mm); determined as a fraction of L_o; typical values for EDL ($0.44L_o$) and for soleus ($0.71L_o$) [13]. Can be used to express muscle shortening velocity (see below).
14. L_o—optimal muscle length (mm); maximal twitch or tetanic responses occur at this length; resting tension/force (g/mN) is used as an index of length.
15. Optimal voltage—the voltage used to activate the muscle to produce maximal twitch or tetanic force responses.
16. Power (e.g., mN × mm/s, mWatts)—determined from force × velocity or work (force × distance)/time).
17. Pulse width (μs)—the duration of each electric pulse delivered by the stimulator (we use a pulse width of 200 μs).
18. Resting force/resting tension—also termed baseline force/tension or just baseline: passive muscle force measured by the force transducer (e.g., ASI servomotor) when the muscle is not electrically activated; typically reported as the initial force representing L_o at which all forces from the electrically activated muscle were obtained.
19. Specific force—active force of a muscle divided by its CSA (g/mm^2; mN/mm^2; kN/m^2).
20. Stress—mechanical term for a force (e.g., active force) applied over a CSA (g/mm^2; mN/mm^2).
21. Tension (g, mN) = force (g, mN).
22. Tetanus (P_o)—maximal force response at some maximal frequency of electrical activation (e.g., 150 Hz; can vary depending on the stimulator used). Tetanic force does not increase despite activation at a frequency above the maximal frequency.
23. Time to peak force or tension (TPF; TPT; ms)—the time for the muscle to rise from resting force/tension to the peak force/tension.
24. Total force/tension (g, mN, g/mm^2, mN/mm^2)—peak force/tension—zero.
25. Twitch (P_t)—maximal force response of a muscle to a single electrical activation pulse (i.e., 1 Hz).
26. Twitch/tetanus ratio—twitch force/maximal tetanic force; ~0.25 but can vary by muscle type, temperature (Head and Arber [14]; Asmussen and Gaunitz [15]).

27. V_{max} (mm/s; L_o/s; L_f/s)—predicted maximum velocity of muscle shortening at zero load; determined from the Hill equation fitting the force–velocity data for a given muscle.

3.6 Stabilizing Protocol

1. After the 10 min equilibration, we do the following to obtain stable responses from the muscle for the experiment.
2. Elicit three twitches at optimal voltage (see below), each 1 min apart. Adjust tension to 1.0 g between each as needed.
3. Elicit three tetani (500 ms stimulus duration) at optimal voltage, each 1 min apart. Adjust tension to 1.0 g between each as needed. We find the tension drops ~0.5 g after the first tetanus, but by the third tetanus after adjusting the resting tension to 1.0 g, the resting tension remains stable. In our lab we have a custom system that includes a stepper motor under computer control that maintains the resting tension at 1.0 g ± 0.05 g. Without the stepper motor, we would manually adjust the resting tension.
4. Allow the muscle to rest quietly for 10 min before performing additional procedures.

3.7 Force–Optimal Voltage Determination

1. Depending on the make and model of the stimulator used to activate the muscles via the platinum electrodes, the optimal voltage to elicit the maximal muscle response may differ. To determine the optimal voltage for a given stimulator, the following procedures can be used. Note, however, that once the optimal voltage has been determined, and is consistent for six to eight muscles, it need only be checked prior to each new study, or periodically (e.g., every 6 months) as desired. The goal is to use the least voltage required to elicit the maximum twitch and tetanic force responses.
2. Maintain the resting tension of the muscle at 1.0 g tension throughout (adjust the tension as needed). Perform twitches then tetani as described below.
3. The muscle should be activated at 1 Hz (i.e., one activating pulse). This can be done with the DMC software. Note that example software protocols to stimulate the muscle are provided by ASI. These can be adapted as needed by the investigator using the software. Load the appropriate stimulation protocol to elicit a twitch.
4. Collect twitches using the DMC software starting with a voltage setting of 5 V assuming the 701C stimulator is being used (depending on the stimulator, the initial voltage may need to be much higher). Perform a quick analysis of the twitch active force (peak force – resting force) manually on paper and record, but also save the twitch file for each voltage for later analysis. Adjust resting tension to 1.0 g after each twitch if needed.

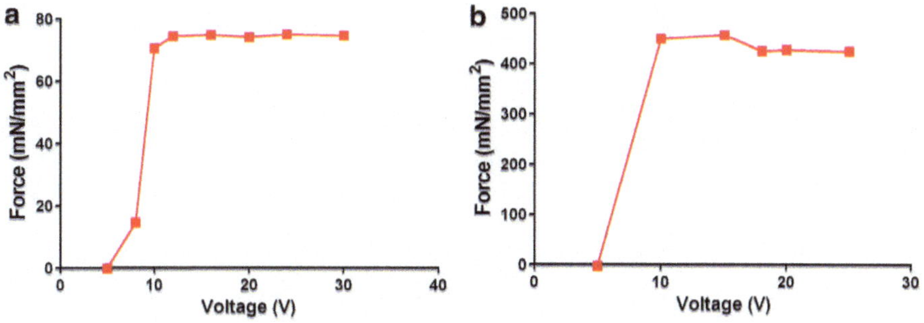

Fig. 4 Example force–voltage plot for EDL twitch (**a**) and tetanic responses (**b**). The optimal voltage for twitches and tetani from these data is 15 V. All force responses were collected at 30 °C

Increase the voltage in 5 V increments. When the maximum force is obtained, record the voltage and go an additional (+5 V) higher to ensure there is no increase in force when voltage is increased. Our experience indicates 5 V increments are suitable, but smaller increments could be used if necessary. Later, a twitch force vs. voltage figure can be plotted (Fig. 4a), the optimal voltage determined, and these data saved to compare to subsequent tests.

5. Test the tetanic response at the twitch optimal voltage. Maintain the muscle at 1.0 g tension for each tetanic stimulation; typically four to five contractions are adequate.
6. Collect tetani (150 Hz), 1 min apart, with voltage set at optimal ± 5 V in (e.g., 20 ± 5 V). Perform quick calculations to determine the tetanic forces elicited, but save each file to plot the tetanic response vs. voltage profile. If the voltage for twitch and tetanus are the same, no further voltage adjustment is needed. If the tetanus requires a higher voltage, use this voltage for the twitches (and all other activation frequencies as well).
7. Repeat the procedure for six to eight muscles. Plot the data and determine the optimal voltage (Fig. 4b). The goal is to use the least voltage required to elicit the maximum twitch and tetanic force responses.

3.8 Length-Tension Relation

1. A fundamental property of a skeletal muscle is the length-tension relation. Briefly, there is an optimal length, termed L_o, at which the muscle will produce the maximum force. At lengths shorter or longer than L_o, the muscle produces less force. This should be established for an $n = 6–8$ samples for each muscle type (e.g., EDL, soleus). Once established, it can be checked prior to each new study, or periodically (e.g., every 6 months) as desired.
2. Muscle resting tension can be used as an index of length to determine L_o. The resting tension used by various labs can vary, likely because of differences in the data collection equip-

ment and stimulators, so it is suggested that it be determined for each set of specific equipment used in a lab.

3. After hanging the muscle in the bath, at a tension of 1.0 g for 10 min, elicit twitches at the optimal voltage every minute for the following resting tension values: 0.5, 0.75, 1.0, 1.25, 1.5, 1.75, 2.0 g. Additional resting tensions can be used, or finer increments if desired. For each twitch response, the resting force for that twitch (measured from the force preceding the twitch response), should be subtracted from the peak twitch force. This difference is called the active force. Record these values on paper, but also save each twitch for subsequent analysis using the DMA. When all the twitches are recorded, plot active force vs. resting force. This plot should reveal the optimal length to elicit maximum twitch active force.
4. There is also the possibility that the maximal tetanic active force will occur at a different resting force. Tetanic active force is peak force minus resting force. To test this possibility, bracket the resting force that elicited optimal twitch active force to test the tetanic responses (e.g., if optimal twitch resting force is 1.0 g, then test tetani at 0.75, 1.0, and 1.25 g); if the tetanic value at 1.25 g is greater than at 1.0 g, then continue to test at higher resting tensions. Use tetani elicited at 150 Hz, 500 ms duration. If the optimal resting forces for the twitch and tetanus are similar (e.g., within 0.5 g), an average resting force can be used. Alternatively, when reporting your methods, the optimal resting force determined for either the twitch or the tetanus can be stated. We typically use the optimal resting force as determined with twitches in our lab (e.g., 1.0 g resting force), for all muscle contractions (including tetani).

3.9 Twitch

1. The force response to a single activating electrical stimulation typically determined at L_o.
2. Clear differences between a fast EDL and slow soleus (Fig. 6a,b).
3. Typical variables: resting force/tension peak force/tension, active force/tension, time to peak force/tension, half relaxation time, contraction time (*see* Common terms for definitions, Subheading 3.6; Fig. 6c) (*see* **Note 8**).

3.10 Tetanus

1. The force response to a maximal electrical activation.
2. Clear differences between a fast EDL and slow soleus (Fig. 7a,b).
3. Typical variables: resting force/tension, peak force/tension, force/tension-time integral.

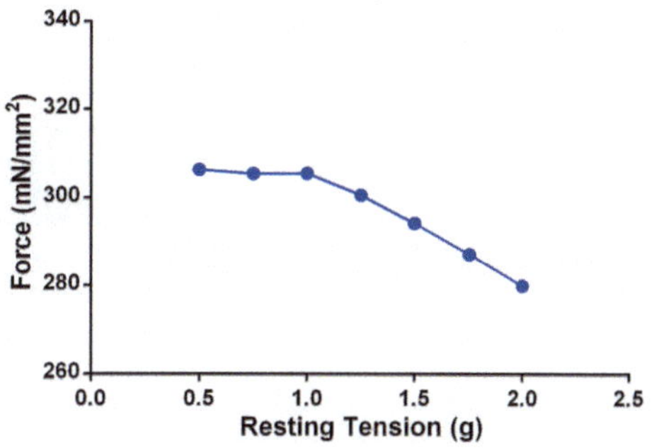

Fig. 5 Example length-tension plot for soleus tetanic responses. The maximum twitch active force value for this muscle was 36.5 mN at 1.0 g tension, but twitch values of 35.2 and 36.0 mN were also obtained at 1.75 and 2.0 g resting tension, respectively (data not shown). Note, however, the tetanic value decreases at resting tensions greater than 1.0 g. Therefore a reasonable optimal resting tension for twitches and tetani could be 0.5–1.0 g; we typically use 1.0 g. All force responses were collected at 30 °C

3.11 Force–Frequency (F-F) Relation

1. A fundamental relationship for muscles in vitro is the force generation capability for a twitch, submaximal activations, and for a maximal tetanus. A fast EDL muscle will have a different relationship compared to a slow muscle, especially when plotted as a percent of the maximum force (Fig. 7d).
2. Following stabilization, the muscle is stimulated at several frequencies from a twitch to a tetanus (e.g., 1, 30, 50, 70, 100, 120, 150 Hz), with a stimulation duration of 1 s and each frequency separated by 1 min. The resting tension is set to 1.0 g after each stimulation.
3. The force responses for each stimulation are saved and subsequently analyzed with the DMA-HT program after all data are collected (e.g., all muscles for that experiment).
4. The DMA-HT will yield a number of variables, but for the *F-F* relation, active force at each frequency will be plotted against frequency. The data can also be plotted relative to CSA (e.g., specific force or stress) vs. frequency. The force responses can also be expressed as a percent of the maximum tetanic force for each specific muscle and plotted against frequency. Different muscle types (EDL and soleus; *see* Fig. 7c,d) or different conditions (e.g., treated vs. untreated) can be plotted by either method to compare the *F-F* relations.
5. The DMA-HT will analyze the force profiles for each muscle collected at each frequency and the summary data table generated can be downloaded to Excel. The data can be plotted in Excel or in another graphing program (e.g., Graphpad Prism) to fit a curve to the points.

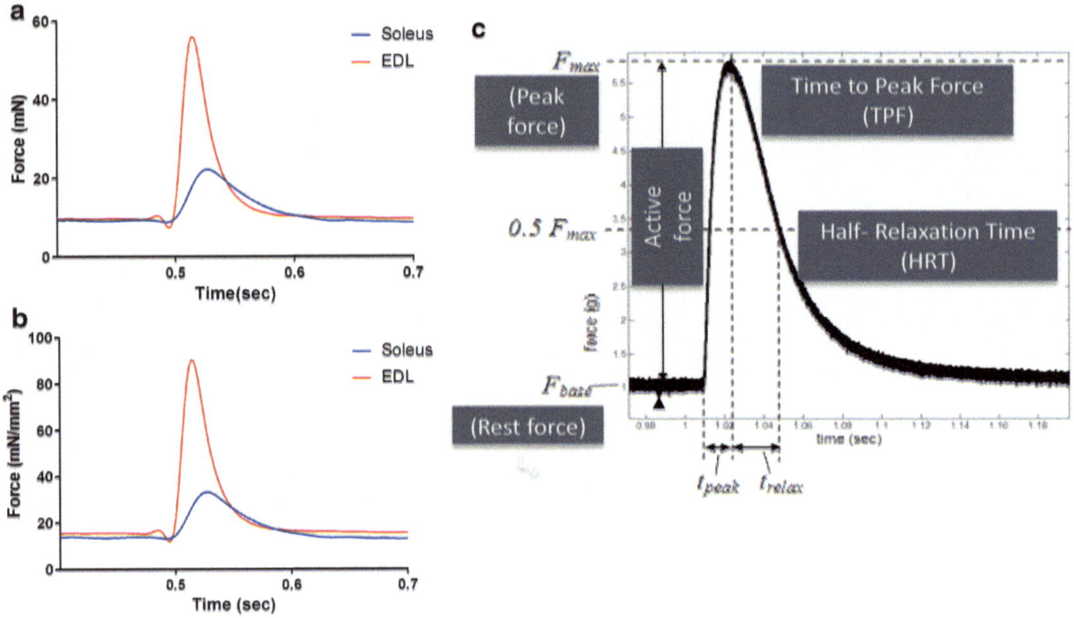

Fig. 6 Examples of EDL and soleus twitches expressed in mN (**a**) and normalized to cross-sectional area expressed in mN/mm^2 (**b**). (**c**) Typical variables determined for a twitch contraction are indicated. F_{base} = Rest force/tension; F_{max} = Peak force/tension. All force responses were collected at 30 °C (*see* **Note 8**)

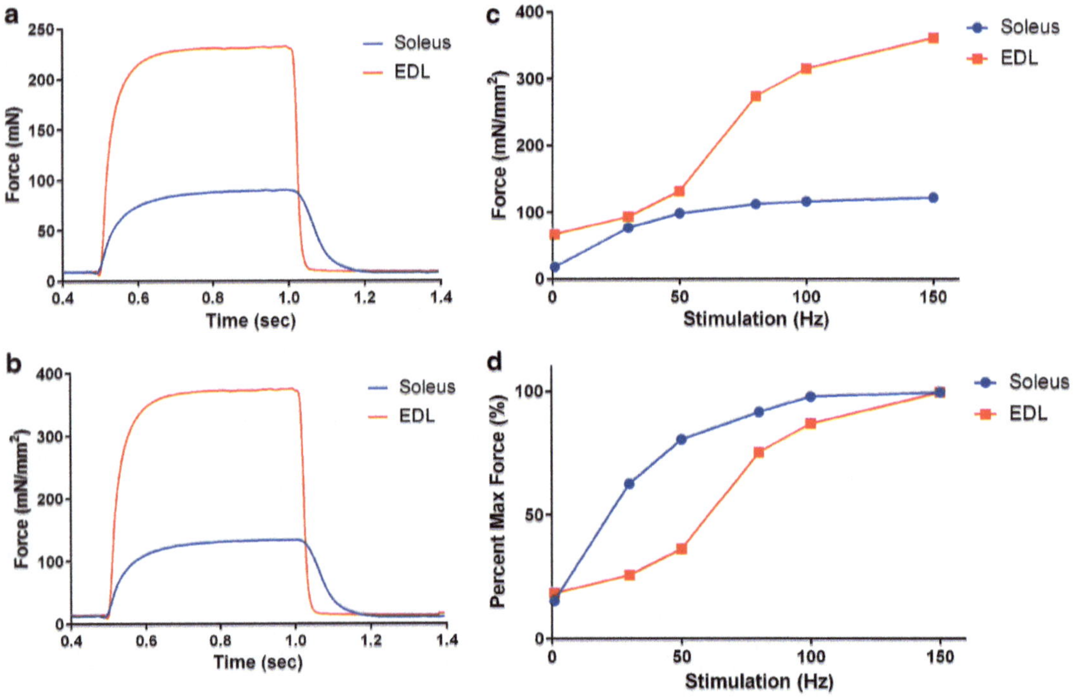

Fig. 7 Examples of EDL and soleus tetani expressed in mN (**a**) and normalized to cross-sectional area expressed in mN/mm^2 (**b**). (**c**) Force–frequency response for an EDL and soleus expressed in mN/mm^2, and as a percent of their respective maximum force responses (**d**). All force responses were collected at 30 °C. Note that mN/mm^2 are units of Specific Force or Stress

6. The data can be plotted as mN, or normalized to CSA (mN/mm^2) vs. frequency. The data can also be plotted relative to each muscle's maximum tetanic response. When comparing a fast EDL to a slow soleus, this approach shows a clear distinction in the frequency response, with the slow soleus reaching a greater percent of its maximum tetanic force at lower frequencies compared to the fast EDL. This response is known as a left-shift for the soleus relative to the EDL *F-F* curve and reflects the slower kinetics of calcium handling in the soleus compared to the EDL.

3.12 Force–Velocity (F-V) Relation

1. Another fundamental property of skeletal muscle is the *F-V* relation.
2. After stabilization of the muscle, the DMC force–velocity protocol is used.
3. The method we use is tetanic afterload [13]. In this method, a load is set on the lever arm as a decimal fraction of the muscle's maximal tetanic force. When the force produced by the muscle exceeds the load on the lever arm, the muscle shortens.
4. We typically use loads of 0.05, 0.1, 0.25, 0.5, 0.75, and 0.90 of the maximal tetanic force for each muscle. The muscle is activated at the frequency for a maximal tetanus (e.g., 150 Hz), with a different load tested every minute. Resting tension is reset to 1.0 g after each measure as needed. Data files for each load are saved and subsequently analyzed in DMA-HT. These data are then copied to a graphing program such as GraphPad Prism.
5. The relationship between load moved and speed of shortening is nonlinear. The resulting plot of load vs. velocity is fit with the Hill equation (*see* description under Common terms, Subheading 3.6). When the constants (a and b) in the Hill equation are determined by curve fitting, then the maximal speed of shortening or V_{max} can be predicted by entering a load of zero in the equation. Note, setting a true load of zero on the arm is technically difficult, therefore we generate a curve and predict V_{max}.
6. There is a distinct difference in the V_{max} for a fast EDL compared to a slow soleus (Fig. 8a), largely because of the different rates at which the myosin ATPase in the sarcomere of each fiber type hydrolyzes ATP (type I slow myosin vs. IIb fast myosin).

3.13 Power Calculation

1. Power is a variable that describes work/time. It is calculated as force times velocity. To generate a power curve, the load (expressed as g) at each velocity (expressed in mm/s) from the force–velocity relation is determined. Power can be expressed in force × velocity units (e.g., mN × mm/s) or in power units (mWatts). The power value at each load is then plotted (Fig. 8b).

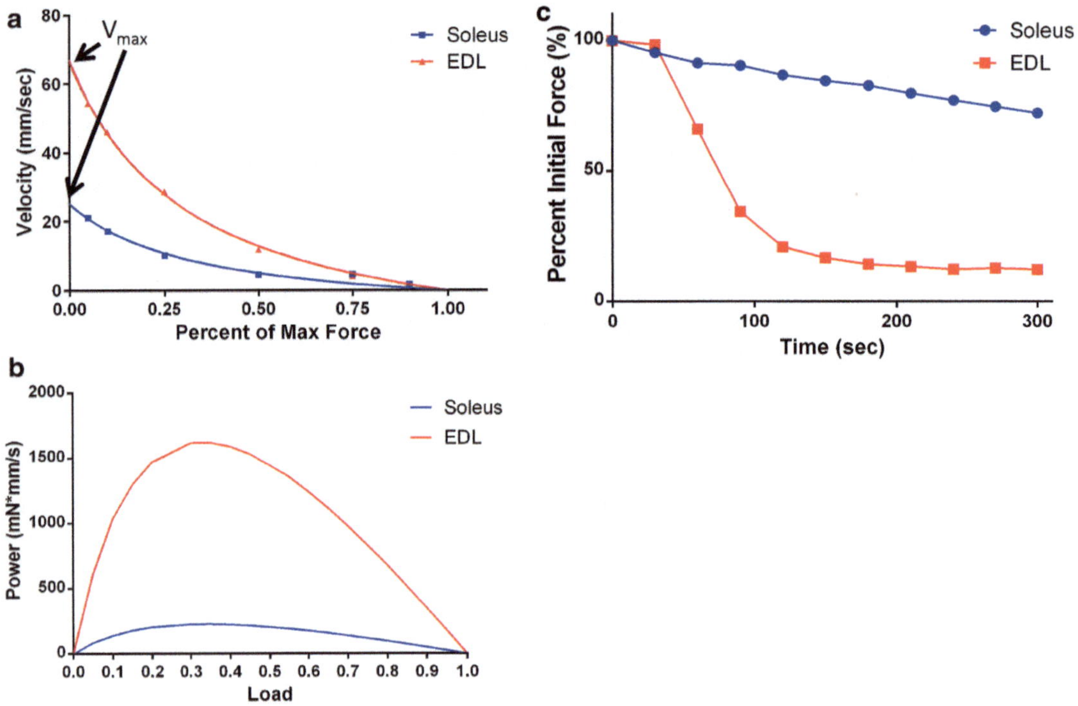

Fig. 8 (**a**) Force–velocity relationships for EDL and soleus. Percent of Max Force on the *x*-axis is the load moved by the muscle during shortening. The *y*-intercept for each curve is called V_{max}, the maximum speed of shortening. V_{max} for EDL was 66.9 mm/s and V_{max} for soleus was 25.2 mm/s. (**b**) Power curves determined from the force–velocity relations shown in Fig. 8. (**c**) Fatigue profiles for EDL and soleus when stimulated for 5 min intermittently (*see* Methods description for fatigue protocol details Subheading 3.14). All data were collected at 30 °C

2. A fast EDL muscle produces more power than a slow soleus because it produces more absolute force and shortens faster compared to the soleus. However, the power profile is similar for both. Note that a maximal tetanic contraction is associated with zero velocity (i.e., maximal isometric contraction), and therefore power = 0. Similarly, at V_{max}, load is zero, so power = 0; hence the dome shape of the curve. Note also that the maximum power occurs at ~0.35–40 of maximum load (*see* Fig. 4a,b [16]).
3. Because movement is dependent on power (i.e., the muscles acting on a joint must produce force and shorten), changes in power can have potentially profound effects on mobility.
4. Power is another example of characterizing the efficacy of a muscle treatment.

3.14 Fatigue

1. Fatigue is the inability of a muscle to produce a desired force.
2. Fatigue is complex with numerous mechanisms [17]. There are many protocols for assessing fatigue in vitro.

3. The DMC protocol we use is intermittent stimulation at submaximal frequencies of 60 Hz for EDL and 100 Hz for soleus. The intermittent pattern we use is 12×1 s stimulations at the selected frequency per minute for a total duration of 5 min. The muscle therefore undergoes 60 submaximal contractions [18].
4. The peak force response for each individual contraction, or at some desired interval (i.e., the value every 30 s) is plotted against time. The units can be absolute (e.g., mN) or normalized to CSA (mN/mm^2). Fatigue data are also commonly plotted as the percent of the initial muscle force (e.g., 100%). The percent of initial force then decreases with an increased number of contractions. Note that the fast EDL fatigues more rapidly than the slow soleus (Fig. 8c), which reflects the predominant oxidative metabolism of the soleus compared to the predominant glycolytic metabolism of the EDL. Another comparison is to determine how long it takes each muscle to reach 75, 50, 25% of initial force.

3.15 Summary

We have described several fundamental in vitro skeletal muscle contractile function assays that can be used to characterize the differences between different muscle types (e.g., slow vs. fast), between a diseased and non-diseased muscle, or importantly, to demonstrate the benefits of a muscle manipulation such as a drug, an overexpressed transgene, or knockout of a specific gene. Characterizing skeletal muscle function in vitro is a powerful approach to demonstrate efficacy of a treatment to rescue diseased muscle and to assess functional regeneration.

4 Notes

1. Details on the in vitro muscle system were obtained from the Aurora Scientific website (http://aurorascientific.com/products/muscle-physiology/. The 300 series Dual-Mode muscle lever systems are used to measure dynamic physical properties of muscle and connective tissues. Dual-Mode means the force and position of the lever arm can be controlled by the Dynamic Muscle Control software, and that the force and position information from the lever arm is reported back to the computer. Some of our equipment differs from that described on the website because it is older or newer.
2. The 800A holds the baths and the 300C Dual-Mode muscle lever system. We have a custom built apparatus that is similar to the 800A (Fig. 1).
3. Radnotti baths are fully water-jacketed to provide stable temperature control, have a separate gas tube attachment for pro-

viding O_2/CO_2 to the muscle, and incorporate quick disconnect fittings.

4. The 604A connects length controllers, force transducers, and stimulators through a National Instruments Analog/Digital card incorporated into a Microsoft Windows computer.
5. Software. In our lab, we use the 615A Dynamic Muscle Control (DMC) and Analysis Software Suite (DMC/DMA-HT (high throughput)) from ASI (most recent versions of the 605A). DMC collects length and force data while also controlling the length and force of the lever arm during the in vitro functional testing. DMA-HT v. 5.30 automates the analysis of single or many files to assess contractile data such as force–frequency, force–velocity, force–time, position–time, and fatigue analysis experimental protocols. Alternative data collection and analysis systems are also commercially available. The methods described herein are for the DMC/DMA software programs.
6. PSS can be made in 1 L volumes and stored at 4 °C for 1 week. Alternatively, Krebs–Henseleit Buffer or Ringer buffer may be used. The most important concern is that there is consistency between experiments in a lab and the buffers are made carefully.
7. A temperature controlled circulating water pump is needed to maintain the PSS in the bath at the desired temperature. The water is circulated between the inner and outer bath walls (i.e., the 'jacket' of the tissue organ bath). We typically use 30 °C. Bath temperature can be checked with a thermometer.
8. Please be aware that the contractile responses reported in Figs. 6–8 were obtained from a specific mouse knock out (unpublished). The EDL vs. soleus responses nicely demonstrate the clear differences between the two types of muscles, but do exaggerate the differences. The responses you may obtain from EDL and soleus muscles will depend on the mouse genotype, age, the equipment, skill of the dissector, bath temperature, among many other factors. To initiate these studies in your own lab, it is suggested that a mouse such as the C57Bl/6J be used to practice the assays to demonstrate the fundamental contractile property differences between the EDL and soleus muscles. A good source for contractile differences in this mouse is [16], and the references therein (e.g., [13]).

References

1. Call JA, Lowe DA (2016) Eccentric contraction-induced muscle injury: reproducible, quantitative, physiological models to impair skeletal muscle's capacity to generate force. In: Kyba M (ed) Skeletal muscle regeneration in the mouse. Springer, New York
2. Blaauw B, Schiaffino S, Reggiani C (2013) Mechanisms modulating skeletal muscle phenotype. Compr Physiol 3(4):1645–1687
3. Schiaffino S, Reggiani C (2011) Fiber types in mammalian skeletal muscles. Physiol Rev 91(4):1447–1531

4. Glaser BW et al (2010) Relative proportions of hybrid fibres are unaffected by 6 weeks of running exercise in mouse skeletal muscles. Exp Physiol 95(1):211–221
5. Luedeke JD et al (2004) Properties of slow- and fast-twitch skeletal muscle from mice with an inherited capacity for hypoxic exercise. Comp Biochem Physiol A Mol Integr Physiol 138(3):373–382
6. Quiat D et al (2011) Concerted regulation of myofiber-specific gene expression and muscle performance by the transcriptional repressor Sox6. Proc Natl Acad Sci U S A 108(25): 10196–10201
7. Lynch GS et al (2001) Force and power output of fast and slow skeletal muscles from mdx mice 6–28 months old. J Physiol 535(Pt 2): 591–600
8. Childers MK et al (2014) Gene therapy prolongs survival and restores function in murine and canine models of myotubular myopathy. Sci Transl Med 6(220):220ra10
9. Filareto A et al (2015) Pax3-induced expansion enables the genetic correction of dystrophic satellite cells. Skelet Muscle 5:36
10. McClung JM et al (2010) Overexpression of antioxidant enzymes in diaphragm muscle does not alter contraction-induced fatigue or recovery. Exp Physiol 95(1):222–231
11. Russell KA et al (2015) Mouse forepaw lumbrical muscles are resistant to age-related declines in force production. Exp Gerontol 65:42–45
12. Mendez J, Keys A (1960) Density and composition of mammalian muscle. Metabolism 9: 184–188
13. Brooks SV, Faulkner JA (1988) Contractile properties of skeletal muscles from young, adult and aged mice. J Physiol 404:71–82
14. Head SI, Arber MB (2013) An active learning mammalian skeletal muscle lab demonstrating contractile and kinetic properties of fast- and slow-twitch muscle. Adv Physiol Educ 37(4): 405–414
15. Asmussen G, Gaunitz U (1989) Temperature effects on isometric contractions of slow and fast twitch muscles of various rodents—dependence on fibre type composition: a comparative study. Biomed Biochim Acta 48(5–6): S536–S541
16. Graber TG et al (2015) C57BL/6 life span study: age-related declines in muscle power production and contractile velocity. Age (Dordr) 37(3):9773
17. Allen DG, Lamb GD, Westerblad H (2008) Skeletal muscle fatigue: cellular mechanisms. Physiol Rev 88(1):287–332
18. Call JA et al (2008) Endurance capacity in maturing mdx mice is markedly enhanced by combined voluntary wheel running and green tea extract. J Appl Physiol (1985) 105(3): 923–932

Chapter 20

In Vivo Assessment of Muscle Contractility in Animal Studies

Shama R. Iyer, Ana P. Valencia, Erick O. Hernández-Ochoa, and Richard M. Lovering

Abstract

In patients with muscle injury or muscle disease, assessment of muscle damage is typically limited to clinical signs, such as tenderness, strength, range of motion, and more recently, imaging studies. Animal models provide unmitigated access to histological samples, which provide a "direct measure" of damage. However, even with unconstrained access to tissue morphology and biochemistry assays, the findings typically do not account for loss of muscle function. Thus, the most comprehensive measure of the overall health of the muscle is assessment of its primary function, which is to produce contractile force. The majority of animal models testing contractile force have been limited to the muscle groups moving the ankle, with advantages and disadvantages depending on the equipment. Here, we describe in vivo methods to measure torque, to produce a reliable muscle injury, and to follow muscle function within the same animal over time. We also describe in vivo methods to measure tension in the leg and thigh muscles.

Key words Skeletal muscle, Lengthening contraction, Contractile function, Torque, Specific force, Injury

1 Introduction

Muscle damage in humans, with muscle injury or muscle disease, is typically assessed with clinical presentations of pain, impaired strength, decreased range of motion, tenderness, and imaging techniques such as MRI and ultrasound. However, muscle damage is difficult to assess because the clinical presentation varies greatly depending on age, gender, underlying genetic mutations and factors that can affect comorbidities, such as diet and level of activity. Furthermore, as the incidence of injury is a random event that is difficult to predict, it is difficult to study injury-induced muscle damage in humans. To obtain biological markers of injury for studying the mechanisms of injury and repair, animal studies have been used [1–9]. These studies have generated much of the data regarding muscle damage and repair, as they provide control

Michael Kyba (ed.), *Skeletal Muscle Regeneration in the Mouse: Methods and Protocols*, Methods in Molecular Biology, vol. 1460, DOI 10.1007/978-1-4939-3810-0_20,

over many of the variables. However, histological markers, which provide a "direct measure" of injury and biological markers, such as serum creatine kinase levels that are typically elevated with muscle injury, do not correlate with loss of function [10]. Thus, contractile force can be used for obtaining a more comprehensive measure of muscle health.

Several animal models exist to measure muscle contractility and to induce muscle injury via eccentric contractions. These were initially limited to "hanging" small, thin muscles in a bath mixture for field stimulation (in vitro), but there is strong interest in using in vivo methods [11], as evidenced by the growing number of publications using such methods [4, 5, 12–15].

Measuring muscle strength in vivo can be performed using one of two systems, with one system measuring muscle torque and the other system measuring muscle tension. The first is a system to measure muscle torque around a joint, which is a method for assessing contractile function without dissecting the muscle. The benefits of this system are that the muscle anatomy and biomechanics are not altered, and that the procedure is not terminal. Thus, one can measure contractility in the same animal over time, and compare results to in vivo imaging, such as MRI. Other advantages are that the nerve is not bypassed for stimulation (such as for in vitro preparations) and that the muscle remains in its normal environment, so effects of inflammation, hormones, or other factors can be studied. However, there are some limitations compared to measuring muscle function in isolated muscles. For example, length changes that occur during isometric or lengthening contractions must be estimated, and the muscle mass cannot be measured until it is harvested (although it can be estimated based on the volume measured via MRI) [16].

The second method to measure muscle function in vivo involves releasing the distal tendon and attaching it to a load cell. One advantage of this model is that the contractility of only one muscle with a known length and mass is measured. This allows one to determine the "specific force" (force per unit of cross-sectional area) of an individual muscle and avoids force transmission from nearby muscles. Although it provides more experimental control when measuring the force of an individual muscle, the experiment becomes less physiological. Surgical release of the muscle can also alter the anatomy and affect force transmission. The experiment is also terminal, so any changes in muscle contractility cannot be monitored over time.

Our custom-designed injury model is based on the same principles used by others to establish contraction-induced injury in animals [7, 9, 17, 18]. Despite the availability of models in the marketplace, there is little instruction beyond the use of the hardware. Our model has specifications in terms of available range of motion and angular velocity that are advantageous [19, 20], but our main goal is to share the methods. Here, we describe the procedures from start to finish for measuring muscle contractility, specifically torque and tension in vivo. We also describe the technique used to induce muscle injury, as susceptibility to injury is a very useful marker in the study of muscle disease.

2 Materials

2.1 Hardware/Software for Muscle Physiology

1. BUD value line cabinet (Newark, 06M4718).
2. Multifunction data acquisition device I/O USB-6221M (National Instruments, 779808-01).
3. Stepper motor controller (Newark, 16M4189).
4. Stepper motor (Newark, 16M4198).
5. Strain gauge amplifier (Honeywell, Sensotec, DV-05).
6. Torque sensor (Honeywell, QWLC-8M).
7. Foot plate and stabilization device (custom designed).
8. Lever arm for quadriceps testing (custom designed).
9. DC amplifier (model P122, Grass Instruments, Warwick, RI).
10. Force Displacement Transducer (model FT03, Grass Instruments, Warwick, RI for in vivo muscle tension).
11. Acquisition Software (PolyVIEW™ 16, Grass Instruments, Warwick, RI).
12. Software for synchronization of contractile activation, joint rotation and torque data collection (LabView version 2013, National Instruments, Austin, TX).

2.2 Muscle Attachment/Harvest

1. Precision vaporizer (Vet Equip, Inc., Pleasanton, CA).
2. Electrodes (J05 Needle Electrode Needles, 36BTP).
3. Micromanipulator (Kite Manipulator, World Precision Instruments Inc.).
4. Warmed mineral oil or DMEM.
5. Sterile ophthalmic cream (Paralube Vet Ointment, PharmaDerm, Floham Park, NJ, if animal is anesthetized for a long period of time).
6. Microdissection tools (if performing tendon release to obtain specific tension).
7. 4.0 Ethicon suture or customized muscle clamp.

3 Methods

All experimental procedures were approved by the University of Maryland Institutional Animal Care & Use Committee. These procedures can be used for mice, rats, and even larger animals such as rabbits. To begin, place the animal supine under inhalation anesthesia (~4–5 % isoflurane for induction in an induction chamber, then ~2 % isoflurane via a nosecone for maintenance) using a precision vaporizer. If the animal is to be anesthetized for a long period of time, apply sterile ophthalmic cream to each eye to protect the corneas from drying. During the procedure, the animal is kept warm by use of a heat lamp. Animals lose body heat while under inhalation anesthesia, so they need to be kept warm. However, the heat lamp should be kept at a safe distance of at least 12 in. away from the animal. Confirm proper anesthesia by lack of a deep tendon reflex (no foot withdrawal in response to strong pinching the foot). Prepare the skin around the leg and ankle by removing hair, and by cleaning with alternating scrubs of Betadine and 70% alcohol to prevent seeding skin bacteria into the soft tissue or bone.

3.1 Ankle (Testing Dorsiflexors)

Similar methods for measuring in vivo dorsiflexion and plantarflexion contractility are also described in Chapter 1 by Call and Lowe [21].

3.1.1 In Vivo Torque

1. A needle (25 G for rat or 27 G for mouse) is manually placed through the proximal tibia by hand. The needle will be used to stabilize the tibia onto the rig (*see* **Notes 1** and **2**).
2. Lock the needle (and consequently, the tibia) into a fixed position with the animal supine on a raised bed such that the tibia lies horizontal to the floor. The ankle should be free to rotate through a full range of motion. We use a custom-made device to secure the needle, and thereby stabilize the leg (Fig. 1a). Other methods for stabilizing the leg can be used instead of the custom-made device, such as a clamp (*see* **Note 3**).
3. Subcutaneous electrodes are used to stimulate the fibular nerve at the outside of the knee where it lies in a superficial position. You should visually confirm isolated dorsiflexion by performing a series of twitches (0.1 ms pulse for the mouse and 1 ms pulse for the rat) before the foot is secured into the foot plate (*see* **Note 4**).
4. Secure the foot onto the custom made footplate with adjustable lever arm using adhesive tape. This footplate is attached to a torque sensor and a stepper motor. Align the axis of rotation of the ankle joint with the axis of motor rotation.
5. The voltage is adjusted to optimize twitch tension. An increase in fused amplitude in response to an increase in voltage confirms that opposing muscles are not being simultaneously stimulated by excessive current.
6. The optimal position of the joint is determined by measuring twitches at different lengths of the muscle. This is done by sequentially rotating the foot plate at 5° increments from start position through the full range of motion. The optimal length

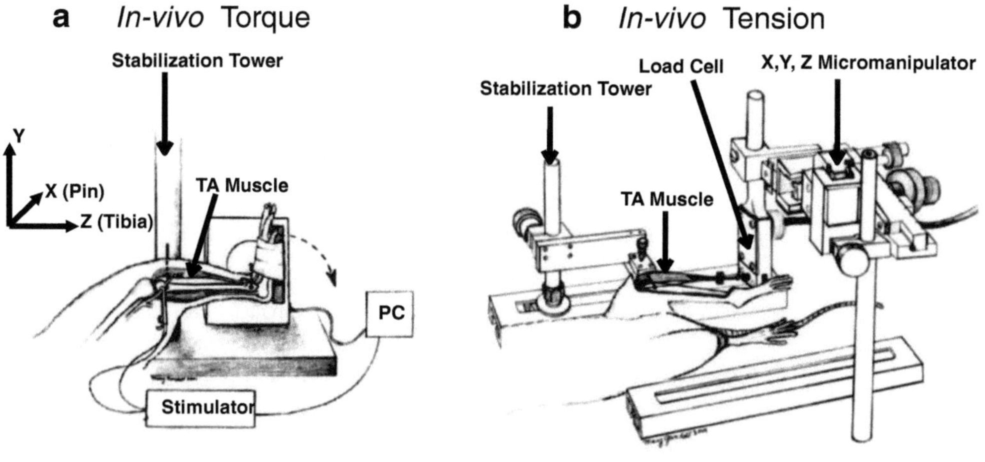

Fig. 1 Testing ankle dorsiflexor muscles. (**a**) In vivo torque apparatus. To measure dorsiflexion isometric torque, the tibia is stabilized and the foot is attached to a footplate that is attached to a torque cell. The ankle dorsiflexors are stimulated by subcutaneous electrodes via the fibular nerve. To induce injury, the ankle dorsiflexors are maximally stimulated while a motor, attached to the footplate, forces the foot into plantar flexion (*dotted arrow*). The magnitude of injury can be regulated by modifying variables such as angular velocity, timing of muscle activation, range of motion, and the number of lengthening contractions. (Image from Lovering et al., J Biomech 2005, used with permission.) (**b**) In vivo tension apparatus. The tibia is stabilized, and the tendon of the primary dorsiflexor, the tibialis anterior (TA), is released and attached to the load cell. The load cell is mounted to the *X,Y,Z* micromanipulator, which is used for aligning the muscle and for adjusting the TA to its resting length. The TA muscle is stimulated by subcutaneous electrodes via the fibular nerve (Image reproduced from Lovering et al., J Vis. Exp. 2011, used with permission)

(L_o or "resting length") is the angle at which peak twitch amplitude occurs in the torque-angle curve (this is typically at the mid-range of motion).

7. The optimal frequency is obtained by measuring the torque with progressively increasing frequency of pulses during a tetanic contractions (typically ~200 ms duration) (*see* **Note 5**).
8. The maximal force-producing capacity is recorded as the "maximal isometric torque" (measured in N mm) with the optimized voltage, the optimized joint angle, and the optimized frequency of pulses. Three separate twitches and tetanic contractions are recorded and saved for further analysis. Representative maximal isometric torque trace is as shown in Fig. 3b.

3.1.2 In Vivo Tension

1. The limb is stabilized, and the nerve is stimulated as described above in Subheading 3.1.1 using the same apparatus (Fig. 1b). All instrumentation is turned on at least 30 min prior to testing for proper calibration and to minimize thermal drift of the

force transducer. In contrast to the in vivo torque measurements in Subheading 3.1.1, this technique involves a surgical procedure (releasing the muscle insertion) and is therefore terminal. Results will no longer be in units of torque (typically N mm), but instead in units of force (typically grams, or Newtons) (*see* **Note 6**).

2. Incise the skin anterior to the ankle, and sever the tendon of the TA muscle as distally as possible (*see* **Note 7**).
3. Carefully tie 4.0 Ethicon silk non-absorbable suture to the tendon and attach the vicryl suture to the load cell (Fig. 1b) (force displacement transducer) via an S-shaped hook (weight = 0.1 g). Alternatively, a custom clamp (weight = 0.5 g) can be used to attach the tendon to the vicryl suture.
4. The load cell is mounted to a micromanipulator (Fig. 1b) so that the muscle can be adjusted to resting length and aligned properly (a straight line of pull between the origin and insertion of the muscle and the load cell).
5. The muscle is protected from cooling by a heat lamp. If the user wishes to expose more than just the tendon, dehydration of the exposed muscle can be minimized by application of mineral oil, as needed (to which extent the TA muscle is released is up to the user, as long as it is performed consistently). Alternatively, a continuous drip of 37 °C culture medium (DMEM) over the muscle can be used.
6. The signals from the load cell (calibrated before each test) are fed via a DC amplifier to an A/D board to be collected and stored by acquisition software.
7. Measure single twitches (rectangular pulse of 1 ms) at different muscle lengths in order to determine resting length (L_o). Muscle resting length is defined as the length at which maximal contractile force is produced.
8. At this length, gradually increase the voltage until maximal force is obtained and then incrementally increase the pulse frequency to establish a force–frequency relationship. A maximally fused tetanic contraction is obtained at approximately 80–150 Hz (200 ms train duration comprising 1 ms pulses) (*see* **Note 5**). A representative force–frequency experiment, with various tension traces generated by increasing frequency stimulation, is shown in Fig. 4a (*see* **Note 8**).
9. Use 150% of the maximum stimulation intensity (pulse voltage) to activate the TA muscle in order to induce maximal contractile tension (P_o). Three separate twitches and tetanic contractions are recorded and saved for further analysis.
10. Maximal tetanic contractions can be performed repeatedly (typically every 0.5–1 s) with the final tension expressed as

percentage of P_o, providing an index of fatigue at any desired point in time. Representative trace of muscle fatigue assessment using repeated maximal tetanic contractions is shown in Fig. 4b.

3.2 Knee (Testing Quadriceps)

3.2.1 In Vivo Torque

1. A needle (25 G for rat or 27 G for mouse) is manually placed through the distal femur (instead of the proximal tibia for testing dorsiflexors) by hand. The needle will be used to stabilize the femur onto the rig (*see* **Notes 1** and **2**).
2. Lock the needle (and consequently, the femur) into a fixed position with the animal supine, such that the femur lies horizontal to the floor. The knee should be free to rotate through a full range of motion. We use a custom-made device to secure the needle and thereby stabilize the thigh (Fig. 2a) (*see* **Note 3**). Subcutaneous electrodes are used to stimulate the femoral nerve anterior to the hip joint, where it lies in a relatively superficial position. You should visually confirm isolated knee extension by performing a series of twitches (0.1 ms pulse width) before the limb is secured (*see* **Note 4**).

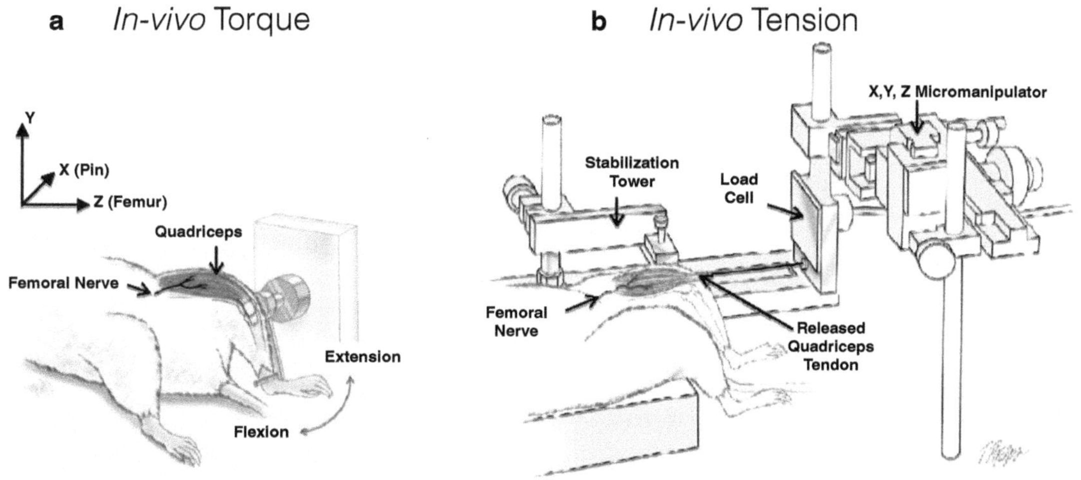

Fig. 2 Testing knee quadriceps muscles. (**a**) In vivo torque apparatus. To measure quadriceps isometric torque, the femur is stabilized, and the distal leg is taped (not shown) to a lever arm, which is attached to a torque cell to measure knee extension. The quadriceps are stimulated by subcutaneous electrodes through the femoral nerve, producing knee extension. To induce injury, the quadriceps muscles are stimulated while the lever arm forces the knee into flexion. As with dorsiflexion, the magnitude of injury can be regulated by modifying angular velocity, timing of muscle activation, range of motion, and the number of lengthening contractions. (Image adapted from Pratt et al., J Physiol. 2012, used with permission.) (**b**) In vivo tension apparatus. The femur is stabilized, and the quadriceps tendon is released and attached to the load cell. The load cell is mounted to the *X*, *Y*, *Z* micromanipulator, which is used for aligning the muscle and for adjusting the quadriceps to their resting length. The quadriceps muscle is stimulated by subcutaneous electrodes through the femoral nerve (Image adapted from Pratt and Lovering, J Biol. Methods 2014, used with permission)

3. Secure the distal leg to a custom-machined adjustable lever arm with adhesive tape. The quadriceps lever arm is attached to a stepper motor and a torque sensor. Align the axis of rotation of the knee joint with the axis of motor rotation.
4. The optimized voltage, optimized joint angle and the optimized frequency of pulses is obtained similar to that of the ankle, described in Subheading 3.1.1. At the optimized parameters, three separate twitches and tetanic contractions are recorded and saved for further analysis.

3.2.2 In Vivo Tension

1. The limb is stabilized, and the nerve is stimulated as described above in Subheading 3.2.1 using the same apparatus. Similar to the measurement of in vivo tension in the TA muscle, all instrumentation is turned on at least 30 min prior to testing.
2. Incise the skin anterior to the knee, and sever the tendon of the quadriceps muscle as distally as possible (*see* **Note 7**).
3. Carefully tie 4.0 Ethicon silk non-absorbable suture to the tendon and attach the vicryl suture to the load cell (Fig. 2b) via an S-shaped hook (weight = 0.1 g). Alternatively, a custom clamp (weight = 0.5 g) can be used to attach the tendon to the vicryl suture. The load cell, DC amplifier, A/D board and the acquisition software are described in Subheading 3.1.2 for measuring in vivo tension in the TA muscle.
4. The load cell is mounted to a micromanipulator (Fig. 2b) so that the muscle can be adjusted to resting length and aligned properly (a straight line of pull between the origin and insertion of the muscle and the load cell).
5. Similar to the TA muscle in Subheading 3.1.2, the quadriceps muscle must be protected from cooling with use of a heat lamp. If the quadriceps muscle is exposed, dehydration must be minimized similar to the TA muscle, using mineral oil or 37 °C culture medium.
6. Similar to the TA muscle in Subheading 3.1.2, the muscle resting length, the optimized voltage and the optimized pulse frequency is measured for the quadriceps muscle.
7. Use 150% of the maximum stimulation intensity (pulse voltage) to activate the quadriceps muscle in order to induce maximal contractile tension (P_o). Three separate twitches and tetanic contractions are recorded and saved for further analysis at the optimized parameters. Index of fatigue can be calculated for the quadriceps muscle similar to the method used for the TA muscle.

3.3 Induction of Muscle Injury

Similar methods for in vivo ankle dorsiflexion and plantarflexion eccentric contraction induced injury are also described in Chapter 1 by Call and Lowe [21].

1. The limb immobilization and the apparatus used for muscle injury induction are identical to the setup used to record maximal isometric torque of the dorsiflexors or quadriceps (Figs. 1a and 2a).
2. We use commercial software to synchronize contractile activation, onset of joint rotation, and torque data collection. For lengthening contractions, stimulation of the muscles occurs while the computer-controlled motor simultaneously moves the foot plate or lever arm against the direction of the muscle contraction, thus leading to a lengthening contraction (also called "eccentric" contraction which, if maximal and repeated, causes injury of the muscle). The specific protocol depends on the desired magnitude of injury and is designed by the investigator. The magnitude of injury, or tissue damage, can be regulated by manipulation of variables such as angular velocity, timing of muscle activation, range of motion, and the number of lengthening contractions.
3. To induce injury, superimpose a lengthening contraction onto a maximal isometric contraction (a contraction during which muscle length is held constant), varying the range of motion, velocity of lengthening, and timing of stimulation as needed (*see* **Notes 9** and **10**). For example, a maximal isometric contraction for a given muscle is obtained before the onset of motor movement and the muscle is immediately lengthened at a selected velocity to approximate normal movement (900°/s) resulting in a high force peak (Fig. 3c). We have shown that activation before and during motor movement, and the degree of lengthening are important factors in obtaining an injury [19]. As reported previously, this model results in injury to muscle [19, 22–26]. The muscle remains stimulated throughout lengthening.
4. After injury, the animal is removed from the apparatus. The pin is removed, the leg is cleaned again with Betadine, and over-the-counter antibiotic cream is applied to the needle placement site. The animal is returned to the cage (placed on a temperature-controlled heating block at 37 °C) and monitored until recovery. If follow-up measurements are desired, a new sterile needle is placed into the same location.

3.4 In Vivo Testing of Other Muscles

1. For testing force in isolated muscles, almost any limb muscle can be tested. We provide two examples here. The device can also be used for measuring in vivo tension of other muscles such as supraspinatus (Fig. 5a), a rotator cuff muscle, and the gastrocnemius (Fig. 5b), a hindlimb muscle.
2. For measuring in vivo tension of supraspinatus muscle, the scapula is stabilized using a clamp. The tendon of the supraspinatus is released and attached to a load cell that is mounted to a micromanipulator, similar to the technique seen in

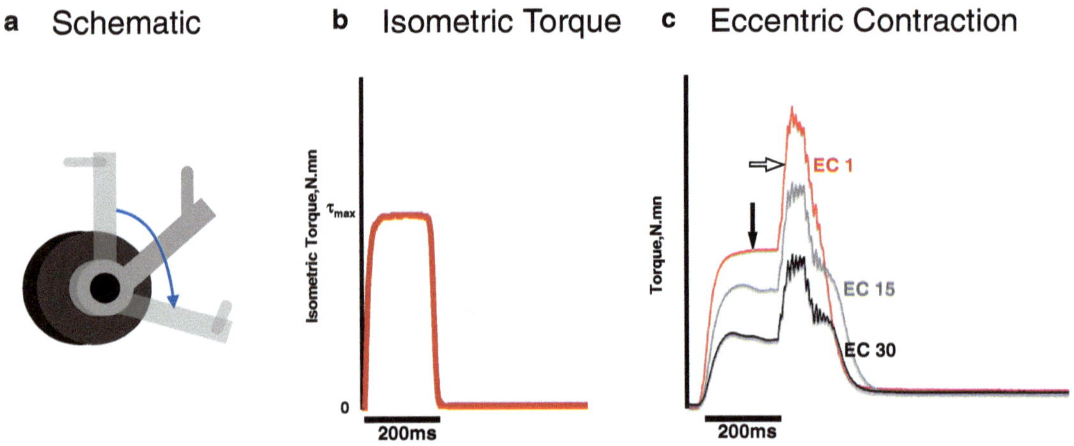

Fig. 3 Representative data from the apparatus used to obtain muscle torque. (**a**) Positioning of the joint. For measurement of isometric torque at resting length (optimal length of the muscle), the lever arm is placed mid-range (solid grey lever arm in schematic). For eccentric contractions used to induce injury, movement typically starts with the lever arm in a position shorter the optimal length of the muscle, so the joint can be moved through a large range of motion (*blue arrow*). (**b**) Isometric Torque. Representative trace data of maximal isometric torque obtained at mid-range (optimal length). (**c**) Eccentric Contractions. Representative data of torque recorded from eccentric contractions (ECs). The muscle is placed at the start position (in this example shorter than optimal length) and stimulated for 200 ms to induce maximal isometric torque (*closed arrow*) before lengthening (*open arrow*) using the lever arm. Repeated maximal ECs result in a quantifiable muscle injury. In this example of 30 ECs, the red tracing represents the first EC and the black tracing represents the last EC

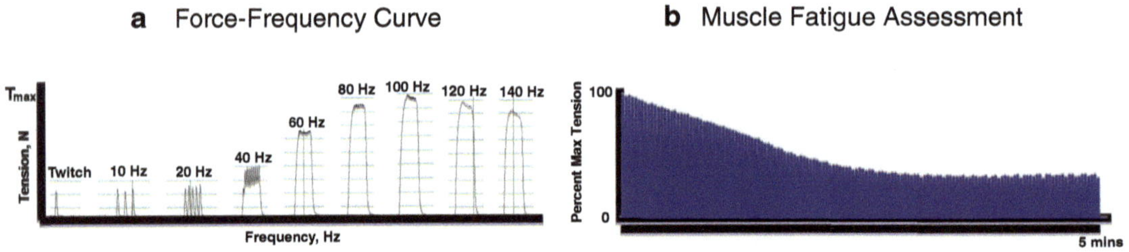

Fig. 4 Representative data from the apparatus used to obtain muscle tension. (**a**) Force Frequency Curve. Representative tracings of maximal tension of a muscle with progressively increasing frequency of pulses during a 200 ms pulse train. The maximal contractile tension typically occurs ~ 80–150 Hz, depending on the muscle, electrode placement, etc. In this example, occurs at the optimal frequency (in this example, 100 Hz). (**b**) Muscle Fatigue Assessment. Representative muscle fatigue assessment, which can be calculated at any point time with repeated maximal tetanic contractions as a percentage of initial maximal contractile tension. In this example, stimulation occurred every 2 s over a period of 5 min. Both these measures (force–frequency curve and muscle fatigue assessment) can also be assessed using the torque apparatus instead of in an isolated muscle, as performed here

Subheading 3.1.2 for the TA muscle and Subheading 3.2.2 for the quadriceps muscle. The suprascapular nerve is stimulated near the suprascapular notch using subcutaneous electrodes. The maximal contractile tension and the muscle fatigue index

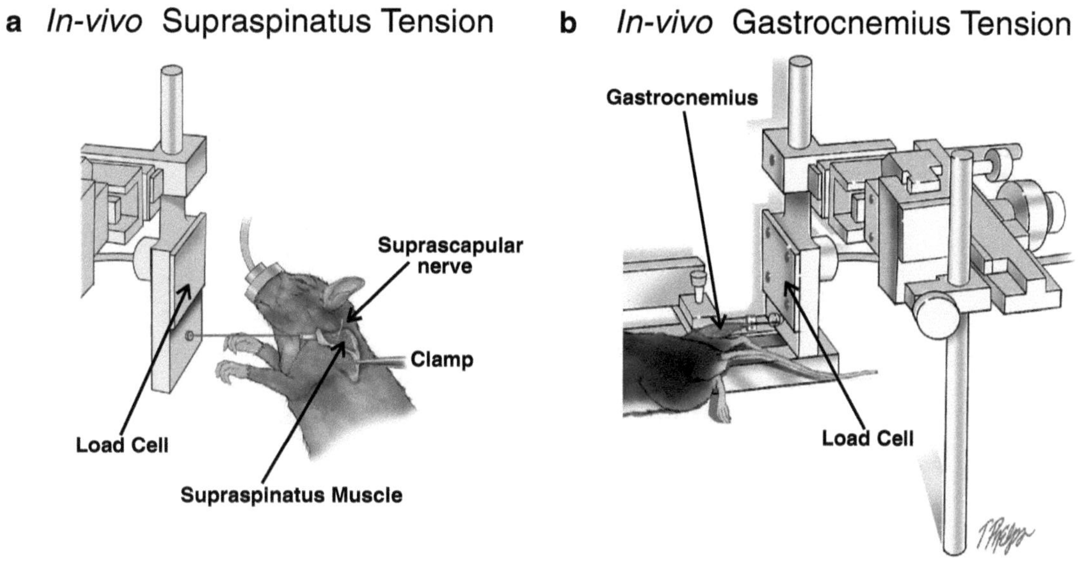

Fig. 5 In vivo testing of other muscles. (**a**) In vivo supraspinatus tension. The scapula is stabilized and supraspinatus tendon is released and attached to the load cell. Similar to the hind limb experiments, the load cell is mounted to the *X*, *Y*, *Z* micromanipulator, which is used for aligning the muscle and for adjusting the supraspinatus to its resting length. The supraspinatus muscle is stimulated by subcutaneous electrodes via the suprascapular nerve. (**b**) In vivo gastrocnemius tension. The tibia is stabilized, and the Achilles tendon is released and attached to the load cell (the soleus muscle can be detached to test isolated gastrocnemius). The load cell is mounted to the *X*, *Y*, *Z* micromanipulator, which is used for aligning the muscle and for adjusting the gastrocnemius to its resting length. The gastrocnemius muscle is stimulated by subcutaneous electrodes via the tibial nerve (not shown)

can be obtained using the same procedures described in Subheadings 3.1.2 and 3.2.2 for the TA muscle and the quadriceps muscles, respectively.

3. For measuring in vivo tension of the gastrocnemius muscle, the tibia is stabilized using similar techniques described in Subheading 3.1.1. The Achilles tendon is released and attached to a load cell that is mounted to a micromanipulator, similar to the technique seen in Subheading 3.1.2 for the TA muscle and Subheading 3.2.2 for the quadriceps muscles. The tibial nerve is stimulated in the popliteal fossa using subcutaneous electrodes. The maximal contractile tension and the muscle fatigue index can be obtained using the same procedures described in Subheadings 3.1.2 and 3.2.2 for the TA muscle and the quadriceps muscles, respectively.

3.5 Harvesting and Storing Muscles

1. At the end of experiments, animals are euthanized per institutional animal care guidelines and the muscles are harvested, weighed, snap frozen in liquid nitrogen, and then stored at −80 °C. This can be performed at any point in time after the in vivo experiments. Muscles are always harvested immediately after the in vivo tension experiments, as

this is a terminal procedure. For detailed morphological studies, the animal is fixed with 4% paraformaldehyde via perfusion through the left ventricle.

4 Notes

1. The device we use to stabilize the limb allows three degrees of freedom for positioning, but this is for convenience; the method used to stabilize the trans-osseous needle is not important, as long as the femur or the tibia is immobilized and the position is accurate. We have tried many methods to stabilize the limb without using a needle, but the trans-osseous needle proved superior in terms of stability. For tests of the same limb over time, in our experience, it is relatively easy to place the tip of the needle under the skin and slide along the bone to find the same location used for earlier procedures.
2. When testing the ankle muscles, the needle should be through the proximal tibia and posterior enough such that the TA muscle is not penetrated by the needle. When testing the quadriceps, the needle should not enter the anterior compartment of the thigh, the quadriceps, or knee joint (he widest aspect of the knee is formed by the femoral epicondyles, which sit just superficial to the condyles and are good landmarks to palpate for needle placement). Needle placement is easy, but can take practice. Correct needle placement can only be confirmed by opening the skin, which can be done during muscle harvesting, whenever the investigator decides to sacrifice the animal.
3. For all testing, the needle can be immobilized in a variety of methods; one could simply use a vise that is positioned correctly. If the needle is placed correctly in the coronal plane (parallel to the floor), the foot will be aligned so the toes are pointing anteriorly and not to the left or to the right.
4. One can obtain dorsiflexion or knee extension with the electrodes in approximately correct position. Be sure to try several locations, and also adjust the electrodes superficial to deep, in order to determine which electrode placement results in maximal torque. This type of trial and error is not necessary after you have an idea of what values animals of a specific gender/age/species generate on your device. The fibular nerve used to activate dorsiflexion is located at the lateral knee and is quite superficial. The femoral nerve is "superficial" to the hip joint and surrounding musculature, but it is comparatively deeper.
5. There are a variety of stimulation parameters in the literature. Very high stimulation frequency (250–300 Hz) is typical for in vitro procedures (removing a muscle from the animal and "hanging it" in a bath for field stimulation). However, for

in vivo procedures (stimulating through the motor nerve), maximal contractile tension occurs at a much lower frequency. In our hands, a maximal fused tetanic contraction in vivo for the TA is obtained at ~80–90 Hz, and in the quadriceps at ~100–150 Hz. The frequency, and voltage, should be optimized for each experiment.

6. Muscle torque is defined mathematically as the cross-product of the force vector and the moment arm vector. For ankle dorsiflexion, the tibia is fixed and the foot is attached to the foot plate. For knee extension, the femur is fixed and the leg is attached to the lever arm. Thus, the length of the moment arm is fixed and the only variable that changes is the magnitude of the muscle force.
7. Muscles can transmit some of their contractile force laterally, such that muscles in a group (e.g., dorsiflexors) can contribute to force measurements even when only one of the muscle (e.g., tibialis anterior) is tied to the load cell for assessment. Thus, for single muscle experiments (where the distal tendon is released for attachment to the load cell), it can be important to release the muscle entirely to its origin to avoid lateral transmission of force from nearby muscles [27]. For the quadriceps, since there are no additional knee extensor muscles, one can release just the tendon. Releasing the quadriceps completely up to its origin is more difficult than with other muscles due to its more complex attachments.
8. For single muscle terminal experiments, once the muscle is attached to the load cell there is a small amount of passive force present when the muscle is stretched to resting length. This passive tension can "drift" after the first few contractions. Having moist suture and performing a series of "warm up" twitches can minimize this drift, but the investigator should adjust resting length as needed.
9. When performing in vivo torque experiments L_o, or resting length, is typically midway through the available range of motion for most joints, but the precise location can be determined by seeing where maximal torque occurs. For inducing injury, the isometric torque that occurs before the onset of lengthening will be lower than the maximal isometric torque if the joint angle start position is not at where L_o was obtained mid-range (Fig. 3a, schematic).
10. As for injury (repeated eccentric contractions), many variables affect the magnitude of injury (operationally defined as a loss in maximal isometric torque). Assuming the velocity is above 100°/s, increasing velocity of lengthening is not likely to matter. Obtaining a full isometric contraction *before* (at least 200 ms long) lengthening occurs, as well as using a large range of

motion (but within physiological limits), will assist in producing a reliable injury that does not require an excessive number of contractions.

Acknowledgements

This work was supported by grants from the National Institutes of Health by grants to APV (T32AG000268-15S1), SRI (AR07592-20), and to RML (1R01AR059179).

References

1. Call JA, Eckhoff MD, Baltgalvis KA, Warren GL, Lowe DA (2011) Adaptive strength gains in dystrophic muscle exposed to repeated bouts of eccentric contraction. J Appl Physiol 111: 1768–1777
2. Lovering RM, De Deyne PG (2004) Contractile function, sarcolemma integrity, and the loss of dystrophin after skeletal muscle eccentric contraction-induced injury. Am J Physiol Cell Physiol 286:C230–C238
3. Ingalls CP, Warren GL, Zhang JZ, Hamilton SL, Armstrong RB (2004) Dihydropyridine and ryanodine receptor binding after eccentric contractions in mouse skeletal muscle. J Appl Physiol 96:1619–1625
4. Rathbone CR, Wenke JC, Warren GL, Armstrong RB (2003) Importance of satellite cells in the strength recovery after eccentric contraction-induced muscle injury. Am J Physiol Regul Integr Comp Physiol 285:R1490–R1495
5. Lieber RL, Shah S, Friden J (2002) Cytoskeletal disruption after eccentric contraction-induced muscle injury. Clin Orthop (403 Suppl): S90–S99
6. Warren GL, Hermann KM, Ingalls CP, Masselli MR, Armstrong RB (2000) Decreased EMG median frequency during a second bout of eccentric contractions. Med Sci Sports Exerc 32:820–829
7. Ingalls CP, Warren GL, Williams JH, Ward CW, Armstrong RB (1998) E-C coupling failure in mouse EDL muscle after in vivo eccentric contractions. J Appl Physiol 85:58–67
8. Lieber RL, Thornell LE, Friden J (1996) Muscle cytoskeletal disruption occurs within the first 15 min of cyclic eccentric contraction. J Appl Physiol 80:278–284
9. Faulkner JA, Jones DA, Round JM (1989) Injury to skeletal muscles of mice by forced lengthening during contractions. Q J Exp Physiol 74:661–670
10. Friden J, Lieber RL (2001) Serum creatine kinase level is a poor predictor of muscle function after injury. Scand J Med Sci Sports 11:126–127
11. Matarese G, La CA, Horvath TL (2012) In vivo veritas, in vitro artificia. Trends Mol Med 18:439–442
12. Best TM, McCabe RP, Corr D, Vanderby R Jr (1998) Evaluation of a new method to create a standardized muscle stretch injury. Med Sci Sports Exerc 30:200–205
13. Barash IA, Mathew L, Ryan AF, Chen J, Lieber RL (2004) Rapid muscle-specific gene expression changes after a single bout of eccentric contractions in the mouse. Am J Physiol Cell Physiol 286:C355–C364
14. Brooks SV, Opiteck JA, Faulkner JA (2001) Conditioning of skeletal muscles in adult and old mice for protection from contraction-induced injury. J Gerontol A Biol Sci Med Sci 56:B163–B171
15. Koh TJ, Brooks SV (2001) Lengthening contractions are not required to induce protection from contraction-induced muscle injury. Am J Physiol Regul Integr Comp Physiol 281: R155–R161
16. Heemskerk AM, Strijkers GJ, Vilanova A, Drost MR, Nicolay K (2005) Determination of mouse skeletal muscle architecture using three-dimensional diffusion tensor imaging. Magn Reson Med 53:1333–1340
17. Lieber RL, Friden J (1988) Selective damage of fast glycolytic muscle fibres with eccentric contraction of the rabbit tibialis anterior. Acta Physiol Scand 133:587–588
18. Warren GL, Lowe DA, Hayes DA, Karwoski CJ, Prior BM, Armstrong RB (1993) Excitation

failure in eccentric contraction-induced injury of mouse soleus muscle. J Physiol 468:487–499

19. Lovering RM, Hakim M, Moorman CT III, De Deyne PG (2005) The contribution of contractile pre-activation to loss of function after a single lengthening contraction. J Biomech 38:1501–1507
20. Dipasquale DM, Bloch RJ, Lovering RM (2011) Determinants of the repeated-bout effect after lengthening contractions. Am J Phys Med Rehabil 90:816–824
21. Call JA, Lowe DA (2016) Eccentric contraction-induced muscle injury: reproducible, quantitative, physiological models to impair skeletal muscle's capacity to generate force
22. Hakim M, Hage W, Lovering RM, Moorman CT III, Curl LA, De Deyne PG (2005) Dexamethasone and recovery of contractile tension after a muscle injury. Clin Orthop Relat Res 439:235–242
23. Lovering RM, Roche JA, Bloch RJ, De Deyne PG (2007) Recovery of function in skeletal muscle following 2 different contraction-induced injuries. Arch Phys Med Rehabil 88:617–625
24. Lovering RM, McMillan AB, Gullapalli RP (2009) Location of myofiber damage in skeletal muscle after lengthening contractions. Muscle Nerve 40:589–594
25. Pratt SJ, Lawlor MW, Shah SB, Lovering RM (2011) An in vivo rodent model of contraction-induced injury in the quadriceps muscle. Injury. 2012;43(6):788–793. doi:10.1016/j.injury.2011.09.015.
26. Pratt SJ, Lovering RM (2014) A stepwise procedure to test contractility and susceptibility to injury for the rodent quadriceps muscle. J Biol Methods. doi:10.14440/jbm.2014.34
27. Huijing PA, Baan GC (2001) Myofascial force transmission causes interaction between adjacent muscles and connective tissue: effects of blunt dissection and compartmental fasciotomy on length force characteristics of rat extensor digitorum longus muscle. Arch Physiol Biochem 109:97–109

Chapter 21

Functional Measurement of Respiratory Muscle Motor Behaviors Using Transdiaphragmatic Pressure

Sarah M. Greising, Carlos B. Mantilla, and Gary C. Sieck

Abstract

The diaphragm muscle must be able to generate sufficient forces to accomplish a range of ventilatory and non-ventilatory behaviors throughout life. Measurements of transdiaphragmatic pressure (Pdi) can be conducted during eupnea, hypoxia (10% O_2)–hypercapnia (5% CO_2), chemical airway stimulation (i.e., sneezing), spontaneously occurring deep breaths (i.e., sighs), sustained airway or tracheal occlusion, and maximal efforts elicited via bilateral phrenic nerve stimulation, representing the full range of motor behaviors available by the diaphragm muscle. We provide detailed methods on the in vivo measurements of Pdi in mice.

Key words Diaphragm muscle, Mouse, Muscle force, Non-ventilatory behavior, Phrenic nerve, Ventilatory behavior

1 Introduction

The diaphragm muscle sustains ventilation throughout the life span in mammals, as the primary muscle of inspiration. Fundamentally being able to measure transdiaphragmatic pressure (Pdi) and respiratory function is imperative. Failure of the respiratory system to sustain ventilation is an ultimate cause of death in many neuromuscular disorders including motor neuron diseases (i.e., amyotrophic lateral sclerosis or spinal muscular atrophy) and muscular dystrophies (i.e., Duchenne muscular dystrophy). The diaphragm muscle must be able to provide sufficient forces to accomplish a range of functions including ventilatory and high force non-ventilatory motor behaviors. Ventilatory behaviors include breathing room air at rest (i.e., eupnea) and breathing during various conditions where the chemical drive for ventilation is stimulated (e.g., hypoxia–hypercapnia, exercise). The diaphragm muscle also performs high force non-ventilatory behaviors such as sneezing, gagging and coughing, which are mostly related to maintenance and clearance of the airway. Each behavior requires varying amounts of force generation. Importantly, measurements

Michael Kyba (ed.), *Skeletal Muscle Regeneration in the Mouse: Methods and Protocols*, Methods in Molecular Biology, vol. 1460, DOI 10.1007/978-1-4939-3810-0_21, © Springer Science+Business Media New York 2016

of diaphragm muscle force generating capacity can be accomplished in vivo with the use of Pdi.

Force generation by the diaphragm muscle, similarly to other skeletal muscles, is accomplished by the recruitment of motor units (Fig. 1), which exhibit diverse contractile and fatigue properties. Activation of the diaphragm motor units follows an orderly recruitment pattern [1], such that ventilatory behaviors recruit slow-twitch and fast-twitch fatigue-resistant motor units while maximum activation would recruit both slow- and fast-twitch motor units, including those that are more fatigable [2–6]. Isometric force is correlated with peak root mean squared (RMS) electromyography (EMG) amplitude. In the diaphragm muscle, we have shown a significant positive correlation between RMS EMG and Pdi across motor behaviors [2, 3]. Thus, RMS EMG can also be used to evaluate diaphragm muscle force [2, 3, 7–10]. In skeletal muscles,

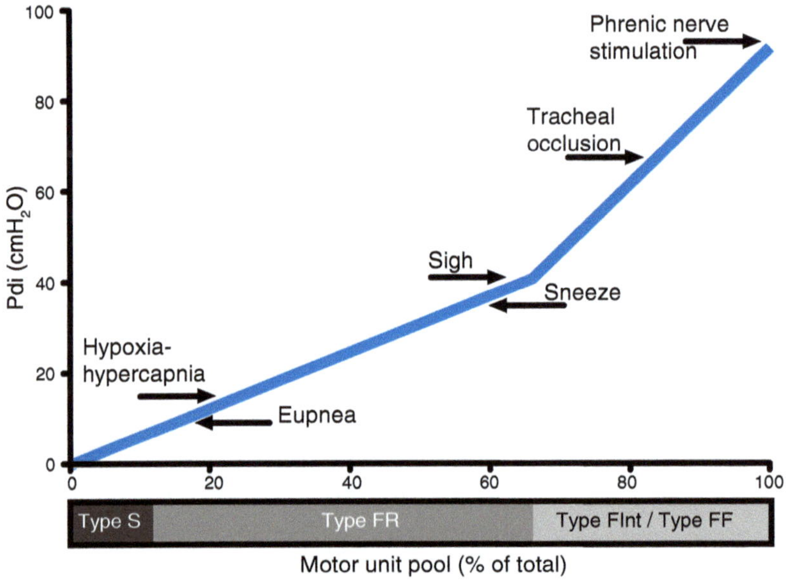

Fig. 1 The diaphragm muscle generates transdiaphragmatic pressures (Pdi; forces) to accomplish a range of motor behaviors from eupnea to high force non-ventilatory behaviors by the recruitment of different motor unit types. This model of motor unit recruitment for the mouse diaphragm muscle indicates that during ventilatory behaviors (eupnea and hypoxia–hypercapnia), type slow-twitch (S) and fast-twitch fatigue resistant (FR) motor units are recruited, while high force non-ventilatory behaviors (tracheal occlusion, and phrenic nerve stimulation) require the recruitment of type fast-twitch fatigue intermediate (FInt) and fast-twitch fatigable (FF) motor units. The Pdi during different motor behaviors are adapted from [14, 15, 24] and the motor unit pool is estimated from fiber type proportions of the mouse diaphragm muscle as determined by myosin heavy chain isoform expression [27, 29], both representing the adult male wild type mouse. This model of motor unit recruitment highlights the importance of measuring Pdi across motor behaviors, not just during eupnea

weakness is best defined as a decrement in maximal force. Accordingly, for the diaphragm muscle, weakness is defined as a decrease in Pdi during maximal activation.

To understand the function of the diaphragm muscle, it is necessary to examine force generation across the full range of motor behaviors (Fig. 1). The percent of maximal Pdi generated during ventilatory behaviors varies across species, but in general, accomplishing eupneic breathing requires ~10–27% of maximal Pdi [2, 11–15]. For this reason, even in conditions in which an entire hemi-diaphragm has been paralyzed by unilateral phrenic denervation, the intact hemi-diaphragm muscle is still able to generate forces required for ventilatory behaviors [3]. Similarly, in old age when the diaphragm muscle is significantly weakened by sarcopenia, ventilatory behaviors are not compromised [15]. In contrast, the ability of the diaphragm muscle to generate high force, non-ventilatory behaviors is compromised, highlighting the importance of examining the full range of behaviors that the diaphragm muscle can accomplish.

Clinically the measurement of Pdi has an important role in understanding the ability of the diaphragm muscle to generate force. Measurements of Pdi have been conducted in a range species including humans [16, 17], sheep [18], dogs [19], cats [11], piglets [20, 21], rabbits [22, 23], hamsters [12, 13], rats [2, 3, 6], and mice [14, 15, 24]. The current position statement of the American Thoracic and the European Respiratory Societies outlines the accepted use of Pdi [25], with the most traditional methodology involving a dual balloon catheter system, with balloons spanning the diaphragm muscle in the thoracic and abdominal cavity. This methodology has been adapted for use with solid-state pressure transducers, which has proven especially useful in smaller animals such as mice [14, 15, 24].

2 Materials

2.1 Technical Equipment and Preparation

1. Connect a pressure transducer (e.g., Millar Instruments MPVS-300; Houston TX) to an analog to digital converter (e.g., ADInstruments PowerLab System 16/35; Colorado Springs, CO) (*see* **Note 1**).
2. Set up LabChart Pro for acquisition of data signal, alternative data acquisition software could be LabView or comparable software. Note there is a freely available reader for any post-hoc analysis (LabChart Reader) available through ADInstruments.
3. Signal should be acquired with at least four channels: two of which will be live data recordings of (1) thoracic (esophageal, Peso) pressure; (2) abdominal (gastric, Pgas) pressure; and two that will represent the data acquired: (3) pressure difference,

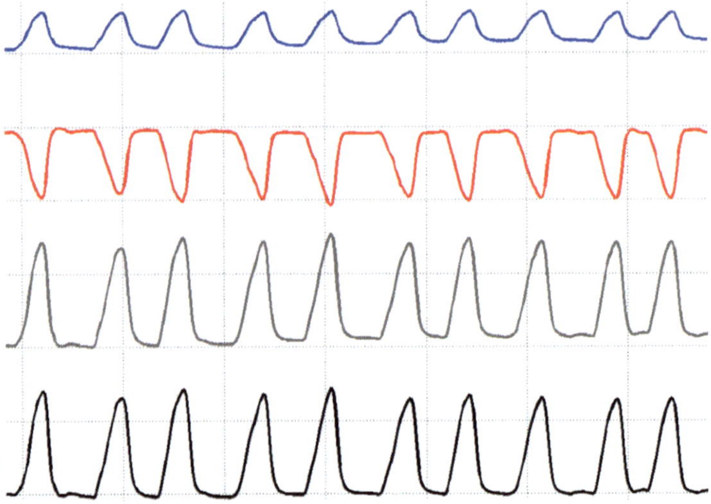

Fig. 2 Measurements of gastric (Pgas) and esophageal (Peso) pressures are used to calculate the transdiaphragmatic pressure (Pdi; Pgas – Peso). The raw pressure signals were band-pass filtered (0.3–30 Hz), to center the signal and remove any possible artifact noise due to cardiac activity. This tracing represents ten inspiratory events during eupnea from an adult male wild type mouse. Note the gridlines are a guide to visualize the effects of filtering on the baseline

i.e., Pdi; and (4) filtered Pdi signal using a 0.3–30 Hz band-pass to center the power of the signal and remove any noise related to cardiac activity (Fig. 2).

4. Equilibrate two Mikro-Tip® solid-state pressure catheters (Millar Instrument; 3.5 F, #SPR-524) in individual syringes filled with saline for at least 30 min.
5. Calibration of the solid-state pressure catheters should follow the manufacture specification. Briefly, a two-point calibration between 0 and 30 cm H_2O using a manometer is sufficient, since the solid-state pressure catheters are highly linear across the physiological range of Peso and Pgas comprising the Pdi measurement. This reliability and linear range is advantageous compared to the more commonly used balloon-tipped system (*see* **Note 2**).

3 Methods

3.1 Mouse Preparation

1. Weigh and anesthetize mice. Consider the use of fentanyl (10 mcg/kg), droperidol (0.2 mg/kg), and diazepam (5 mg/kg). This anesthetic regimen is consistent with limited instrumentation of the animal and has minimal impact on ventilation or parameters of skeletal muscle function [26].

2. Maintain mouse on a heating pad, consider using thermometer and pulse oximetry (e.g., MouseOX, Starr Life Science Corp., Oakmont, PA) to monitor animal vital signs throughout the procedure. Note that stressing the animal to maintain thermoregulation may result in changes in ventilation so it is critical to maintain body temperature.
3. To constrain abdominal movements and maintain near isometric conditions, bind the abdomen of the mouse using a thick woven gauze or elastic bandage. Taking care to make sure the binding is below the base of the rib cage, and covering the body of the mouse.
4. Position and stabilize the mouse supine on a surgical board. The use of a magnetic board (e.g., Braintree Scientific, ACD 014) will allow for precise controlled retraction and positioning of the mouse. Note alternative positioning, such as prone, is possible but should be standardized across the experiment, since gravity and posture may affect Pdi.
5. Secure the upper airway of the mouse open at the mouth by the use of a rubber band or 4-0 suture attached to the incisors.
6. Tracheal cannulation (Fig. 3): begin by making a vertical midline incision ~25 mm distal to the jaw, near the hyoid bone to just proximal of the sternum. Retract the skin on both the right and left side. Bluntly split and dissect away the submaxillary gland to expose the trachea and larynx. Using a dissecting microscope carefully split and dissect away the sternohyoid and sternothyroid muscle layers surrounding the trachea. Place two pieces of 5-0 silk suture, each ~10 cm long under the trachea at the level of the cricoid cartilage. Cut the trachea approximately two rings just below the cricoid cartilage. Carefully advance the cannula into the trachea, then using the two previously placed sutures, tie two basic square knots to secure the cannula in place. Take care to tie the knots between cartilage rings on the smooth muscle portion of the trachea. Trim excess suture. In general, a 19 G blunt tipped metal cannula (e.g., Fishman #Z512119) is sufficient for mice, but very small (<15 g) or frail/old mice may require a smaller cannula such as a 21 G (e.g., Fishman #Z512121). Note the use of thin wall cannulas is beneficial (*see* **Note 3**).
7. Exposure of phrenic nerves (Fig. 4): bilaterally expose the phrenic nerves for maximal stimulation during behavioral testing. Following placement of tracheal cannula, bluntly dissect the sternocleidomastoid muscle and retract carefully. Using the dissecting microscope, take care to avoid the jugular veins, carotid arteries and nerve plexus (ansa cervicalis) while dissecting deeply to reach the brachial plexus and phrenic

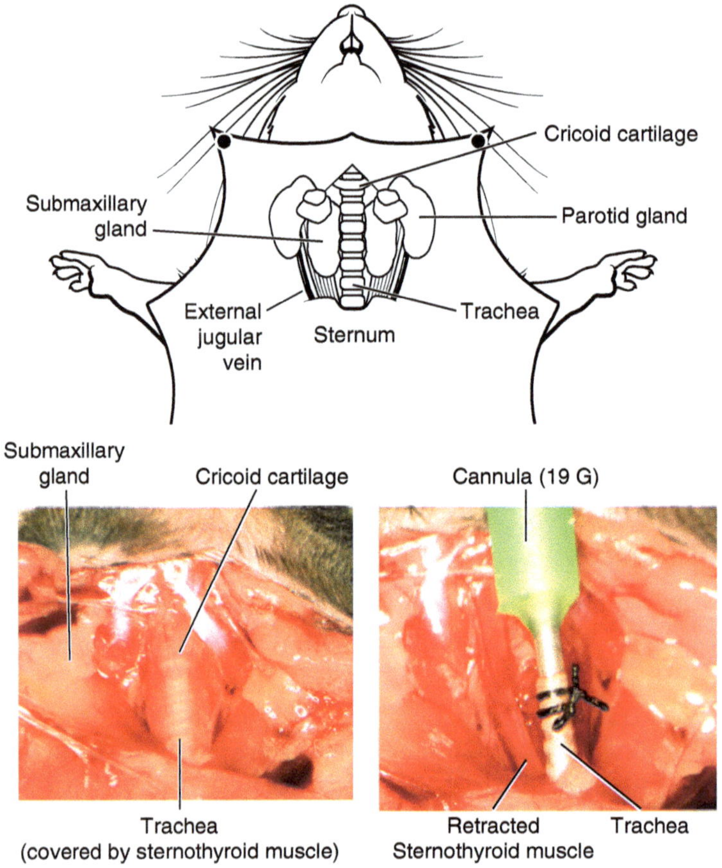

Fig. 3 Anatomical schematic with the structures and landmarks necessary for the tracheal cannulation procedure in the mouse. Note this view is observed following the vertical midline incision, retraction of the skin and blunt dissection of the submaxillary gland to expose the trachea. The photographs highlight the same view; with a view of the sternothyroid muscle overlying the trachea, which needs to be retracted to expose the trachea, and tracheostomy followed by cannulation with a 19 G blunt tipped cannula. Note in this view the sternothyroid muscle is retracted and sutures are placed securing the cannula between cartilage rings

nerve. Using the trachea as a reference, the phrenic nerve will be dorsal, lateral and parallel. Also running parallel will be the vagus nerve, which has a much larger diameter and is located more medial between the phrenic nerve and the trachea, next to the carotid and jugular vessels. The brachial plexus will run at roughly a 45° angle from the trachea. Note in some species there may be an accessory phrenic nerve (from lower cervical roots) that joins the main phrenic nerve lower in the neck and which may require additional dissection (*see* **Note 3**).

3.2 Pdi Measurement of Motor Behaviors

1. Insert solid-state pressure catheters, one at a time, into the esophagus through the mouth of the mouse (Fig. 5). Adjust placement of one of the catheters while monitoring pressure until it becomes positive (Pgas indicating gastric placement in abdominal cavity). Advance the second catheter to obtain the

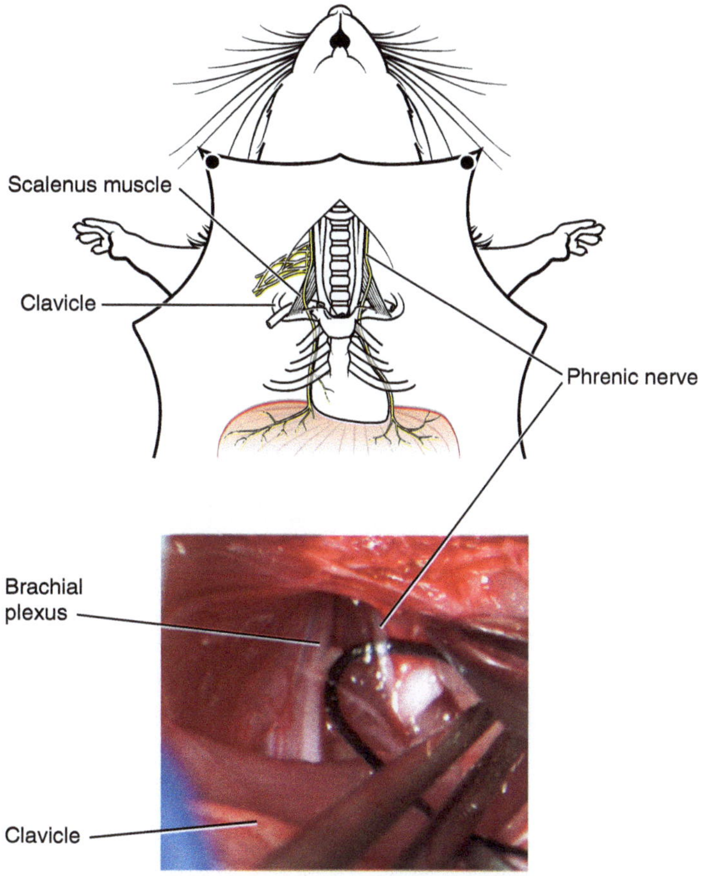

Fig. 4 Anatomical schematic showing the structures and landmarks related to phrenic nerve within the neck of the mouse. The phrenic nerve innervates the diaphragm muscle on the corresponding side of the muscle. Following dissection into the anterior aspect of neck, the brachial plexus and phrenic nerve's run on both side of the trachea. Using the trachea as a reference, the phrenic nerve will be dorsal, lateral and parallel. The photographs highlight a zoomed in view; with the phrenic nerve and brachial plexus on the lateral side of the trachea

maximum negative pressure (Peso indicating esophageal placement in thoracic cavity). While inserting catheters, it is useful to use a blunt forceps to hold tongue, which will aid in advancing the catheters. Note there is no need to use lubrication for insertion of the catheters. Correct placement of the catheters can be confirmed post mortem.

2. Once catheters are placed spanning the diaphragm muscle, begin recording while animals breathe room air (eupnea)—recommended duration of at least 5 min.
3. To chemically stimulate breathing, expose the animals (in a chamber) to a hypoxia and/or hypercapnia gas mixture containing: 10% O_2, 5–7% CO_2, balance N_2—recommended

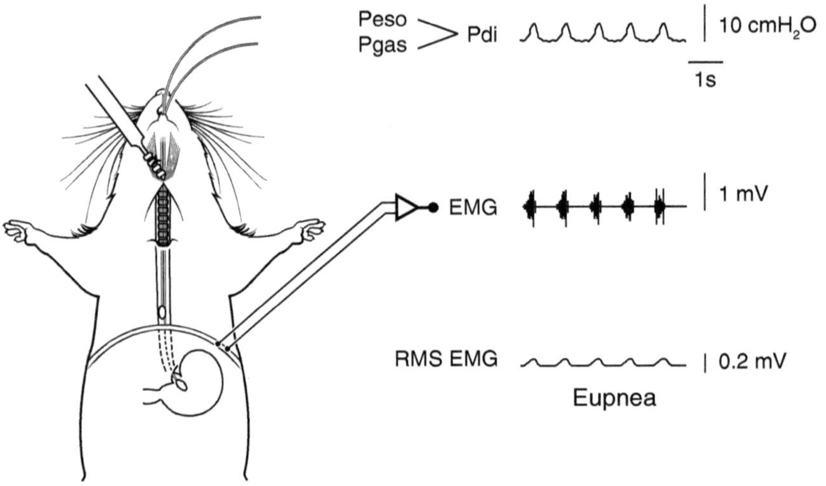

Fig. 5 Anatomical schematic showing placement of the solid-state pressure catheters for measurement of gastric (Pgas) and thoracic esophageal (Peso) pressures in mice—note the pair of catheters is advanced into the stomach (Pgas) and esophagus in the thoracic cavity (Peso) through the mouth. The difference between Pgas and Peso is used to determine transdiaphragmatic pressure (Pdi; see raw tracing). The Pdi amplitude is significantly correlated to peak root mean squared (RMS) of diaphragm muscle electromyography (EMG) activity [2, 3]. Raw tracings are provided for Pdi, diaphragm EMG, and RMS EMG

duration of at least 5 min of acclimatization followed by 2 min for data analysis.

4. Spontaneously occurring sighs can be noted during ventilatory behaviors, in general the amplitude for a sigh will be at least two times that of eupnea in the mouse (*see* Fig. 6). This criterion for sighs will allow for proper identification during analysis.

5. Inspiratory efforts against an occluded trachea provide a stronger neural drive for diaphragm activation. To accomplish tracheal occlusion the top of the tracheal cannula is completely obstructed for 15 s. Typically, this time will allow analysis ~10 escalating efforts, and the five maximal efforts are used for analysis. As an alternative, if tracheal cannulation is not conducted (e.g., with survival procedures), obstruction of the upper airway can be achieved by completely occluding the mouth and nose.

6. Maximal bilateral phrenic nerve stimulation provides a measure of maximal Pdi for referencing other efforts. When stimulating the phrenic nerves ensure that the electrodes make good contact with the nerve by freeing the area connective tissue, and by placing a few drops of mineral oil in the exposed area to aid in electrical isolation. Use straight bipolar electrodes on each nerve (e.g., FHC #PBSD08075; Bowdoin ME) and stimulate the nerve up to three times using a Grass stimulator together with a stimulation isolation unit (e.g., Grass Technologies, Warwick, RI) using the following settings deter-

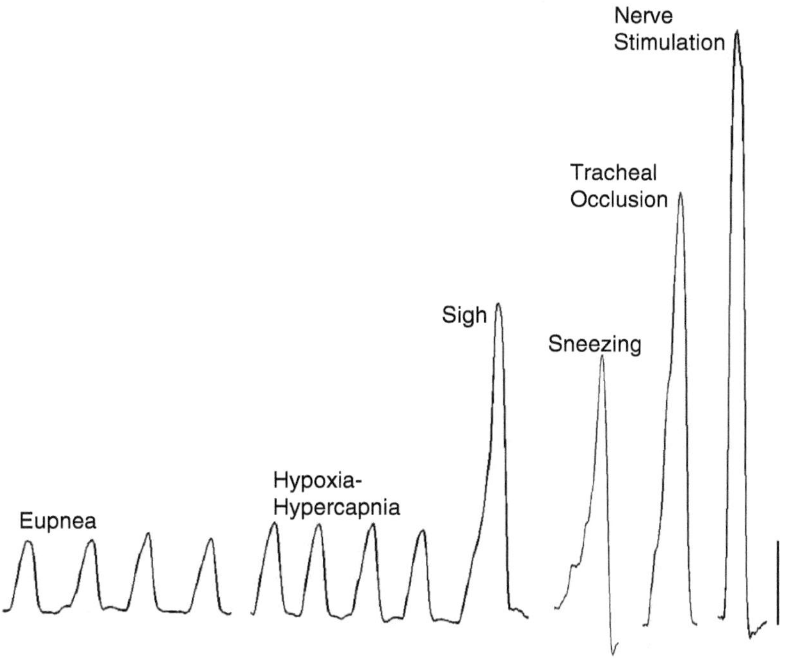

Fig. 6 Representative transdiaphragmatic pressure (Pdi) tracings during various motor behaviors in an adult mouse. Increasing Pdi (force) is found from eupnea, hypoxia–hypercapnia, sighs, sneezing, tracheal occlusion, and maximal phrenic nerve stimulation in mice. Bar represents 10 cm H_2O

mined in separate studies [27]: 150 Hz (maximal tetanic force), 0.5 ms pulse duration, in 300 ms duration trains at supramaximal intensity (determined by increasing stimulus current until maximal Pdi response is elicited). Allow at least 1 min between each stimulation train. Recommend analysis of only the maximal amplitude of the three evoked responses.

7. Sneezing induced with capsaicin (Sigma; M2028), using a 10 μl Hamilton syringe to inject 30 μM (30% alcohol) into both nostrils, ~5 μl per side. Allow 5 min of recording and analyze any sneezing that is seen. Note, while the sneeze response is robust and consistent in the rat, not all mice will sneeze and the response can be more variable. In general the amplitude for a sneeze will be at about two times that of eupnea. Alternative concentrations of capsaicin may be considered [28]. Note that you must clean the syringe thoroughly with water and 70% alcohol after use.

8. Allow at least 5 min between each behavior to allow for Pdi amplitudes to return to eupneic values. Consider completing behaviors in the order presented; (1) eupnea; (2) hypoxia–hypercapnia; (3) sighs; (4) tracheal occlusion; (5) maximal phrenic nerve stimulation; and (6) sneezing.

3.3 Analytical Procedures

1. The primary outcome is the Pdi amplitude across motor behaviors (Fig. 6) (*see* **Note 4**).
2. Analysis of Pdi amplitude across motor behaviors can be done through various analytical platforms such as LabChart, MATLAB, R, C, Java, or Python.
3. Additional parameters of analysis may include amplitude duration, inspiratory time, expiratory time, and estimates of drive [24]. Additionally, respiratory frequency and frequency of naturally occurring sighs can be determined during ventilatory behaviors.

4 Notes

1. All specific equipment and product numbers listed is what our lab uses for Pdi; many aspects can be conducted with comparable equipment or platforms.
2. All cleaning and storage of the solid state pressure catheters should follow the manufacturer specification.
3. The Pdi procedure can be conducted during survival procedures and thus longitudinally in the same animal, with the omission of tracheal cannulation and maximal bilateral nerve stimulation.
4. Appropriate inclusion and exclusion criteria based on Pdi amplitude should be determined a priori and based on maximal values from phrenic nerve stimulation.

Acknowledgments

Any procedures conducted on animals in the development of these methods were conducted following institutional protocols and animal care guidelines, in compliance with National Institute of Health Guidelines.

This work was supported by grants from National Institute of Health R01-AG-044615 and R01-HL-096750 (CBM and GCS), T32-HL105355 (SMG), and the Mayo Clinic.

References

1. Henneman E (1957) Relation between size of neurons and their susceptibility to discharge. Science 126(3287):1345–1347
2. Mantilla CB, Seven YB, Zhan WZ, Sieck GC (2010) Diaphragm motor unit recruitment in rats. Respir Physiol Neurobiol 173(1):101–106
3. Gill LC, Mantilla CB, Sieck GC (2015) Impact of unilateral denervation on transdiaphragmatic pressure. Respir Physiol Neurobiol 210:14–21. doi:10.1016/j.resp.2015.01.013
4. Sieck GC, Ferreira LF, Reid MB, Mantilla CB (2013) Mechanical properties of respiratory muscles. Compr Physiol 3(4):1553–1567. doi:10.1002/cphy.c130003
5. Mantilla CB, Sieck GC (2013) Impact of diaphragm muscle fiber atrophy on neuromotor

control. Respir Physiol Neurobiol 189(2):411–418. doi:10.1016/j.resp.2013.06.025

6. Mantilla CB, Sieck GC (2011) Phrenic motor unit recruitment during ventilatory and non-ventilatory behaviors. Respir Physiol Neurobiol 179(1):57–63. doi:10.1016/j.resp.2011.06.028, S1569-9048(11)00241-2 [pii]
7. Mantilla CB, Greising SM, Zhan WZ, Seven YB, Sieck GC (2013) Prolonged C2 spinal hemisection-induced inactivity reduces diaphragm muscle specific force with modest, selective atrophy of type IIx and/or IIb fibers. J Appl Physiol 114(3):380–386. doi:10.1152/japplphysiol.01122.2012
8. Gransee HM, Zhan WZ, Sieck GC, Mantilla CB (2013) Targeted delivery of TrkB receptor to phrenic motoneurons enhances functional recovery of rhythmic phrenic activity after cervical spinal hemisection. PLoS One 8(5):e64755. doi:10.1371/journal.pone.0064755
9. Gransee HM, Zhan WZ, Sieck GC, Mantilla CB (2015) Localized delivery of brain-derived neurotrophic factor-expressing mesenchymal stem cells enhances functional recovery following cervical spinal cord injury. J Neurotrauma 32(3):185–193. doi:10.1089/neu.2014.3464
10. Mantilla CB, Gransee HM, Zhan WZ, Sieck GC (2013) Motoneuron BDNF/TrkB signaling enhances functional recovery after cervical spinal cord injury. Exp Neurol 247C:101–109. doi:10.1016/j.expneurol.2013.04.002
11. Sieck GC, Fournier M (1989) Diaphragm motor unit recruitment during ventilatory and nonventilatory behaviors. J Appl Physiol 66(6):2539–2545
12. Sieck GC (1994) Physiological effects of diaphragm muscle denervation and disuse. Clin Chest Med 15(4):641–659
13. Sieck GC (1991) Neural control of the inspiratory pump. NIPS 6:260–264
14. Greising SM, Sieck DC, Sieck GC, Mantilla CB (2013) Novel method for transdiaphragmatic pressure measurements in mice. Respir Physiol Neurobiol 188(1):56–59. doi:10.1016/j.resp.2013.04.018
15. Greising SM, Mantilla CB, Medina-Martinez JS, Stowe JM, Sieck GC (2015) Functional impact of diaphragm muscle sarcopenia in both male and female mice. Am J Physiol Lung Cell Mol Physiol 309(1):L46–L52. doi:10.1152/ajplung.00064.2015
16. Polkey MI, Harris ML, Hughes PD, Hamnegard CH, Lyons D, Green M, Moxham J (1997) The contractile properties of the elderly human diaphragm. Am J Respir Crit Care Med 155(5):1560–1564
17. Tolep K, Higgins N, Muza S, Criner G, Kelsen SG (1995) Comparison of diaphragm strength between healthy adult elderly and young men. Am J Respir Crit Care Med 152(2):677–682
18. Bazzy AR, Haddad GG (1984) Diaphragmatic fatigue in unanesthetized adult sheep. J Appl Physiol 57(1):182–190
19. Hubmayr RD, Sprung J, Nelson S (1990) Determinants of transdiaphragmatic pressure in dogs. J Appl Physiol 69(6):2050–2056
20. Watchko JF, Mayock DE, Standaert A, Woodrum DE (1986) Postnatal changes in transdiaphragmatic pressure in piglets. Pediatr Res 20:658–661
21. Howell S, Maarek JM, Fournier M, Sullivan K, Zhan WZ, Sieck GC (1995) Congestive heart failure: differential adaptation of the diaphragm and latissimus dorsi. J Appl Physiol 79(2): 389–397
22. Sassoon CS, Gruer SE, Sieck GC (1996) Temporal relationships of ventilatory failure, pump failure, and diaphragm fatigue. J Appl Physiol 81(1):238–245
23. Sassoon CS, Caiozzo VJ, Manka A, Sieck GC (2002) Altered diaphragm contractile properties with controlled mechanical ventilation. J Appl Physiol 92(6):2585–2595
24. Medina-Martinez JS, Greising SM, Sieck GC, Mantilla CB (2015) Semi-automated assessment of transdiaphragmatic pressure variability across motor behaviors. Respir Physiol Neurobiol 215:73–81. doi:10.1016/j.resp.2015.05.009
25. American Thoracic Society/European Respiratory Society (2002) ATS/ERS Statement on respiratory muscle testing. Am J Respir Crit Care Med 166(4):518–624
26. Ingalls CP, Warren GL, Lowe DA, Boorstein DB, Armstrong RB (1996) Differential effects of anesthetics on in vivo skeletal muscle contractile function in the mouse. J Appl Physiol 80(1):332–340
27. Greising SM, Mantilla CB, Gorman BA, Ermilov LG, Sieck GC (2013) Diaphragm muscle sarcopenia in aging mice. Exp Gerontol 48(9):881–887. doi:10.1016/j.exger.2013.06.001
28. Costanzo MT, Yost RA, Davenport PW (2014) Standardized method for solubility and storage of capsaicin-based solutions for cough induction. Cough 10:6. doi:10.1186/1745-9974-10-6
29. Greising SM, Medina-Martínez JS, Vasdev AK, Sieck GC, Mantilla CB (2015) Analysis of muscle fiber clustering in the diaphragm muscle of sarcopenic mice. Muscle Nerve 52(1):76–82. doi:10.1002/mus.24641

Chapter 22

Assessment of the Contractile Properties of Permeabilized Skeletal Muscle Fibers

Dennis R. Claflin, Stuart M. Roche, Jonathan P. Gumucio, Christopher L. Mendias, and Susan V. Brooks

Abstract

Permeabilized individual skeletal muscle fibers offer the opportunity to evaluate contractile behavior in a system that is greatly simplified, yet physiologically relevant. Here we describe the steps required to prepare, permeabilize and preserve small samples of skeletal muscle. We then detail the procedures used to isolate individual fiber segments and attach them to an experimental apparatus for the purpose of controlling activation and measuring force generation. We also describe our technique for estimating the cross-sectional area of fiber segments. The area measurement is necessary for normalizing the absolute force to obtain specific force, a measure of the intrinsic force-generating capability of the contractile system.

Key words Muscle physiology, Skeletal muscle, Single muscle fiber, Permeabilized muscle fiber, Cross-sectional area, Contractility, Isometric force, Specific force

1 Introduction

In vivo muscle force generation is the end result of a sequence that includes motor nerve action potentials, neuromuscular transmission, muscle fiber action potentials, release of intracellular calcium, and activation of a system of regulatory proteins that allow the force-producing activity of contractile proteins. A key attribute of the permeabilized fiber preparation is that it eliminates most of the steps required for initiation of contractile activity in vivo, with only the regulatory and contractile functions associated with the myofibrillar apparatus remaining. Because the cell membrane barrier is no longer functional, the concentration of activating calcium can be controlled by the investigator, resulting in a simplified system that allows functional assessment of the isolated regulatory and contractile structures in their native configuration. Measurements of force generated by permeabilized skeletal muscle fibers can thus be used in experimental strategies designed to identify sources of muscle weakness observed in vivo. Examples include characterization

Michael Kyba (ed.), *Skeletal Muscle Regeneration in the Mouse: Methods and Protocols*, Methods in Molecular Biology, vol. 1460, DOI 10.1007/978-1-4939-3810-0_22, © Springer Science+Business Media New York 2016

of the force generating capacity of fibers from myostatin deficient mice [1] and investigation of the cause of persistent muscle weakness exhibited following chronic rotator cuff tears [2, 3].

In this chapter, we describe in detail the techniques used to measure the maximum force generating capacity of single fibers from chemically permeabilized skeletal muscle samples. Specifically, we describe the steps required to extract an individual fiber segment (referred to herein as a "fiber") from a pre-permeabilized bundle of fibers and then suspend the fiber in relaxing solution, attached at one end to a force-transducer and at the other end to a length-controlling servomotor. The techniques for adjusting fiber length to obtain optimal force production and for maximally activating the fiber by introducing a high concentration of calcium ions will then be discussed. Finally, we describe the approach used to estimate fiber cross-sectional area, which is required for normalization of absolute force to produce specific force, a measure of the intrinsic force-generating capability of the fiber. A video article that demonstrates several of the steps described here is also available [4].

2 Materials

2.1 Preparation and Storage of Fiber Bundles

1. Dissecting, permeabilizing, and storage solutions: potassium propionate (*see* **Note 1**), imidazole, EGTA, $MgCl_2$, Na_2H_2ATP, Brij 58, glycerol, KOH for adjusting pH.
2. Forceps (Dumont #5).
3. Microdissecting scissors.
4. 50 mm polystyrene culture dish plated with 6 mm of silicone elastomer (Sylgard 184).
5. Stainless steel insect pins, 100 μm diameter.
6. Stereomicroscope.
7. Polystyrene test tube with cap (5 mL).
8. 0.5 mL microcentrifuge tubes + caps with O-rings.

2.2 Testing Fibers

1. Relaxing, preactivating, and activating solutions: HEPES, MgO, EGTA (ethyleneglycol-bis-(β-aminoethyl ether)-*N*,*N*,*N′*,*N′*-tetraacetic acid), HDTA (1,6-diaminohexane-*N*,*N*,*N′*,*N′*-tetraacetic acid), $CaCO_3$, NaN_3, Na_2H_2ATP, creatine phosphate, KOH for adjusting pH.
2. Fiber chamber system (Aurora Scientific, Model 802D).
3. Force transducer (Aurora Scientific, Model 403A).
4. Servo motor for fiber length control (Aurora Scientific, Model 312C).
5. Vibration isolation table.

6. Stereomicroscope equipped with digital camera.
7. Monofilament nylon suture size USP 10-0, non-sterile (Ashaway Line & Twine).
8. Laser diode.
9. Computer equipped with data acquisition hardware, software.
10. Computer software for analyzing acquired force and length recordings.
11. Computer software for analyzing fiber images.

3 Methods

3.1 Preparation and Storage of Fiber Bundles

1. Dissecting solution: combine potassium propionate (125 mM), imidazole (20 mM), EGTA (5 mM), $MgCl_2$ (2 mM), and Na_2H_2ATP (2 mM) in deionized water. Adjust pH to 7.0 with KOH.
2. Permeabilizing solution: add Brij 58 (0.5 % w/v) to dissecting solution.
3. Storage solution: identical to dissecting solution, but instead of water, the compounds are dissolved in a mixture of 50% deionized water and 50% glycerol (*see* **Note 2**). Adjust pH to 7.0 with KOH.
4. Skeletal muscle tissue samples are typically obtained by open biopsy [5] or needle biopsy [6]. The fresh tissue is quickly transferred to a container of chilled (4 °C) dissecting solution, agitated gently to remove adherent blood, and then placed in a dissecting dish (50 mm polystyrene culture dish plated with 6 mm of Sylgard 184, a silicone elastomer) filled with fresh dissecting solution.
5. The dissecting dish is then placed on the transillumination stage of a stereomicroscope and stainless steel insect pins (100 μm diameter) are passed through peripheral regions of the biopsy and into the silicone base of the dissecting dish to stabilize the tissue during dissection.
6. The goal of the dissection is to divide the gross sample into multiple small bundles of fibers that are arranged in parallel. The bundles are approximately 4 mm long by 1 mm diameter. Dissection into bundles is accomplished by gently grasping the sample with fine forceps in one hand and cutting parallel to the long axis of the fibers using the microdissection scissors in the other hand.
7. Long strips of parallel fibers that result from longitudinal dissection are transected into 4 mm lengths using microdissection scissors. These steps are illustrated in Fig. 1a and b (*see* **Note 3**).

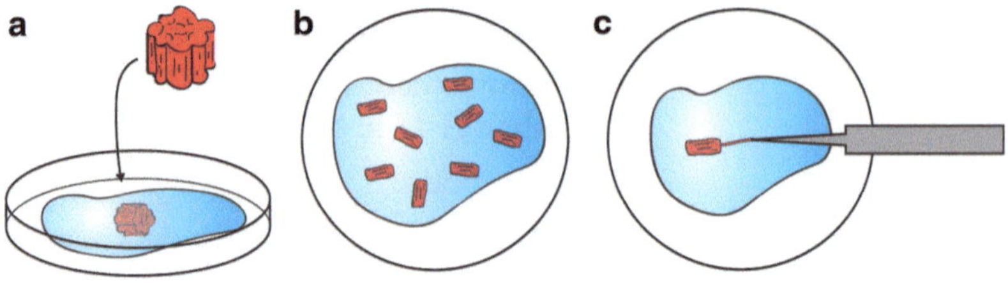

Fig. 1 Preparation of fiber bundles and extraction of individual fibers. (**a**) Fresh tissue is placed in a dissecting dish and (**b**) divided into small bundles of parallel fibers in preparation for permeabilization and freezer storage. (**c**) On the day of an experiment, a bundle of permeabilized fibers is removed from freezer storage and pinned to a dissection dish. Individual fibers are extracted from the bundle by pulling gently on one end of a fiber with the pulling force applied along the long axis of the fiber. Reproduced from Roche et al. 2015 [4] with permission from JoVE

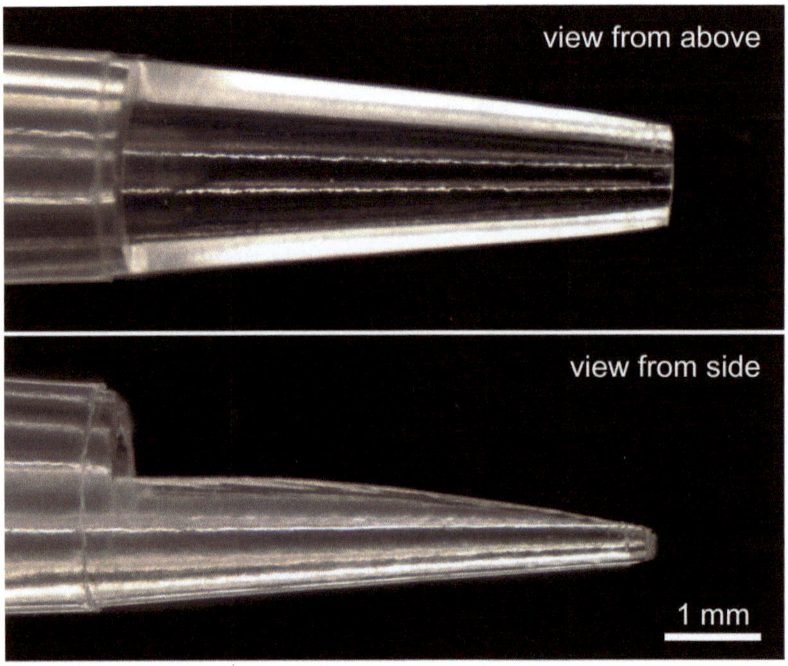

Fig. 2 Fiber transfer scoop formed by trimming a 100 μL disposable pipette tip. Reproduced from Roche et al. 2015 [4] with permission from JoVE

8. Immediately following isolation, fiber bundles are incubated in permeabilizing solution at 0 °C for 30 min.
9. After incubation, bundles are transferred to a 10 mL beaker of fresh dissecting solution and agitated for ≈1 min to remove adherent permeabilizing solution, and then placed in a 5 mL polystyrene test tube filled with cold (0 °C) storage solution.

10. Following overnight (12–16 h) storage at 4 °C, individual fiber bundles are aliquoted into labeled 0.5 mL microcentrifuge tubes filled with storage solution and stored at −80 °C (*see* **Note 4**).

3.2 Extraction of Individual Fibers

1. On the day of an experiment, a microcentrifuge tube containing a bundle of fibers is removed from the freezer and placed on ice for approximately 15 min.
2. The thawed bundle is then removed with forceps and immersed in dissecting solution in a dissecting dish and placed on the transillumination stage of a stereomicroscope.
3. Stainless steel insect pins (100 μm diameter) are passed through the periphery of the bundle at all four corners to secure it to the silicone elastomer at the bottom of the dish.
4. Extraction of an individual permeabilized fiber from the bundle is accomplished by grasping the end of a fiber with fine forceps and gently pulling with a steady force along the long axis of the fiber until it slides free from the bundle (Fig. 1c) (*see* **Notes 5** and **6**).
5. After extraction, the fiber is transferred to the experimental chamber system in a solution-filled "scoop" fashioned from a 100 μL disposable pipette tip (Fig. 2) and the dissection dish is stored on ice until a fresh fiber is needed (*see* **Notes** 7 and **8**).

3.3 Solutions for Fiber Experiments

Three solutions are used during experiments in which the functional properties of maximally activated permeabilized fibers are assessed: relaxing solution, preactivating solution, and activating solution. The solutions described here are adapted from Moisescu and Thieleczek [7] and are prepared as follows (*see* **Note 9**).

1. Add, in order, the ingredients listed in Table 1 to deionized water (begin with approximately 80% of the desired final volume) (*see* **Notes 10** and **11**).

Table 1
Compounds that are combined with water and maintained at 70–80 °C for 30 min as the initial step in the preparation of relaxing, preactivating, and activating solutions

	Relaxing	Preactivating	Activating
HEPES (acid)	90.0	90.0	90.0
MgO	10.3	8.5	8.12
EGTA (acid)	52.0		52.0
HDTA (acid)		50.0	
$CaCO_3$			50.0

Table entries are mM concentrations in the final solution

Table 2
Final compositions of the completed relaxing, preactivating, and activating solutions

Constituent	Relaxing solution	Activating solution	Preactivating solution
HEPES	90.0	90.0	90.0
Mg (total)	10.30	8.12	8.50
Mg^{2+}	1.00	1.00	1.00
EGTA	50.0	50.0	0.10
Ca (total)	None added	50.0	None added
HDTA	0.0	0.0	50.0
ATP	8.00	8.00	8.00
CrP	10.0	10.0	10.0
NaN_3	1.00	1.00	1.00
Na (total)	37	37	37
K (total)	125	125	125
pCa	>9	≈4.5	>8

Table entries are mM concentrations in the final solution with the exception of the final entry, which is given as pCa ($= -\log_{10}[Ca^{2+}]$)

2. Heat to between 70 °C and 80 °C for 30 min, stirring throughout (*see* **Note 12**).
3. Cool to room temperature, then add 1 mM NaN_3, which serves as a preservative (*see* **Note 13**).
4. Adjust pH to approximately 7.0 using 8 N KOH (*see* **Notes 14** and **15**).
5. Add Na_2H_2ATP (8 mM) and Na_2CrP (10 mM) (*see* **Notes 16** and **17**).
6. For the ***preactivating solution only***, add 1 part relaxing solution in 500 parts final solution (e.g., for a 1000 mL batch of preactivating solution, add 2 mL relaxing solution before performing **step** 7).
7. Adjust pH to 7.10 with KOH while maintaining solution at the temperature to be used during experiments, and then bring solution to final volume with deionized water (*see* **Note 18**). Final compositions of relaxing, preactivating, and activating solutions are given in Table 2.
8. Aliquot solutions into 5 mL cryovials and store at −80 °C for up to two years (*see* **Note 19**).

3.4 Fiber Mounting

Prepare the fiber chamber system by filling chamber 1 with relaxing solution, chamber 2 with preactivating solution, and chamber 3 with activating solution. Adjust the temperature control system to the desired experimental temperature. Transport an extracted fiber to chamber 1 as described in Subheading 3.2 (*see* Fig. 4b). Chamber 1 is unique in that is has a right-angle prism embedded in one sidewall, allowing visualization and imaging of the fiber from the side after mounting. The mounting procedure begins with the formation of loops from USP 10-0 monofilament nylon suture (Fig. 3) and then proceeds as described in the numbered steps, below. The mounting procedure is performed using fine forceps with the aid of a stereomicroscope attached to a boom stand and positioned over the chamber. The chamber system is mounted to a vibration isolation table to minimize contamination of force recordings by environmental vibrations.

1. Two suture loops are threaded onto both the servomotor and force transducer attachment extensions (Fig. 4a).
2. The two outermost suture loops are first tightened over the fiber to secure it to the apparatus (Fig. 4c). These ties hold the fiber in place while the more critical innermost loops are positioned and tightened.
3. After trimming excess suture from the ends of the outermost ties with microdissection scissors, the final two suture loops are positioned over the fiber near the ends of the attachment points extending from the force transducer and servomotor. These innermost loops are then tightened over the fiber, securing it to the apparatus and defining the length of the fiber that is contributing to measurements made during subsequent experimental manipulations (Fig. 4d). Excess suture is then trimmed from the ends of the outer ties using microdissection scissors (*see* **Note 20**).
4. With the fiber "just taut", the *x*-*y*-*z* micromanipulators that control the servomotor position are adjusted with the goal of

Fig. 3 Loop of USP 10-0 monofilament nylon suture used to attach the fiber to the experimental apparatus. Note that the knot type is double-overhand, not single-overhand. Reproduced from Roche et al. 2015 [4] with permission from JoVE

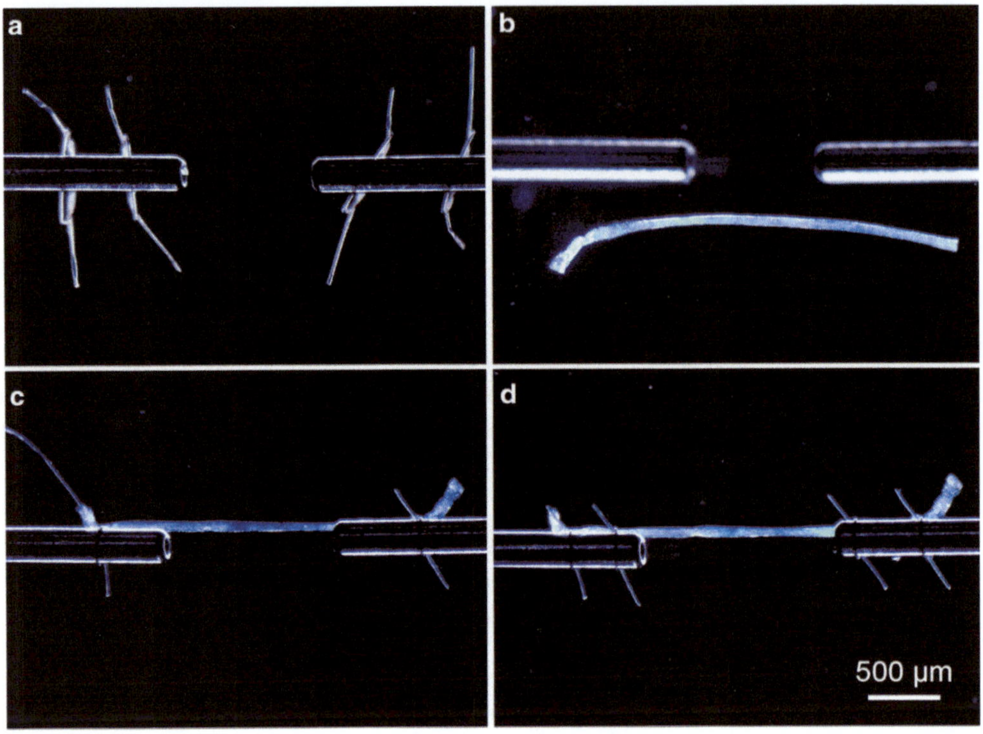

Fig. 4 Sequence for attaching fiber to force transducer and servomotor. (**a**) Thread two suture loops onto the force transducer and servomotor attachment extensions. (**b**) Transfer fiber to chamber using transfer scoop (*see* Fig. 2). (**c**) Secure fiber using two outermost suture ties. (**d**) Secure fiber using two innermost suture ties. Reproduced from Roche et al. 2015 [4] with permission from JoVE

making sure the fiber force vector is parallel to the force-sensing axis of the force transducer. These adjustments are made while viewing the fiber from above and from the side (via the wall-mounted prism), making sure the long axis of the fiber is parallel to both the floor and the sides of the chamber.

5. Next, fiber sarcomere length is adjusted by monitoring the interference pattern that results from passing light from a laser through the fiber [8]. The fiber is modeled as a one-dimensional diffraction grating [9] with periodicity equal to that of fiber sarcomeres due to optical density variations arising from sarcomere structures. The goal of this step is to achieve optimal overlap of sarcomeric thick and thin filaments, which results in maximum force development [10]. A translucent target screen must be constructed and placed a distance "*D*" from the fiber (*see* Fig. 5). The laser interference pattern emerges from the fiber and appears on the target screen. Only the undiffracted beam ("center") and first-order interference maxima are shown in Fig. 5. Higher order maxima are faint and difficult to detect. The target screen is constructed based upon the diffraction equation [8], which results in the following expression for

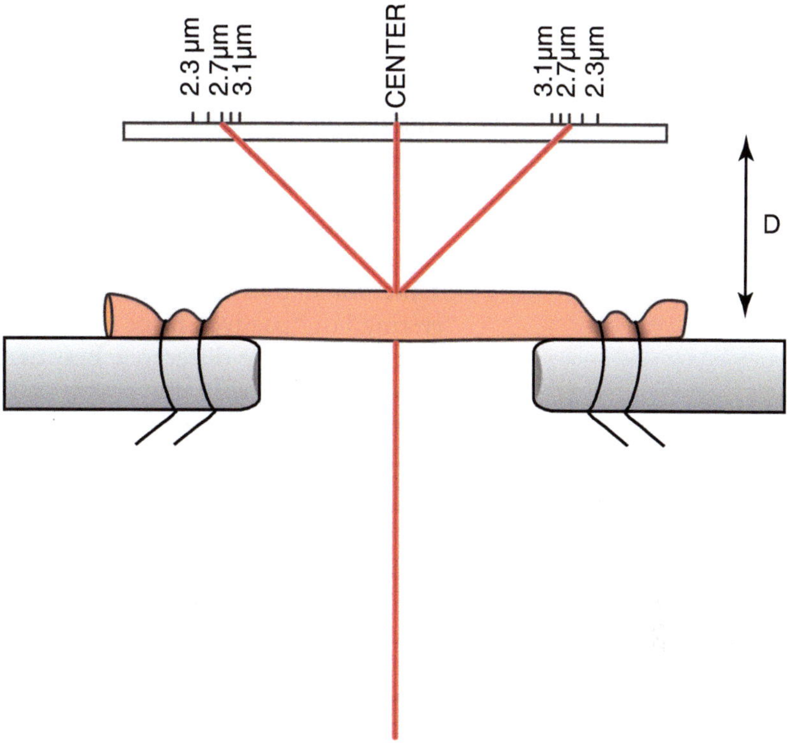

Fig. 5 Laser light diffracted by the fiber is used to adjust sarcomere length. The laser target screen is calibrated for a specific laser wavelength and "*D*", the distance from fiber to screen (see text for explanation). Reproduced from Roche et al. 2015 [4] with permission from JoVE

"*T*", the distance between the undiffracted beam and the first-order maxima at the target.

$$T = D \times \tan\left[\sin^{-1}\left(\frac{\lambda}{\mathrm{SL}}\right)\right]$$

Note that *T* depends on the distance from fiber to target (*D*), the wavelength of the laser light (λ) and sarcomere length (SL). Markings corresponding to sarcomere lengths of 2.1 µm through 2.9 µm are typically indicated on the target and fiber length is adjusted using the micromanipulator attached to the servomotor until the first-order interference maxima appear on the target markings corresponding to the desired sarcomere length (*see* **Notes 20–25**). "Fiber length" (L_f) is defined as the length of the fiber, measured between the two innermost ties, after adjustment of optimal sarcomere length.

3.5 Fiber Activation

1. Use the chamber system controller to remove the fiber from relaxing solution (chamber 1) and immerse it in preactivating solution (chamber 2). Allow to equilibrate for 3 min (*see* **Note 26**).

2. While in preactivating solution, start the software that records the servomotor position and force transducer output, then use the servomotor control software to apply a rapid ("step") reduction in fiber length followed 30 ms later by a rapid return to the original length (*see* **Note 27**). This maneuver, termed herein as a "find-force-zero" movement, causes the fiber to briefly become "slack", and the absence of tension results in an accurate indication, in the experimental record, of the output of the force transducer when subjected to zero force. The "resting" or "passive" force required to maintain the fiber at L_f is measured as the difference between the force level just prior to the find-force-zero movement and the force transducer zero revealed by the movement.
3. After the 3 min incubation in preactivating solution, while continuing to record servomotor and force transducer outputs, use the chamber system controller to immerse the fiber in activating solution (chamber 3). Allow force to reach a maximum, steady level and then apply another find-force-zero movement with the servomotor. This establishes the zero force level corresponding to chamber 3 in the experimental record (*see* **Note 28**). Force will rapidly regenerate at the completion of the find-force-zero movement. After force returns to a maximum, steady level, the chamber system controller is used to immerse the fiber once again in chamber 1 where the activating calcium is removed and the fiber relaxes. Maximum *total* isometric force is measured as the difference between force level established by the find-force-zero movement in the activating chamber and the maximum force reached during the activation. Maximum *active* isometric force is the difference between maximum total isometric force and the passive force measured when the fiber was in chamber 2. Figure 6 illustrates typical recordings of force and servomotor position during the determination of maximum isometric force (*see* **Note 29**).

3.6 Estimating Fiber Cross-Sectional Area

Maximum isometric *specific* force ("specific force") is a measure of the intrinsic force-generating capacity of the contractile filament system of a fiber. Specific force is obtained by dividing the absolute force generated by the cross-sectional area of the fiber, thereby neutralizing the effects of fiber size. Valid estimates of specific force therefore require accurate estimates of fiber cross-sectional area.

1. While maintaining the length of the fiber at L_f (i.e., at optimal sarcomere length), capture a "top-view" image using a camera mounted to an imaging port on the stereomicroscope (*see* **Note 30**).
2. Move the stereomicroscope in the direction perpendicular to the long axis of the fiber until the side view of the fiber, as

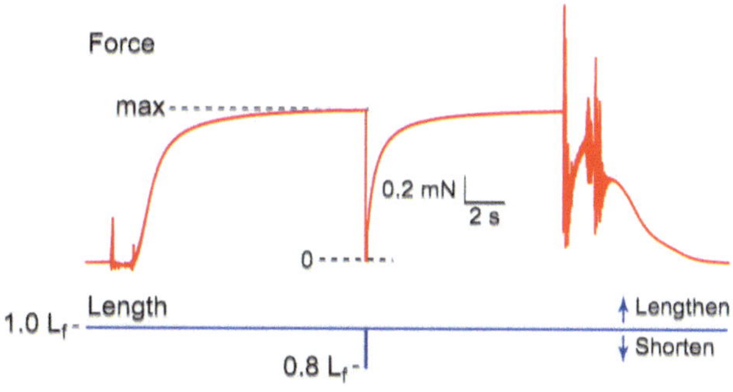

Fig. 6 Representative recording of maximum isometric force generation. Upon immersion in a high [Ca^{2+}] solution, force increases rapidly (*top trace*) to a steady level ("max"). The servomotor is then controlled to introduce a large, brief reduction in fiber length followed by return to original length (*bottom trace*). The brief shortening serves to indicate the output of the force transducer when the fiber is slack and force is zero. The large fluctuations near the beginning and end of the force recording are caused by the stepper motors that move the chamber system and by the force transducer extension passing through the liquid-air interface during solution changes. They are not indicative of fiber tension during those times

viewed through the prism embedded in the chamber wall, is centered in the field of view. Refocus and capture a "side-view" image.

3. Using image analysis software, measure the top-view and side-view diameters at several locations along the fiber. For a typical fiber with $L_f \approx 1.5$ mm, we sample five diameter pairs separated by approximately 100 μm as illustrated in Fig. 7.
4. Fiber cross-sectional area is estimated by assuming an elliptical fiber cross section, computing the area at each measurement location as $\pi \times a \times b$ (where a is one-half the top-view diameter and b is one-half the side-view diameter). The cross-sectional areas for all measurement locations are averaged to give a global estimate of fiber cross-sectional area.
5. Upon completion of all measurements, the fiber is removed and, if desired, stored for subsequent analysis of myosin isoform composition. The chambers are emptied and refilled with fresh solutions in preparation for analysis of the next fiber.

4 Notes

1. Potassium propionate is available from TCI America, all other reagents are available from Sigma-Aldrich.
2. Glycerol is quite viscous and therefore difficult to measure by volume. Instead we put the beaker on a scale and add glycerol by mass: 100 mL ≈ 126 g.

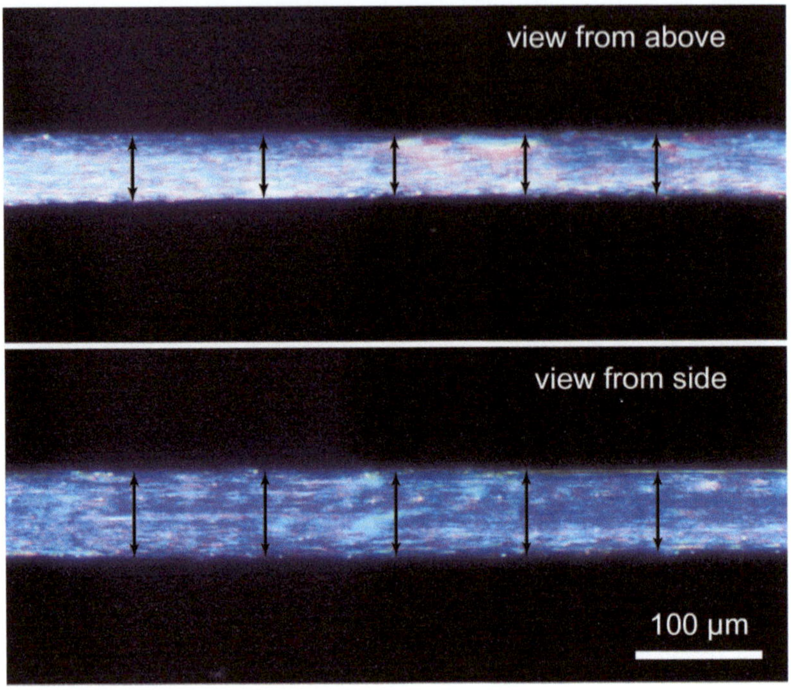

Fig. 7 Estimation of fiber cross-sectional area. Fiber cross-sectional area is estimated by measuring the "top-view" and "side-view" diameters at several locations (indicated by *double-headed arrows*), computing the cross-sectional area at each location under the assumption of an elliptical cross section, and then averaging the individual estimates to obtain global fiber cross-sectional area (see text for explanation). Reproduced from Roche et al. 2015 [4] with permission from JoVE

3. The fiber bundle isolations are performed at room temperature and should therefore be completed as quickly as possible to prevent deterioration of the tissue.
4. Fibers extracted from bundles stored at −80 °C have been tested after storage durations of up to 12 months with no apparent functional deficits as assessed by measurements of specific force.
5. Fibers from the periphery of the bundle are avoided as they are more likely to have been damaged during bundle preparation than fibers nearer the core of the bundle.
6. Several fibers are withdrawn from the bundle and discarded before proceeding to pull fibers that will be used in experiments. This step "unpacks" the bundle, allowing subsequent fibers to be extracted using less force which, in turn, reduces the likelihood that a fiber will be damaged by excessive stretch.
7. The fiber transfer scoop is made by using a razor blade to trim the distal 5 mm of a 100 μL disposable pipette tip as illustrated in Fig. 2.

8. A single bundle is used as the source of fresh fibers for several hours, allowing as many as 6–8 fiber experiments to be performed from one bundle. After a bundle has been thawed once, it is not reused, although we have not tested the function of refrozen fibers.
9. Creatine kinase, which catalyzes transfer of a phosphate group from CrP (creatine phosphate) to ADP, is not included in the solution formulas specified here because sufficient endogenous creatine kinase remains within the permeabilized fibers to maintain ATP levels during measurement of maximum isometric force at cool temperatures [11]. Experiments at higher temperatures or during which fast fibers are required to shorten at high speeds during activation might require supplemental exogenous creatine kinase to maintain sufficient ATP levels [12].
10. HDTA, an "EGTA-like" compound without the Ca^{2+}-binding properties of EGTA, is available from TCI America. All other reagents are available from Sigma-Aldrich.
11. $CaCO_3$ is used as the source of calcium in the activating solution and MgO is used as the source of magnesium in all solutions to avoid having high concentrations of Cl^- in the final solutions, as would be the case if $CaCl_2$ and $MgCl_2$ were used.
12. The heating step is required to remove, by way of CO_2 effervescence, the carbonic acid (H_2CO_3) formed when $CaCO_3$ combines with EGTA in the activating solution. The relaxing and preactivating solutions are also heated to ensure consistent treatment of all solutions. The heat also helps to dissolve the MgO.
13. The NaN_3 is added from 100 mM stock solution made by dissolving NaN_3 in 0.1 N KOH.
14. Concentrated KOH is used for this step to avoid the need to add large volumes. For example, a 1000 mL batch of solution requires approximately 8.5 mL of 8 N KOH to reach a pH of 7.0.
15. The KOH that is used to adjust pH is the sole source of potassium in the final solutions.
16. When weighing the ATP and CrP compounds, water content must be accounted for and can be found by consulting the "Certificate of Analysis" for the lot supplied by the manufacturer.
17. The sodium content of the final solutions (37 mM) comes from Na_2H_2ATP, Na_2CrP, and NaN_3.
18. A common experimental temperature for permeabilized fiber experiments is 15 °C. For convenience, pH can be adjusted at a warmer temperature while accounting for the relationship

between pH and temperature. Example: adjusting pH to 7.03 at 20 °C will result in a pH of 7.10 at 15 °C.

19. Storage for up to 2 years has been verified to result in no apparent deficiency in the effectiveness of the solutions for experiments in which maximum isometric force is being assessed.
20. Making the suture ties too loose will result in slippage of the fiber during force generation, whereas tying too tight will result in damage at the tie, resulting in tearing or breaking of the fiber at the tie. Practice is required to develop a feel for the level of tension required during the tightening of the loops that will avoid these undesirable outcomes.
21. Example calculation for the laser target: for a sarcomere length of 2.5 μm, laser wavelength of $\lambda = 635$ nm, and distance from fiber to target of $D = 102$ mm, the distance from the zero-order and first-order laser maxima at the target is 26.8 mm.
22. Optimal sarcomere length varies both within and among species. Experiments on fibers from rodent limb muscles are typically performed at a sarcomere length of 2.5 μm [13–15] whereas optimal length for human fibers is closer to 2.7 μm [16, 17].
23. A fiber that remains "slack" (not straight, no resting tension) at a fiber length that should correspond to optimal sarcomere length has likely been damaged due to excessive stretch during bundle preparation or fiber extraction and should be discarded.
24. Attaching the laser target to the objective of the stereomicroscope ensures that, when the fiber is in focus, the fiber-to-target distance remains fixed regardless of the position of the fiber in the chamber (assuming that a parfocal adjustment has been made to the stereomicroscope eyepieces prior to use).
25. Take care to fill chamber 1 in such a way that the top surface of the liquid is flat. A concave or convex surface geometry can affect, by refraction, the diffracted laser light that passes through the liquid-air interface on the way to the laser target.
26. The preactivating solution is weakly buffered for calcium. This step greatly reduces the intramyofibrillar EGTA concentration (*see* Table 2), resulting in physiologically rapid activation and force generation when the fiber is subsequently immersed in activating solution.
27. While 30 ms is typical, the amount of time spent at the shorter length is not critical as long as the force reported by the force transducer has sufficient time to reach a steady-state minimum level before the fiber is returned to original length.
28. In general, each chamber will have a unique force transducer output that corresponds to zero force, making it necessary to establish chamber-specific zero levels.

29. It is common practice to activate the fiber a second time for the determination of maximum isometric force, with the first activation serving as a "warm up" to make sure the suture attachments of the fiber to the experimental apparatus are "set" (tight, secure).
30. The optical gain must be determined by capturing an image of a stage micrometer (calibration slide) at the same stereomicroscope zoom setting that is used for capturing fiber top-view and side-view images.

Acknowledgements

This work was supported by the following funding sources: R01-AG050676, R01-AR063649, F31-AR065931.

References

1. Mendias CL, Kayupov E, Bradley JR, Brooks SV, Claflin DR (2011) Decreased specific force and power production of muscle fibers from myostatin-deficient mice are associated with a suppression of protein degradation. J Appl Physiol 111:185–191
2. Gumucio JP, Davis ME, Bradley JR, Stafford PL, Schiffman CJ, Lynch EB, Claflin DR, Bedi A, Mendias CL (2012) Rotator cuff tear reduces muscle fiber specific force production and induces macrophage accumulation and autophagy. J Orthop Res 30:1963–1970
3. Mendias CL, Roche SM, Harning JA, Davis ME, Lynch EB, Sibilsky Enselman ER, Jacobson JA, Claflin DR, Calve S, Bedi A (2015) Reduced muscle fiber force production and disrupted myofibril architecture in patients with chronic rotator cuff tears. J Shoulder Elbow Surg 24:111–119
4. Roche SM, Gumucio JP, Brooks SV, Mendias CL, Claflin DR (2015) Measurement of maximum isometric force generated by permeabilized skeletal muscle fibers. J Vis Exp 100:e52695
5. Wood LK, Kayupov E, Gumucio JP, Mendias CL, Claflin DR, Brooks SV (2014) Intrinsic stiffness of extracellular matrix increases with age in skeletal muscles of mice. J Appl Physiol 117:363–369
6. Claflin DR, Larkin LM, Cederna PS, Horowitz JF, Alexander NB, Cole NM, Galecki AT, Chen S, Nyquist LV, Carlson BM, Faulkner JA, Ashton-Miller JA (2011) Effects of high- and low-velocity resistance training on the contractile properties of skeletal muscle fibers from young and older humans. J Appl Physiol 111:1021–1030
7. Moisescu DG, Thieleczek R (1978) Calcium and strontium concentration changes within skinned muscle preparations following a change in the external bathing solution. J Physiol 275:241–262
8. Cleworth DR, Edman KA (1972) Changes in sarcomere length during isometric tension development in frog skeletal muscle. J Physiol 227:1–17
9. Kawai M, Kuntz ID (1973) Optical diffraction studies of muscle fibers. Biophys J 13:857–876
10. Gordon AM, Huxley AF, Julian FJ (1966) The variation in isometric tension with sarcomere length in vertebrate muscle fibres. J Physiol 184:170–192
11. Moss RL (1982) The effect of calcium on the maximum velocity of shortening in skinned skeletal muscle fibres of the rabbit. J Muscle Res Cell Motil 3:295–311
12. Chase PB, Kushmerick MJ (1995) Effect of physiological ADP concentrations on contraction of single skinned fibers from rabbit fast and slow muscles. Am J Physiol 268:C480–C489
13. Edman KA (2005) Contractile properties of mouse single muscle fibers, a comparison with amphibian muscle fibers. J Exp Biol 208:1905–1913
14. Phillips SK, Woledge RC (1992) A comparison of isometric force, maximum power and isometric heat rate as a function of sarcomere length in mouse skeletal muscle. Pflugers Arch 420:578–583

15. Stephenson DG, Williams DA (1982) Effects of sarcomere length on the force-pCa relation in fast- and slow-twitch skinned muscle fibres from the rat. J Physiol 333:637–653
16. Walker SM, Schrodt GR (1974) I segment lengths and thin filament periods in skeletal muscle fibers of the Rhesus monkey and the human. Anat Rec 178:63–81
17. Gollapudi SK, Lin DC (2009) Experimental determination of sarcomere force-length relationship in type-I human skeletal muscle fibers. J Biomech 42:2011–2016

Chapter 23

Analysis of Aerobic Respiration in Intact Skeletal Muscle Tissue by Microplate-Based Respirometry

Jonathan Shintaku and Denis C. Guttridge

Abstract

Mitochondrial function is a key component of skeletal muscle health, and its dysfunction has been associated with a wide variety of diseases. Microplate-based respirometry measures aerobic respiration of live cells through extracellular changes in oxygen concentration. Here, we describe a methodology to measure aerobic respiration of intact murine skeletal muscle tissue. The tissues are not cultured, permeabilized, or enzymatically dissociated to single fibers, so there is minimal experimental manipulation affecting the samples prior to acquiring measurements.

Key words Skeletal muscle, Mitochondria, Aerobic respiration, Oxygen consumption rate, Seahorse

1 Introduction

Mitochondria have a central role in the ability of skeletal muscle to facilitate movement and regulate glucose homeostasis. Accordingly, mitochondrial dysfunction is involved in the pathogenesis of diseases such as diabetes [1, 2], sarcopenia [3–5], inclusion body myositis [6], and various mitochondrial myopathies [7–9]. Methods to accurately and reliably measure aerobic respiration can support studies examining how mitochondrial function is regulated in health and disease.

The Seahorse Bioscience XF^{e}24 utilizes solid state sensor probes to measure the concentration of oxygen surrounding live cells. The rate of oxygen utilization by the cells in indicative of aerobic respiration. This system is integrated into a microplate format for analyzing multiple samples simultaneously. Its most well established use has been the analysis of tissue culture cells [10]. Following baseline measurements, cells can then be treated with drugs or substrates to further dissect specific aspects of their metabolic phenotype [11].

Measuring respiration from tissues, while advantageous for studying animal disease models, has been more technically

Michael Kyba (ed.), *Skeletal Muscle Regeneration in the Mouse: Methods and Protocols*, Methods in Molecular Biology, vol. 1460, DOI 10.1007/978-1-4939-3810-0_23, © Springer Science+Business Media New York 2016

challenging. Isolating mitochondria from tissues is the most straightforward way to do this, but they can undergo damage during the isolation process and differential centrifugation can bias the analyzed population [12]. In addition, mitochondrial function has been linked to interactions with the intracellular environment [13–15]. Alternatively, researchers have used permeabilized skeletal muscle single fibers or permeabilized tissues. Both of these strategies keep intracellular structures intact and enable the use of a myriad of substrates, inhibitors, and stimulants to dissect mitochondrial function [16, 17]. However, tissue permeabilization, while allowing compounds into cells, also releases the cytoplasm and any pertinent cytosolic factors from cells. This strategy can yield vast amounts of data from muscle tissue, as long as cytoplasmic components are not pertinent to the conclusions being drawn.

Here, we present a methodology for measuring the oxygen consumption rate (OCR) of intact, non-permeabilized skeletal muscle tissue using the Seahorse Bioscience XF^{e}24. The protocol we describe here does not offer the range of data that could be obtained from drug and substrate treatments, though our preliminary data suggest that treatment with the uncoupling agent carbonyl cyanide-p-trifluoromethoxyphenylhydrazone (FCCP) is a feasible and useful tool for analyzing muscle tissue respiration. This protocol provides a measure of basal respiration from muscle tissues that retain more of their relevant physiological traits. We have, as an example, compared soleus and extensor digitorum longus (EDL) muscles and observed an expectedly higher OCR in the soleus (Fig. 1), where there are proportionally more slow fibers and mitochondria. In addition, this methodology can be applied to muscles of disease models and muscles electroporated with gene expression plasmids. As mentioned above, aging has been associated with a decline in mitochondrial function. This decline has been measured experimentally through enzymatic staining of muscle cross sections [18], gene expression

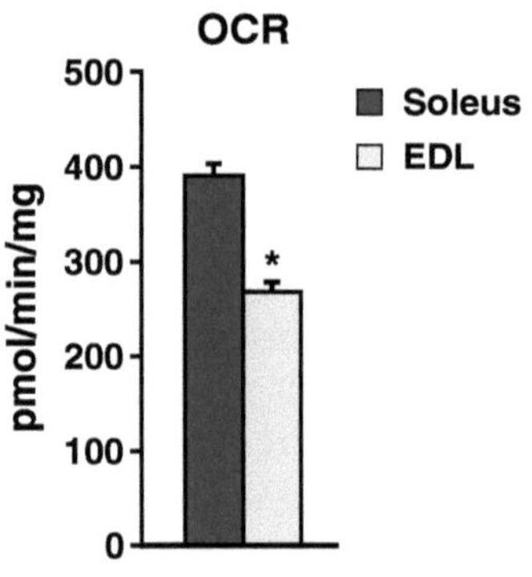

Fig. 1 OCR measurements reflect differences between slow and fast muscles. OCR was measured from soleus and EDL muscles of 8-week-old mice (*see* **Note 5**)

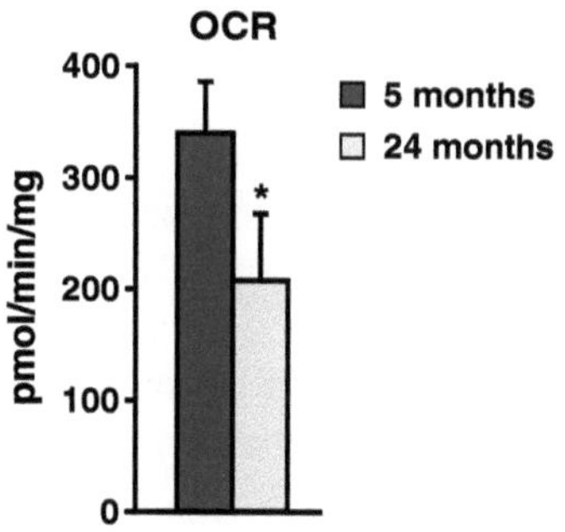

Fig. 2 Aging reduces skeletal muscle OCR. The OCR from TA muscles of 5-month and 24-month-old mice were measured and normalized to dry tissue weight

profiling [19], and vO_2-max measurements [5]. Here, we have compared tibialis anterior (TA) muscles of 5-month and 24-month-old mice and found a significant reduction in OCR associated with aging (Fig. 2). In summary, this methodology can discern differences in the basal respiration rate of healthy, diseased, and genetically modified intact skeletal muscle tissues.

2 Materials

2.1 Oxygen Consumption Rate Measurements

1. XF^e24 Extracellular Flux Analyzer (Seahorse Bioscience).
2. XF^e24 Assay Kit sensor plate.
3. XF Calibrant Solution.
4. XF24 Islet Capture Microplate with capture screens.
5. XF Islet Capture Screen Insert Tool.

2.2 Tissue Preparation

1. Dissection instruments: scissors, forceps, needles, tray.
2. 70% ethanol.
3. Razor blade.
4. 60 mm petri dishes.
5. Complete Assay Medium: XF Assay Medium, 2 mM l-glutamine, 1 mM sodium pyruvate, 10 mM glucose. For working solution, combine 98 mL of XF Assay Medium, 1 mL of 200 mM l-glutamine, 1 mL of 100 mM sodium pyruvate, and 0.180 g of glucose. Adjust pH to 7.4 at 37 °C with sodium hydroxide (*see* **Notes 1–3**).

2.3 Normalization

1. 60 °C oven.
2. Analytical balance.

3 Methods

3.1 XF^e24 Assay Kit Sensor Plate Preparation

1. One day before the experiment, prepare an extracellular flux assay sensor plate by adding 1 mL of calibrant solution to the bottom of each well, placing the sensors into the wells, and storing overnight in a 37 °C non-CO_2 incubator. Keep the spacer between the sensors and plate until the calibration (Subheading 3.6, **step 2**).
2. Check that the XF^{e}24 Analyzer is on and set to 37 °C.

3.2 Tissue Preparation

1. Aliquot 100 μL of Complete Assay Medium into each well of an Islet Capture Microplate and place the plate in the 37 °C non-CO_2 incubator.
2. Prepare a 60 mm petri dish for each tissue sample by adding 10 mL of Complete Assay Medium and labeling. Place petri dishes in the 37 °C non-CO_2 incubator.
3. Prepare one additional 60 mm petri dish with 10 mL Complete Assay Medium and the islet capture screens and place in the 37 °C non-CO_2 incubator.
4. Place remaining Complete Assay Medium in the 37 °C non-CO_2 incubator.
5. Dissect muscles and place each in its own 60 mm petri dish containing warm Complete Assay Medium. Keep the dishes at 37 °C while dissecting the other muscle samples.
6. For each muscle, remove tendons with a razor, then cut muscle sections measuring approximately 2 × 2 × 1 mm. As you cut each piece, place it back in the warm medium (*see* **Notes 4** and **5**).

3.3 Assay Plate Setup

1. Add one tissue piece per microplate well, keeping indicated wells empty for background measurements (Fig. 3).
2. Slide a screen into each well using forceps, being careful that the tissue is centered under the screen and there is no trapped air underneath (*see* **Note 6**). Screens should also be added to background wells.

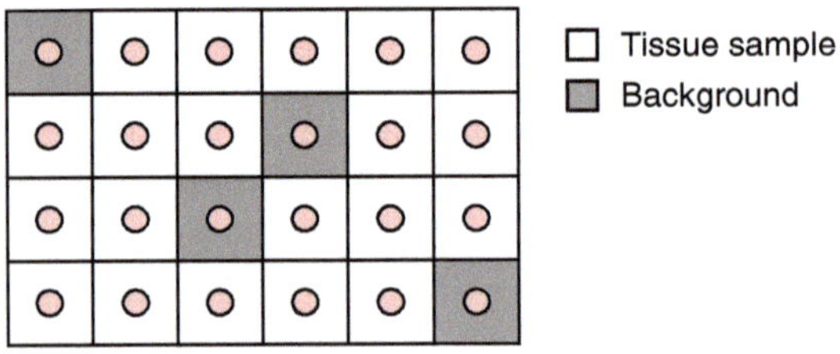

Fig. 3 Assay plate loading scheme. Wells labeled as "background" should contain medium and a screen, but no tissue. This will serve as a control for temperature fluctuations throughout an experiment and across the plate

3. Press down on the screen with the XF Islet Capture Screen Insert Tool so that it is firmly seeded at the bottom of the well.
4. Add 600 μL of Complete Assay Medium to each well.
5. Place in the 37 °C non-CO_2 incubator for 1 h.

3.4 XFe24 Template Setup

1. Design a new blank template.
2. Under the Group Definitions tab, click on the Assay Conditions screen. Define a separate "cell type" for each sample.
3. Create a Well Group for each "cell type." Pair a "cell type" to each Well Group.
4. On the Plate Map screen, assign each Well Group to its corresponding wells.
5. Under the Instrument Protocol tab, edit the measurement details. Set measurement procedure to MIX 3 min, WAIT 5 min, and READ 2 min. Repeat basal measurement at least four times (*see* **Note 7**).
6. Under the Review and Run tab, be sure the Calibrate and Equilibrate boxes are checked.

3.5 XFe24 Measurements

1. Click on Start Run.
2. Start the protocol by calibrating the sensors. Remove the spacer and place the sensors back in the wells containing calibration buffer. When prompted, load the sensors with wells onto the tray that slides out of the Analyzer.
3. When the 1 h tissue incubation is complete, prompt the Analyzer to proceed and replace the calibration buffer plate with the tissue samples.

3.6 Normalization

1. After measurements are completed, remove screens and transfer the tissues to labeled Eppendorf tubes.
2. Incubate uncapped tubes in a 60 °C oven for 48 h.
3. Weigh dried tissues.
4. Calculate OCR per mg of dry tissue for each sample.

4 Notes

1. Adjusting the pH of the Complete Assay Medium should be done at 37 °C and in equilibrium with ambient air, so that the adjusted pH will remain constant during the experiment. A glass beaker or conical tube stabilized in a 37 °C water bath works for this.
2. Seahorse Assay Medium does not contain any buffering agent like bicarbonate, so the pH should be adjusted in small increments.

3. Prepared Complete Assay Medium can be filter-sterilized and stored up to 1 week at 4 °C without having to check the pH again. For longer storage, check the pH before using.
4. Skeletal muscle fiber types are not evenly distributed. In general, deep fibers tend to be more oxidative than more superficial fibers. Therefore, care should be taken to take similar representative samples when comparing tissue samples, by comparing either total cross sections or similar anatomical regions. We have found it useful to analyze relatively small muscles such as the tibialis anterior or soleus when this is possible and appropriate for the experimental design.
5. When developing this protocol, one concern was whether slicing through fibers would disturb OCR measurements. When comparing fully intact EDL muscles versus EDL muscles cut in half, measurements normalized to dry tissue weight were nearly identical.
6. The screens are "U" or cup shaped and should be placed into the wells with the "U" open end facing up. If it is hard to determine which way they are facing, the screens will be substantially easier to pick up with the forceps if they are oriented the right way.
7. Repeated measurements should be fairly consistent. If they are progressively declining, you might try increasing the MIX and WAIT times. In addition, increase the number of repeated measures to see whether the readings stabilize after the tissues equilibrate with the medium. We have found that incubating the plate and tissues less than an hour prior to measurements gives declining OCR values over time, until they eventually stabilize. These early values are unreliable, so care should be taken to allow tissues to equilibrate. We have observed stable OCR values up to 3 hours after dissection.

References

1. Kelley DE, He J, Menshikova EV, Ritov VB (2002) Dysfunction of mitochondria in human skeletal muscle in type 2 diabetes. Diabetes 51(10):2944–2950
2. Lindroos MM, Majamaa K, Tura A, Mari A, Kalliokoski KK, Taittonen MT, Iozzo P, Nuutila P (2009) m.3243A>G mutation in mitochondrial DNA leads to decreased insulin sensitivity in skeletal muscle and to progressive beta-cell dysfunction. Diabetes 58(3):543–549. doi:10.2337/db08-0981
3. Taylor DJ, Kemp GJ, Thompson CH, Radda GK (1997) Ageing: effects on oxidative function of skeletal muscle in vivo. Mol Cell Biochem 174(1-2):321–324
4. Conley KE, Jubrias SA, Esselman PC (2000) Oxidative capacity and ageing in human muscle. J Physiol 526(Pt 1):203–210
5. Short KR, Bigelow ML, Kahl J, Singh R, Coenen-Schimke J, Raghavakaimal S, Nair KS (2005) Decline in skeletal muscle mitochondrial function with aging in humans. Proc Natl Acad Sci U S A 102(15):5618–5623. doi:10.1073/pnas.0501559102
6. Boncompagni S, Moussa CE, Levy E, Pezone MJ, Lopez JR, Protasi F, Shtifman A (2012) Mitochondrial dysfunction in skeletal muscle of amyloid precursor protein-overexpressing mice. J Biol Chem 287(24):20534–20544. doi:10.1074/jbc.M112.359588

7. Lalani SR, Vladutiu GD, Plunkett K, Lotze TE, Adesina AM, Scaglia F (2005) Isolated mitochondrial myopathy associated with muscle coenzyme Q10 deficiency. Arch Neurol 62(2):317–320. doi:10.1001/archneur.62.2.317
8. Holt IJ, Harding AE, Morgan-Hughes JA (1988) Deletions of muscle mitochondrial DNA in patients with mitochondrial myopathies. Nature 331(6158):717–719. doi:10.1038/331717a0
9. Nakamura S, Sato T, Hirawake H, Kobayashi R, Fukuda Y, Kawamura J, Ujike H, Horai S (1990) In situ hybridization of muscle mitochondrial mRNA in mitochondrial myopathies. Acta Neuropathol 81(1):1–6
10. Watanabe M, Houten SM, Mataki C, Christoffolete MA, Kim BW, Sato H, Messaddeq N, Harney JW, Ezaki O, Kodama T, Schoonjans K, Bianco AC, Auwerx J (2006) Bile acids induce energy expenditure by promoting intracellular thyroid hormone activation. Nature 439(7075):484–489. doi:10.1038/nature04330
11. Wu M, Neilson A, Swift AL, Moran R, Tamagnine J, Parslow D, Armistead S, Lemire K, Orrell J, Teich J, Chomicz S, Ferrick DA (2007) Multiparameter metabolic analysis reveals a close link between attenuated mitochondrial bioenergetic function and enhanced glycolysis dependency in human tumor cells. Am J Physiol Cell Physiol 292(1):C125–C136. doi:10.1152/ajpcell.00247.2006
12. Piper HM, Sezer O, Schleyer M, Schwartz P, Hutter JF, Spieckermann PG (1985) Development of ischemia-induced damage in defined mitochondrial subpopulations. J Mol Cell Cardiol 17(9):885–896
13. Milner DJ, Mavroidis M, Weisleder N, Capetanaki Y (2000) Desmin cytoskeleton linked to muscle mitochondrial distribution and respiratory function. J Cell Biol 150(6): 1283–1298
14. Kay L, Nicolay K, Wieringa B, Saks V, Wallimann T (2000) Direct evidence for the control of mitochondrial respiration by mitochondrial creatine kinase in oxidative muscle cells in situ. J Biol Chem 275(10):6937–6944
15. Saks VA, Belikova YO, Kuznetsov AV (1991) In vivo regulation of mitochondrial respiration in cardiomyocytes: specific restrictions for intracellular diffusion of ADP. Biochim Biophys Acta 1074(2):302–311
16. Veksler VI, Kuznetsov AV, Sharov VG, Kapelko VI, Saks VA (1987) Mitochondrial respiratory parameters in cardiac tissue: a novel method of assessment by using saponin-skinned fibers. Biochim Biophys Acta 892(2):191–196
17. Kuznetsov AV, Veksler V, Gellerich FN, Saks V, Margreiter R, Kunz WS (2008) Analysis of mitochondrial function in situ in permeabilized muscle fibers, tissues and cells. Nat Protoc 3(6):965–976. doi:10.1038/nprot.2008.61
18. Wenz T, Rossi SG, Rotundo RL, Spiegelman BM, Moraes CT (2009) Increased muscle PGC-1alpha expression protects from sarcopenia and metabolic disease during aging. Proc Natl Acad Sci U S A 106(48):20405–20410. doi:10.1073/pnas.0911570106
19. McCarroll SA, Murphy CT, Zou S, Pletcher SD, Chin CS, Jan YN, Kenyon C, Bargmann CI, Li H (2004) Comparing genomic expression patterns across species identifies shared transcriptional profile in aging. Nat Genet 36(2):197–204. doi:10.1038/ng1291

Index

Michael Kyba (ed.), *Skeletal Muscle Regeneration in the Mouse: Methods and Protocols*, Methods in Molecular Biology, vol. 1460, DOI 10.1007/978-1-4939-3810-0, © Springer Science+Business Media New York 2016

MIX
Papier aus verantwortungsvollen Quellen
Paper from responsible sources
FSC® C105338

If you have any concerns about our products,
you can contact us on
ProductSafety@springernature.com

In case Publisher is established outside the EU,
the EU authorized representative is:
Springer Nature Customer Service Center GmbH
Europaplatz 3, 69115 Heidelberg, Germany

Printed by Libri Plureos GmbH
in Hamburg, Germany